Photovoltaic and
Photoelectrochemical
Solar Energy Conversion

NATO ADVANCED STUDY INSTITUTES SERIES

A series of edited volumes comprising multifaceted studies of contemporary scientific issues by some of the best scientific minds in the world, assembled in cooperation with NATO Scientific Affairs Division.

Series B. Physics

Recent Volumes in this Series

This series is published by an international board of publishers in conjunction with NATO Scientific Affairs Division

A Life Sciences	Plenum Publishing Corporation
B Physics	London and New York
C Mathematical and Physical Sciences	D. Reidel Publishing Company Dordrecht, Boston, and London
D Behavioral and Social Sciences	Sijthoff & Noordhoff International Publishers
E Applied Sciences	Alphen aan den Rijn, The Netherlands, and Germantown, U.S.A.

NATO Advanced Study Institute on Photovoltaic and
" Photoelectrochemical Solar Energy Conversion (1980: Ghent,
Belgium)

Photovoltaic and Photoelectrochemical Solar Energy Conversion

Edited by

F. Cardon
W. P. Gomes
and

W. Dekeyser

Laboratory for Crystallography and the Study of Solids
and
Laboratory for Physical Chemistry
State University of Gent
Gent, Belgium

PLENUM PRESS • NEW YORK AND LONDON
Published in cooperation with NATO Scientific Affairs Division

Library of Congress Cataloging in Publication Data

NATO Advanced Study Institute on Photovoltaic and Photoelectrochemical Solar
    Energy Conversion (1980: Ghent, Belgium)
    Photovoltaic and photoelectrochemical solar energy conversion.
    (NATO advanced study institutes series. Series B, Physics; v. 69)
    "Proceedings of a NATO Advanced Study Institute on Photovoltaic and
Photoelectrochemical Solar Energy Conversion held August 25 – September 5, 1980,
at Gent, Belgium."—T. p. verso.
    "Published in cooperation with NATO Scientific Affairs Division."
    Includes bibliographies and indexes.
    1. Solar cells—Congresses. 2. Photovoltaic power generation—Congresses. I.
Cardon, Felix. II. Gomes, Walter. III. Dekeyser, W. (Willy) IV. North Atlantic Treaty
Organization. Division of Scientific Affairs. V. Title. VI. Series.
TK2960.N38 1980   621.31'244   81-10666
ISBN 0-306-40800-7   AACR2

Proceedings of a NATO Advanced Study Institute on Photovoltaic and Photoelectrochemical Solar Energy Conversion held August 25-September 5, 1980, at Gent, Belgium

© 1981 Plenum Press, New York
A Division of Plenum Publishing Corporation
233 Spring Street, New York, N.Y. 10013

PREFACE

In recent years there has been an increasing interest in
systems which enable the conversion of solar energy into electrical or chemical energy. Many types of systems have been proposed
and studied experimentally, the fundamentals of which extend from
solid state physics to photo- and electrochemistry. For most
of the systems considered excitation of an electron by absorption
of a photon is followed by charge separation at an interface. It
follows that the different fields involved (photovoltaics, photoelectrochemistry, photogalvanics, etc.) have several essential
aspects in common.

It was the main purpose with the NATO Advanced Study Institute held at Gent, Belgium, from August 25 to September 5, 1980,
to bring together research workers specializing in one of these
fields in order to enable them not only to extend their knowledge
into their own field but also to promote the interdisciplinary
exchange of ideas. The scope of the A.S.I. has been limited to
systems which have not or have hardly reached the stage of practical development. As a consequence, no lectures on economical
aspects of solar energy conversion have been included.

The topics covered in this volume are the fundamentals of
recombination in solar cells (P. Landsberg), theoretical and
experimental aspects of heterojunctions and semiconductor/metal
Schottky barriers (J.J. Loferski, W.H. Bloss and W.G. Townsend),
photoelectrochemical cells (H. Gerischer and A.J. Nozik), photo-

galvanic cells (W.J. Albery) and finally, surfactant assemblies (M. Grätzel).

The editors are grateful to the lecturers of this Institute for providing extended lecture notes. In most cases this involved a considerable writing task in order to cover the subject in a comprehensive way.

The support of the NATO Science Committee is gratefully acknowledged. Our gratitude also goes to the authors and editors of books and periodicals who granted permission to reproduce figures, diagrams, or other material.

Finally we also wish to thank Plenum Press for providing the necessary support for the publication of these Proceedings.

F. CARDON

W.P. GOMES

W. DEKEYSER

Gent, December 1980

CONTENTS

The Iron Thionine Photogalvanic Cell   313
W. John Albery

RECOMBINATION IN SOLAR CELLS: THEORETICAL ASPECTS

Peter T. Landsberg

University of Southampton

Southampton SO9 5NH, UK

1. INTRODUCTION

The study of the effect of recombination processes on solar
cells becomes a difficult subject as soon as one goes into it in
detail.   Experiments measure currents and voltage, possibly
light absorption, capacitance and temperature dependences as well.
However, what holds for bulk materials, namely that it is
difficult to identify experimentally the dominant recombination
mechanisms, is even truer in devices.   The situation is not much
better in a theoretical approach; for, as soon as it is made
reasonably realistic one loses the advantage of simple formulae.
To get over these difficulties, while still offering the reader an
introduction to the different ways of approaching the subject, the
lectures have been divided into three parts.

Part I (§2,3,4) deals with the macroscopic aspects of solar
cells, and specifically with the directions of the various
currents; we see also how to discuss maximum power.   Relevant to
the title of these lectures in this Part are particularly the
directions of the recombination currents and how recombination in
the junction region increases the ideality factor (§4).   Part II
(§5-10) deals with recombination more microscopically in deriving
steady-state recombination rates (§6), which are essential parts
of the junction current (§5), and can be evaluated by simple
integrations if a simple model of the p-n junction is adopted
(§7-9).   Numerical results are given in figs. 8.1 and 8.2 and
in (§13).   In Part III (§11-13) we deal somewhat less mathematic-
ally, but still systematically, with the theory of the Schottky

barrier MIS solar cell in support of lectures given on this topic
in other parts of this course.   From the theory (which is not as
fully expounded in Part III as in the appropriate approximate
theory in Part II) recombination current densities can be inferred
(§13).

The importance of recombination effects in solar cell
performance and methods of treating them should be clear from
these three Parts.

PART I

2.   CONVENTIONS USUALLY MADE FOR p–n JUNCTIONS AND SOLAR CELLS

In most discussions of solar cells certain conventions are
adopted.   They are:-

(2.1)   One uses band diagrams in which the vertical axis repre-
        sents the electrostatic potential energy of an electron,
        expressed as a potential or voltage.   Thus if ϕ is the
        electrostatic potential, the vertical axis represents
        $-e\phi/e = -\phi$ where $e = |e|$ is the electronic charge.
        The horizontal axis is the x-axis and represents the
        direction of current flow, from right to left or from left
        to right.   Such theories are in this sense one-dimensional,
        and require extension if this assumption is to be discarded.

(2.2)   The radiation will here be assumed to be incident from the
        left, i.e. from $x < 0$.   For cells which receive light
        from both sides this convention has to be discarded.   It
        is in any case not essential.

(2.3)   There are now two broad classes of cells, depending on the
        sign of the electric field in the right-hand portion of
        the active region.   If it is positive we shall call it a
        "class p" type of cell, and this includes a n–on–p
        junction cell and a p-type Schottky barrier type of cell.
        If the field in the right-hand portion is negative we shall
        call it a "class n" type cell.   Such cells include a
        p–on–n junction cells and a n-type Schottky barrier cells.

3.   THREE LAWS OF PHOTOVOLTAICS

The <u>first law of photovoltaics</u> must deal with the direction
of the various currents which flow inside the cell under various
conditions.   They are given in table 3.1.

Table 3.1

Directions of various conventional currents inside the cell

General specification or direction	Directions using conventions (2.1,2,3)	Currents
Towards low values of $(-\phi)$	class p: right to left class n: left to right	Forward current in the dark under applied voltage; Recombination current.
Towards high values of $(-\phi)$	class p: left to right class n: right to left	Light-induced current; Generation current; Reverse current in the dark under applied voltage.

To see the reason for the entries in table 1, consider first a cell in the dark under an applied voltage.   For a forward current one must reduce the main internal field

$$E(x) = -\phi' = d(-\phi)/dx \qquad\qquad (3.1)$$

e.g. the field at the p-n junction.   This means one must raise the low $(-\phi)$-values or lower the high $(-\phi)$-values.   Thus the positive terminal of the external battery must be connected to the end with the high $(-\phi)$-values and the negative terminal to the end with the low $(-\phi)$-values.   This yields a forward current from right to left for a class p cell (Fig.3.1) and a forward current from left to right for a class n cell (Fig.3.2).   A reverse current increases the field and so flows in the opposite direction in the cell.

Now replace the battery by a load and illuminate the cell. The light-induced current is due to the photo-generated carriers and follows the internal field.   Thus the photo-holes move up the $(-\phi)$-curve, i.e. towards high $(-\phi)$-values, the photo-electrons move in the opposite direction and so both combine to give a conventional current which flows towards high $(-\phi)$-values. It remains to check that the potential developed corresponds to a forward potential.   For this purpose let $\rho(x)$ be the space charge density and let x = w be a plane where the electric field

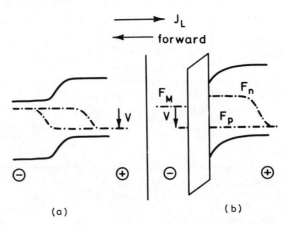

Fig. 3.1   Class p cells: (a) A n-p junction;   (b) a metal –
            insulator – p-type semiconductor cell. The polarity
            shown indicates an externally applied forward
            potential.

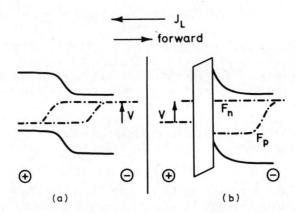

Fig. 3.2   Class n cells: (a) A p-n junction;   (b) a metal –
            insulator – n-type semiconductor cell. The polarity
            shown indicates an externally applied forward
            potential.

E(x) is negligibly small.   Integrating Poisson's equation (ε is the dielectric constant, ρ is the space charge density)

$$-\phi''(x) = 4\pi\rho(x)/\varepsilon \qquad (3.2)$$

from a general x to x = w in the active layer (i.e. ignoring back contact fields, etc.)

$$E(x) = -\phi'(x) = -(4\pi/\varepsilon)\int_x^w \rho(x)dx . \qquad (3.3)$$

For the class p cell $\rho(x)$ is negative, and the majority carriers are holes as in a p-type semiconductor with ionised (negatively charged) acceptors.   The quasi-Fermi level for holes on the right is therefore lowered by the photo-generated holes. Since it is this Fermi level which determines the voltage developed by illumination, its lowering corresponds to the case when the positive terminal of the battery is connected to the right-hand side, i.e. to the case of a forward potential applied externally. Similarly, for a class n cell, $E(x) < 0$.   Thus $\rho(x)$ is positive as for an n-type semiconductor with ionised donors. Electrons are the majority carriers and the photo-generated electrons raise the quasi-Fermi level on the right.   This Fermi level determines the voltage developed by illumination, and it corresponds to a forward potential applied externally (Fig.3.3).

These considerations help one to understand the apparently paradoxical property of solar cells, namely, that they develop a forward potential under illumination, while the current delivered flows through the cell in a direction which corresponds to the reverse direction if an external voltage were applied.

Finally, the thermal equilibrium situation of zero current can be regarded as due to a balance between a recombination and a generation current I_r and I_g .   For a type p cell a forward potential corresponds to a positive potential being applied to a p-layer.   This tends to repel holes and attract electrons, and this increased intermingling of current carriers of different types adds to the recombination current while leaving the generation current at roughly its thermal equilibrium value. Similarly for type n cells (Fig.3.4) a forward current results from the application of a negative potential to an n-layer. This again adds to the recombination current.   Thus in both cases the recombination current can be thought of as having the same direction as the forward current.   The generation current flows in the opposite direction in order to be able to balance the recombination current in equilibrium.   As expected the light-induced current and the generation current flow therefore in the same direction.

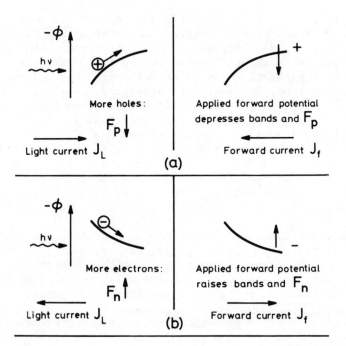

Fig. 3.3   Incident light causes a forward potential.   (a) Class
            p cell; (b) Class n cell. J_L flows towards high
            values of $(-\phi)$, the forward current flows in the
            opposite direction.

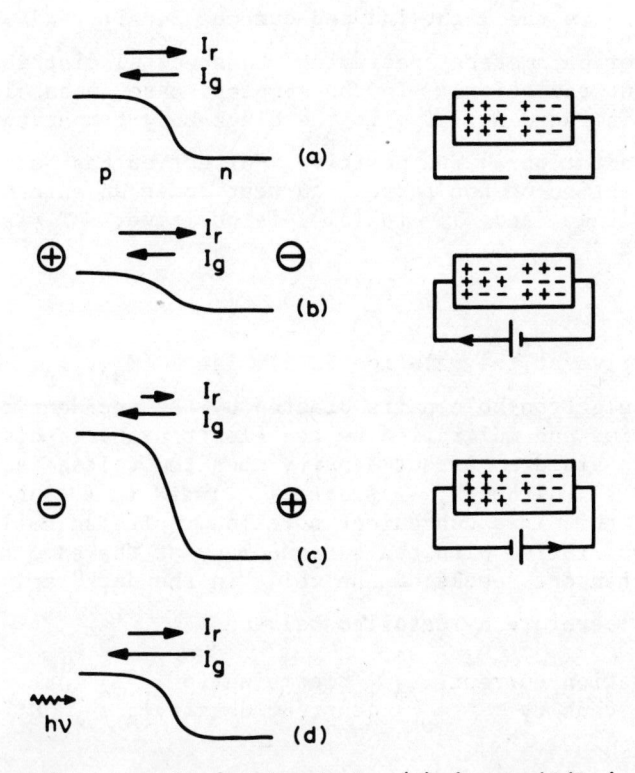

Fig. 3.4   Class n cell in the dark: (a) in equilibrium, i.e.
open-circuit, condition; (b) under forward bias;
(c) under reverse bias; (d) Illuminated class n cell
without bias.

The <u>second law of photovoltaics</u> states that the current-density J drawn from the cell and the voltage V across the cell at absolute temperature T are approximately related by

$$J(E_G,V,T,T_r) = J_o(E_G,T)\left[\exp\frac{eV}{\alpha kT} - 1\right] - J_L(E_G,T_r) . \quad (3.4)$$

Here E_G is a characteristic energy gap (e.g. an average gap if there are several gaps) of the solar cell, α is its ideality factor[+], J_L is the light-induced current density and T_r stands for a set of parameters specifying the spectral distribution of the incident radiation. In the simplest case, when black-body radiation is incident, T_r is the black-body temperature of the incident radiation. The positive x-direction has been taken to be the direction of the forward current under an externally applied voltage, and J_L in (3.4) is positive. T is the cell temperature.

Dark current

To arrive at the relation (3,4), let $Q(E_G,V,T_r)^*$ be the number of electron-hole pairs created by the incident radiation per unit time and multiplied by the electron charge divided by the area to yield a current density when the voltage across the cell is V. Although V is itself related to the incident radiation, this is a convenient notation. If the cell is in thermal equilibrium with the surroundings at the same temperature $T_r = T$, then one speaks of the cell "in the dark" and one has $V = 0$. Therefore by detailed balance

$$\left(\begin{array}{c}\text{generation current}\\ \text{density}\end{array}\right)_o = \left(\begin{array}{c}\text{recombination}\\ \text{current density}\end{array}\right)_o = Q(E_G,0,T) . \quad (3.5)$$

The suffix 0 refers to thermal equilibrium. In this case recombination and generation are in balance with each other and with black-body radiation at temperature T.

The recombination current density in the dark (i.e. at temperature T) at voltage V across the cell, can be obtained by assuming the recombination rate at a plane x in the active region to be proportional to the product of the carrier concentrations. For non-degenerate semiconductors assuming quasi-Fermi

---

*Q depends also on the cell temperature T, but this is not shown explicitly.
+A derivation of α for the transition current density due to traps is given in section 8.

levels $F_n(x)$, $F_p(x)$ to exist, and denoting the intrinsic carrier density by n_i , one has an integral over the region considered :

$$\text{(i)} \qquad Q\ (E_G,0,T) = C \int \left[n(x)p(x) \right]_o dx = C \int n_i^2(x) dx, \qquad (3.6)$$

where C is a recombination constant, and

$$\text{(ii)} \qquad \frac{n(x)p(x)}{\left[n(x)p(x) \right]_o} = \exp \frac{F_n(x) - F_p(x)}{kT} \left[= \exp \frac{eV}{kT} \right] \qquad (3.7)$$

If the Fermi levels have a constant separation eV in the transition region in the usual way, one finds the last expression in (3.7) provided recombination through traps or surface effects do not upset it. If the separation is not quite constant the contributions to $Q\ (E_G,V,T)$ which arise from different slabs of thickness dx are different, and an ideality factor $\alpha \neq 1$ may be expected. Hence one finds from (3.6) and (3.7)

$$Q\ (E_G,V,T) = C \int n(x)\ p(x)\ dx = \exp(eV/kT) \int \left[n(x)p(x) \right]_o dx$$

i.e.
$$\frac{Q\ (E_G,V,T)}{Q\ (E_G,0,T)} = \exp \frac{eV}{\alpha kT} \left[\sim \frac{\int e^{\frac{1}{kT}(F_n - F_p)} np\ dx}{\int n_o\ p_o\ dx} \right]$$

Thus the current density in the dark is

recombination current density − generation current density

$$\begin{bmatrix} \text{carriers have to overcome} \\ \text{field, V-dependent} \end{bmatrix} \qquad \begin{bmatrix} \text{carriers aided by field,} \\ \text{V-independent,}\ Q\ (E_G,0,T) \end{bmatrix}$$

$$= Q\ (E_G,0,T) \left[\exp \frac{eV}{\alpha kT} - 1 \right].$$

If traps are important one may find two terms

$$J = J_1 \left(\exp \frac{eV}{kT} - 1 \right) + J_2 \left(\exp \frac{eV}{2kT} - 1 \right) \qquad (3.4a)$$

instead of a single term with an ideality factor (see sections 8 and 9). The reason is that for a midgap localized level a contribution is expected which behaves as the second term[1].

Current with Illumination

If illumination specified by the parameters T_r is applied, the generation current density is changed to $Q\ (E_G,V,T_r)$. One finds

$$J(E_G,V,T,T_r) = Q(E_G,0,T) \exp \frac{eV}{\alpha kT} - Q(E_G,V,T_r)$$

$$= Q(E_G,0,T) \left[\exp \frac{eV}{\alpha kT} - 1\right] - \left[Q(E_G,V,T_r) - Q(E_G,0,T)\right]$$

This is essentially (3.4) with the interpretation

$$J_o(E_G,T) \equiv Q(E_G,0,T) = \left\{\begin{array}{l}\text{reverse saturation current} \\ \text{in the dark}\end{array}\right\} \qquad (3.8)$$

$$J_L(E_G,T_r) \equiv Q(E_G,V,T_r) - Q(E_G,0,T) \qquad (3.9)$$

A dependence of $J_L(E_G,T_r)$ on voltage V, which (3.9) suggests, implies a correction to (3.4): one cannot just take the dark characteristic J versus V and shift it by a voltage-independent quantity J_L, though this so-called "shift theorem" is often a good approximation[2]. This simple deduction of (3.8) and (3.9) goes back to 1975[3].

A series resistance R_s is always built into a solar cell through metallised contacts. Its effect is to drop the voltage across the load when it is drawing a current AJ from

$$V \text{ to } V - AJR_s. \qquad (3.10)$$

For an operating cell J in (3.4) is negative and for a good cell both J_o and R_s have to be kept as small as possible. The conditions for maximum power from such a cell are readily obtained (see section 4).

The third law of photovoltaics is that solar cell efficiencies are reduced if the cell temperature is raised; they are increased by concentration until the efficiency is pulled down again by the potential drop across the built-in series resistance due to the large currents produced.

To understand the temperature effect, note that by (3.4) and (3.9) the short-circuit current is, ignoring signs,

$$|J_{sc}| = J_L(E_G,T_r) = Q(E_G,V,T_r) - Q(E_G,0,T). \qquad (3.11)$$

The open-circuit voltage is given from equations (3.4), (3.8) and (3.9) by

$$\exp \frac{eV_{oc}}{\alpha kT} = \frac{Q\ (E_G, V, T_r)}{Q\ (E_G, 0, T)} \ . \tag{3.12}$$

Since the equilibrium pair creation-recombination traffic goes up with cell temperature T, one sees from (3.11) and (3.12) that $|J_{sc}|$ and V_{oc} drop.* This confirms the tendency to lower efficiencies.

The temperature effect can also be understood from the increase of n_i with temperature, which increases the reverse saturation current by virtue of equations (3.6) and (3.8). Similarly, concentration of illuminating light increases the pair creation rate $Q(E_G, V, T_r)$ and this raises $|J_{sc}|$ and V_{oc} by (3.11) and (3.12). The effect of the series resistance (3.10) is here ignored, but it comes in as a limiting condition for the larger concentration factors.

These three laws of photovoltaics and the discussion of the directions of the currents is based on [5].

4. MAXIMUM POWER, RECOMBINATION AND THE IDEALITY FACTOR

It was shown in 1957[1] that while normally $\alpha = 1$, the recombination in the space charge region can lead to $\alpha = 2$. We shall therefore, without further enquiry, study (3.4) in the generalised form (3.4a)

$$J = J_o(e^\theta - 1) + J_o'(e^{\theta'} - 1) - J_L \qquad \left(\theta \equiv \frac{eV}{\alpha KT}, \ \theta' \equiv \frac{eV}{\alpha' kT} \right) \tag{4.1}$$

J_o and J_o' are two coefficients. We see at once that the shortcircuit current ($\theta = \theta' = 0$) is

$$J_{sc} = -J_L . \tag{4.2}$$

The open-circuit voltage ($J = 0$) is given by

$$J_o e^{\theta_{oc}} + J_o' e^{\theta'_{oc}} = J_o + J_o' - J_{sc} \tag{4.3}$$

so that

$$J = J_o(e^\theta - e^{\theta_{oc}}) + J_o'(e^{\theta'} - e^{\theta'_{oc}}), \tag{4.4}$$

---

*In fact, $V_{oc} \to 0$ as $T \to T_r$.

and

$$V \frac{dJ}{dV} = \theta J_o e^{\theta} + \theta' J'_o e^{\theta'} .$$ (4.5)

Given θ_{oc}, θ'_{oc}, J_o, J'_o one can find the optimal working voltage, V_m say; hence θ_m and θ'_m can be found. The argument is as follows.

The maximum power output occurs at a voltage V_m such that $d(VJ)/dV = 0$. This condition is

$$(1+\theta_m)J_o e^{\theta_m} + (1+\theta'_m)J'_o e^{\theta'_m} = J_o e^{\theta_{oc}} + J'_o e^{\theta'_{oc}}.$$ (4.6)

For given $J_o, J'_o, \alpha, \alpha', J_L$ and T, one finds that (4.3) determines V_{oc}. By (4.6) this in turn determines V_m.

Given, then, V_m and V_{oc} the maximum power can be determined using (4.4) and (4.3), from

$$F_m \equiv \frac{P_m}{J_{sc} V_{oc}} = \frac{V_m}{V_{oc}} \frac{J_o \left(e^{\theta_m} - e^{\theta_{oc}} \right) + J'_o \left(e^{\theta'_m} - e^{\theta'_{oc}} \right)}{J_o \left(1 - e^{\theta_{oc}} \right) + J'_o \left(1 - e^{\theta'_{oc}} \right)}$$

$$= \frac{\theta_m}{\theta_{oc}} \frac{J_o \left(e^{\theta_{oc}} - e^{\theta_m} \right) + J'_o \left(e^{\theta'_{oc}} - e^{\theta'_m} \right)}{J_o \left(e^{\theta_{oc}} - 1 \right) + J'_o \left(e^{\theta'_{oc}} - 1 \right)}$$ (4.7)

This is the so-called "fill factor" which is a product of two factors, each smaller than unity. This expression is in the literature only for the case $J'_o = 0$, when it is[4]

$$F_m = \frac{\theta_m}{\theta_{oc}} \frac{e^{\theta_{oc}} - e^{\theta_m}}{e^{\theta_{oc}} - 1} .$$ (4.8)

As α is increased there is often a tendency for the maximum power output to decrease, particularly when recombination in the transition region mops up photo-excited carriers which are then not available for the generation of electricity. On the other

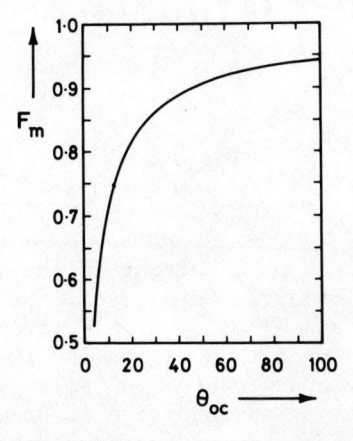

Fig. 4.1   The optimal fill factor F_m as a function of $eV_{oc}/\alpha kT$ for a solar cell satisfying equation (4.1) with $J'_o = o$ [4].

hand if V_{oc} is increased, and hence θ_{oc} is increased, F_m improves (Fig.4.1).

Typical values are

$$F_m = 0.82 \text{ for } \alpha=1, \text{ T=300K, } V_{oc} = 0.55V.$$

$$F_m = 0.72 \text{ for } \alpha=2, \text{ T=300K, } V_{oc} = 0.55V.$$

It has also been confirmed by direct experiment that life times tend to be greater, and recombination traffic to be smaller, for cells having lower α-values. Low α-values are therefore preferable[4].

We now take a closer look at a typical junction to show that recombination is in fact a crucial element in the operation of solar cells.

PART II

5. JUNCTION CURRENTS AS RECOMBINATION CURRENTS

Let the semiconductor pass current in the x-direction only so that a one-dimensional model is being used. Consider any narrow region lying between x and $x+\Delta x$. Then electron and hole current densities flow in and out of this region and there is also a recombination current density $eU\Delta x$ (where $e \equiv |e|$). The illustration fig. 5.1 of the charge conservation condition is drawn with the current directions appropriate for a p-n junction under forward bias. One sees that

$$J_h(x) = eU\Delta x + J_h(x+\Delta x)$$

$$J_e(x+\Delta x) = J_e(x) + eU\Delta x$$

whence

$$\frac{dJ_e(x)}{dx} = -\frac{dJ_h(x)}{dx} = eU. \tag{5.1}$$

The total current density at any plane x is

$$J(x) = J_e(x) + J_h(x) \tag{5.2}$$

With the notation $x = -a$, $x = c$ established in Fig. 5.1, integration of (5.1) yields

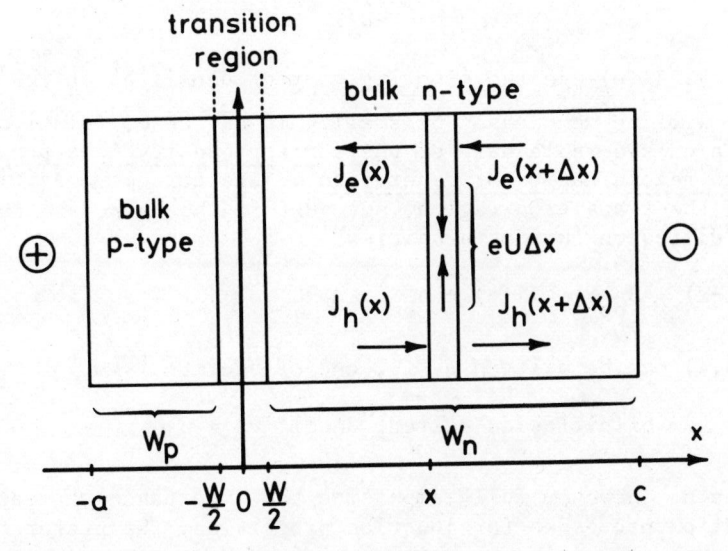

Fig. 5.1   Main x-coordinates of a p-n junction. Charge
conservation diagram for forward bias.

$$J_h(-a) - J_h(c) = e \int_{-a}^{c} U \, dx \qquad\qquad (5.3)$$

Let the region of transition from p- to n-type be as shown in Fig. 5.1 Then the last term in (5.3) can be split into recombination currents for the bulk p-type, the transition region and the bulk n-type. Putting (5.3) into (5.2) we have, for instance at x = -a,

$$J = J_e(-a) + J_h(c) + J_{bp} + J_{tr} + J_{bn} \qquad\qquad (5.4)$$

where

$$J_{bp} \equiv e \int_{-a}^{-W/2} U \, dx \;,\; J_{tr} \equiv e \int_{-W/2}^{W/2} U \, dx \;,\; J_{bn} \equiv e \int_{W/2}^{c} U dx \quad (5.5)$$

Now $J_e(-a)$, $J_h(c)$ are the minority current densities at the contacts and they will in many cases be small. In any case, relation (5.4) reveals a considerable part of the <u>current density in a p-n junction as due to recombination or generation</u>[6]. The current densities at the edges of the transition region, but just in the bulk, are the so-called "diffusion current densities".

$$J_e(-a) + J_{bp} \;,\; J_h(c) + J_{bn} \;.$$

Hence (5.4) can be written as ([1], end of section III.)

$$J = \text{sum of diffusion current densities} + J_{tr}.$$

This does not however fully emphasize the importance of recombination - generation processes for junction properties. We prefer to retain the term "recombination current densities" here.

The above considerations hold whenever the bulk regions and the transition region can be defined by (possibly voltage-dependent) planes. To proceed further, however, one must know, or assume, something about the variation of quasi-Fermi levels and electrical potential within the different regions. The accurate determination of these quantities, by the solution of Poisson's equation and the drift-diffusion-recombination equations, poses a complicated problem, leading to much computation. It is usual instead to base the theory on simple assumptions concerning the shapes of the quasi-Fermi levels and band edges, and this will be done here in section 7.

It should be pointed out, however, that general integrations of the basic equations can be done numerically. This then leads to

interesting results, which have in the case of metal-insulator-semiconductor solar cells been worked out[7]. For p-n junction see de
Mari's paper[8].

6. STEADY-STATE RECOMBINATION RATES AT A GIVEN PLANE X

Consider a non-degenerate semiconductor containing recombination
centres of one type, each of which can trap only one electron. We
shall consider the types of transition shown in Fig.6.1. If forward
and reverse processes were counted separately, there would be
eighteen processes. With the notation of the figure and by means of
the mass-action laws, the transition rates per unit volume may be
determined for any given plane x as follows:

Unavoidable processes (i.e. not involving traps) (*):

$$U(a) = B^S(np - n_i^2), \quad U(b) = B_1 n(np - n_i^2) ,$$

$$U(c) = B_2 p(np - n_i^2). \tag{6.1}$$

Electron-trapping processes:

$$U(d) = T_1^S(np_t - n_1 n_t), \quad U(e) = T_1 n(np_t - n_1 n_t),$$

$$U(f) = T_2 p(np_t - n_1 n_t). \tag{6.2}$$

Hole-trapping processes:

$$U(g) = T_2^S(pn_t - p_1 p_t), \quad U(h) = T_4 p(pn_t - p_1 p_t)$$

$$U(k) = T_3 n(pn_t - p_1 p_t). \tag{6.3}$$

Here n_t is the concentration of trapped electrons, and p_t is the
concentration of empty traps. If N is the concentration of traps,
then

$$N = n_t + p_t \tag{6.4}$$

n_1, p_1 are respectively the electron and hole concentrations when

---

* Band-band processes are unavoidable in the sense that they cannot
be eliminated by purification[8a].

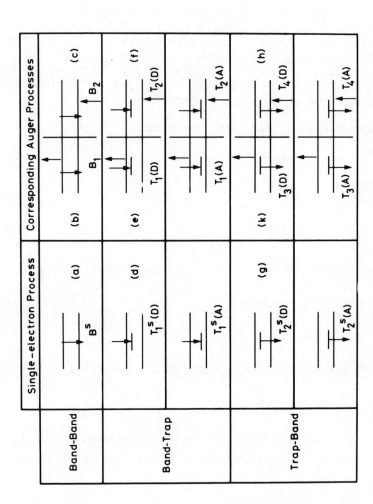

Fig. 6.1 Recombination processes considered here, showing reaction constants.

the Fermi level is at the trap level. Mass-action constants have
been introduced on the following basis: The superscript S indicates
a single-electron process, and its absence indicates a two-electron
process. The letter B is a reminder that only bands participate in
the process, while the letter T shows that traps are involved. The
evaluation of these mass-action constants will not be discussed.

The processes in Fig. 6.1 have been arranged to show how Auger
processes can compete with single-electron processes (for example
with single-electron radiative processes). The trapping processes
are analogous for donors and acceptors, which are shown separately.

Summation of the rates (6.1) yields, for the total unavoidable
recombination rate per unit volume,

$$F(np - n_i^2) \tag{6.5}$$

Similarly, the total electron-trapping rate and the hole-trapping
rate per unit volume are, by (6.4)

$$U_e = G\left[nN - (n + n_1)n_t\right] \tag{6.6}$$

and

$$U_h = H\left[(p + p_1)n_t - p_1N\right] \tag{6.7}$$

Here

$$\left.\begin{aligned}
F &\equiv B^S + B_1 n + B_2 p \\
G &\equiv T_1^S + T_1 n + T_2 p \\
H &\equiv T_2^S + T_3 n + T_4 p
\end{aligned}\right\} \tag{6.8}$$

As attention will be confined to the steady state, one must have
$U_e = U_h$ in order that the number of trapped electrons be constant.
Hence the steady-state concentration of trapped electrons, and the
steady-state recombination rate due to the traps are given by[6]

$$n_t = \frac{(Gn + Hp_1)N}{G(n + n_1) + H(p + p_1)} \tag{6.9}$$

$$U_t = \frac{(np - n_i^2)GHN}{G(n + n_1) + H(p + p_1)}$$
(6.10)

Equation (6.10) can be put into a form familiar to that of the simple Shockley-Read theory, by means of the definition

$$\tau_n^* \equiv 1/NG, \quad \tau_p^* \equiv 1/NH$$
(6.11)

$$U_t = \frac{(np - n_i^2)}{\tau_p^*(n + n_1) + \tau_n^*(p + p_1)}$$
(6.12)

The lifetimes τ_{no} and τ_{po} of the Shockley-Read theory are seen to be replaced by concentration-dependent quantities. Thus if the analysis of recombination data on the basis of a Shockley-Read model is found to yield a concentration-dependent τ_{no} or τ_{po}, this would be evidence in favour of the occurrence of one or more of the Auger processes of Fig.6.1 e,f,h,k.

Although τ_n^* and τ_p^* cannot always be regarded as true steady-state lifetimes

$$\tau_n \equiv \frac{\delta n}{U_t}, \quad \tau_p \equiv \frac{\delta p}{U_t}$$
(6.13)

they go over into such quantities under the following special conditions. If $p \gg n$, p_1, n_1, and if n_p is the equilibrium electron concentration in such a region, then the minority carrier lifetime is

$$\tau_n = \frac{n - n_p}{U_t} \doteq \frac{\tau_n^* p(n - n_p)}{np - n_i^2} \doteq \tau_n^*$$
(6.14)

A similar relation holds for τ_p. This justifies our notation for τ_n^* and τ_p^*.

Note that the total steady-state recombination rate per unit volume is

$$U = (np - n_i^2)\left[F + \frac{GHN}{G(n + n_1) + H(p + p_1)}\right]$$
(6.15)

This expression will be used throughout the rest of these lectures

There are many published mass-action treatments in which single-electron processes and various types of Auger processes have been treated separately. But strictly speaking these processes should be treated jointly as above.

The recombination rate per unit volume (6.15) applies to any plane x of the device. The x-dependence of the quantities n, p, F, G, H, n_1, p_1 must of course be known for the purposes of the recombination current densities (5.5). The calculation from first principles is complicated and can be done only numerically. Thus it is desirable to make simplifying assumptions.

7. JUNCTION MODEL AND SPACE-DEPENDENCES

As an example of a recombination current density calculation consider the transition region of a p-n junction, fig. 7.1. Band edges and quasi-Fermi levels are

$$E_c \equiv kT\eta_c \quad , \quad E_v \equiv kT\eta_v \quad , \quad F_n \equiv kT\gamma_n \quad , \quad F_p \equiv kT\gamma_p \quad , \qquad (7.1)$$

all quantities being functions of position (x). Furthermore, let

$$N_c \equiv 2\left(\frac{2\pi m_c kT}{h^2}\right)^{3/2} \quad , \quad N_v \equiv \left(\frac{m_v}{m_c}\right)^{3/2} N_c \qquad (7.2)$$

Then the volume density of electrons and holes in a non-degenerate semiconductor at uniform temperature are

$$n(x) = N_c \exp(\gamma_n - \eta_c) \quad , \quad p(x) = N_v \exp(\eta_v - \gamma_p) \qquad (7.3)$$

The intrinsic Fermi level $E_i(x) = kT\eta_i(x)$ is defined by $n(x) = p(x)$, so that

$$E_i(x) = \frac{1}{2}\{E_c(x) + E_v(x) + kT\ln(N_v/N_c)\} \qquad (7.4)$$

(This level lies half-way in the forbidden gap if $m_c = m_v$.) If $F_n = F_p = E_i$ the corresponding carrier concentration is from (7,3,4)

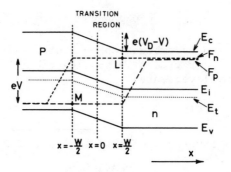

Fig. 7.1   Model of a p-n junction incorporating assumptions
concerning quasi-Fermi levels and electrostatic
potential. Forward voltage implies V > 0 and will
be developed by a solar cell under illumination. When
the forward voltage is due to an external voltage, the
left-hand side is connected to the positive terminal,
and the right-hand side to the negative, terminal of a
battery. E_t denotes a trapping level.

$$n_i(x) = N_c \exp\left[\gamma_i(x) - \eta_c(x)\right]$$

$$= N_c \exp\{\tfrac{1}{2}[\eta_c(x) + \eta_v(x)]\}\left(\frac{N_v}{N_c}\right)^{\frac{1}{2}} \exp(-\eta_c)$$

Hence for $\gamma_n(x) = \gamma_p(x) \left\{\equiv \gamma_i(x)\right\}$

$$n(x) = p(x) = (N_c N_v)^{\frac{1}{2}} \exp(-\eta_G/2) = \text{independent of } x$$

$$= \text{independent of voltage across junction } (\equiv n_i) \tag{7.5}$$

The concentrations $n(x)$, $p(x)$ can be expressed in terms of n_i using the fact that

$$\gamma_i(0) = \gamma_i(-\tfrac{W}{2}) - \tfrac{1}{2}(\eta_D - \eta) = \gamma_i(\tfrac{W}{2}) + \tfrac{1}{2}(\eta_D - \eta) \tag{7.6}$$

as is clear from the assumed potential and quasi-Fermi level variation proposed in Fig. 7.1. Hence for the transition region $-W/2 < x < W/2$

$$\frac{n(x)}{n_i} = \frac{\exp\left[\gamma_n - \eta_c(x)\right]}{\exp\left[\gamma_i(x) - \eta_c(x)\right]} = \exp\left[\gamma_n - \gamma_i(x)\right] \tag{7.7}$$

$$\frac{p(x)}{n_i} = \frac{\exp\left[\eta_v(x) - \gamma_p\right]}{\exp\left[\eta_v(x) - \gamma_i(x)\right]} = \exp\left[\gamma_i(x) - \gamma_p\right] \tag{7.8}$$

Thus from point L one sees that the electron concentration in the bulk n-type region has become constant for $x \geqslant W/2$ at a value n_n say, and from point M one sees that the hole concentration in the bulk p-type semiconductor has become constant for $x < -W/2$ at a value p_p say. By (7.7,8) these values are given by

$$\frac{n_n}{n_i} = e^{\gamma_n - \gamma_i(W/2)} \quad , \quad \frac{p_p}{n_i} = e^{\gamma_i(-W/2) - \gamma_p} \tag{7.9}$$

It follows from (7.6) (see Fig. 7.1) that

$$\gamma_i(\tfrac{W}{2}) + \gamma_i(-\tfrac{W}{2}) = 2\gamma_i(0) \quad ,$$

whence

$$\delta \equiv \frac{1}{2} \ln \frac{n_n}{p_p} = \frac{1}{2} (\gamma_n + \gamma_p) - \gamma_i(0) \quad \text{(in the transition} \qquad (7.10)$$
$$\text{region)}$$

The asymmetry of the doping of the junction is measured by δ in the sense that $\delta = 0$ for a symmetrically doped junction.

From Fig. 7.1 note that

$$\left. \begin{array}{l} \gamma_n - \frac{1}{2}\eta - \gamma_i(0) = \gamma_n - \frac{1}{2}(\gamma_n - \gamma_p) - \gamma_i(0) = \delta \text{ by } (7.10) \\[2mm] \gamma_p + \frac{1}{2}\eta - \gamma_i(0) = \gamma_p + \frac{1}{2}(\gamma_n - \gamma_p) - \gamma_i(0) = \delta \end{array} \right\} (7.11)$$

It follows from (7.9), (7.6) and (7.11) that

$$\frac{n_n}{n_i} = e^{\gamma_n - \gamma_i(W/2)} = e^{\gamma_n - \gamma_i(0) + \frac{1}{2}(\eta_D - \eta)} = e^{\frac{1}{2}\eta_D + \delta} \qquad (7.12)$$

$$\frac{p_p}{n_i} = \frac{p_p}{n_n} \frac{n_n}{n_i} = e^{-2\delta} \, e^{\frac{1}{2}\eta_D + \delta} = e^{\frac{1}{2}\eta_D - \delta} \qquad (7.13)$$

Again from Fig. 7.1 and (7.11)

$$\gamma_i(x) = \gamma_i(0) - \theta\frac{x}{W} \quad (\theta \equiv \eta_D - \eta) \qquad (7.14)$$

$$= \gamma_n - \frac{1}{2}\eta - \delta - \frac{\theta x}{W} = \gamma_p + \frac{1}{2}\eta - \delta - \frac{\theta x}{W} \qquad (7.15)$$

Use of (7.15) in (7.7,8) yields

$$n(x) = n_i \exp\left[\tfrac{1}{2}\eta + \delta + \theta x/W\right] \qquad\qquad (7.16)$$

$$p(x) = n_i \exp\left[\tfrac{1}{2}\eta - \delta - \theta x/W\right] \qquad\qquad (7.17)$$

$$n(x)p(x) - n_i^2 = n_i^2(\exp \eta - 1) \qquad\qquad (7.18)$$

$$n + p = 2n_i \exp(\tfrac{1}{2}\eta)\cosh(\delta + \theta x/W) \qquad\qquad (7.19)$$

$$n_1 + p_1 = 2n_i \cosh\left[\eta_t(x) - \gamma_i(x)\right] = \text{independent of } x \qquad (7.20)$$

The main assumptions made so far are: (a) Non-degenerate bands, (b) Quasi-Fermi levels constant inside the transition region and majority quasi-Fermi levels parallel to the band edges in the bulk regions, (c) Constant small field outside the transition region and constant field inside it.

The first part of (b) is used for example in regarding η as a constant in the transition region. The second part of (b) is used in regarding n_n and p_p as the valid n- and p- concentrations in the bulk regions. The second part of (c) is used in the linear change in band edges and in E_i in (7.14). The first part of (c) implies that the potential drop across the bulk regions is small compared with the applied voltage, so that V (Fig.7.1) is effectively the applied voltage in a normal diode or the light-induced voltage in a solar cell, or the voltage across the device. Fig. 7.2 gives slightly more realistic diagrams. In these lectures we do not take account of the voltage drops across the bulk regions, as the principles of operation are not affected by this simplification.

The requirement of minority carrier status in the bulk regions may be obtained from the following table which is obtainable from (7.16,17). One sees that the conditions $n(-W/2) < p_p$, $p(W/2) < n_n$ translate into a maximum voltage V as follows

$$V < V_D \mp \frac{kT}{e}\ln\frac{n_n}{p_p} \; .$$

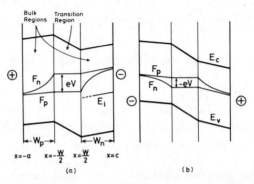

Fig. 7.2   More realistic distribution of potentials in a p-n
           junction. The slopes are exaggerated. (a) Forward
           bias; (b) Reverse bias.

Table 7.1 Concentrations at the edges of the transition region.

	$x = -W/2$	$x = +W/2$
$\ell n \dfrac{n(x)}{n_i}$	$\ell n \dfrac{n_p}{n_i} = \eta + \delta - \tfrac{1}{2}\eta_D$	$\ell n \dfrac{n_n}{n_i} = \delta + \tfrac{1}{2}\eta_D \qquad (7.11)$
$\ell n \dfrac{p(x)}{n_i}$	$\ell n \dfrac{p_p}{n_i} = -\delta + \tfrac{1}{2}\eta_D \qquad (7.12)$	$\ell n \dfrac{p_n}{n_i} = \eta - \delta - \tfrac{1}{2}\eta_D$

8. TRANSITION REGION RECOMBINATION CURRENT DENSITY

Using (6.15) in (5.5), the band–band recombination current density due to the transition region is

$$J_{tr}(bb) = e \int_{-W/2}^{W/2} (B^S + B_1 n + B_2 p)\, (np - n_i^2)\, dx \qquad (8.1)$$

$$= e n_i^2 (e^\eta - 1) \int_{-W/2}^{W/2} \left\{ B^S + n_i e^{\frac{1}{2}\eta} \left[B_1 e^{\delta + \theta x/W} + B_2 e^{-\delta - \theta x/W} \right] \right\} dx$$

$$= e n_i^2 W (e^\eta - 1) \left\{ B^S + \frac{2}{\theta} n_i e^{\frac{1}{2}\eta} \left[B_1 e^\delta + B_2 e^{-\delta} \right] \sinh \tfrac{1}{2}\theta \right\}$$

$$= e n_i^2 W (e^\eta - 1) \left\{ B^S + \frac{\exp\left[\tfrac{1}{2}(\eta_D - \eta)\right]}{(\eta_D - \eta)/2} \left[B_1 n_p + B_2 p_n \right] \sinh\left(\frac{\eta_D - \eta}{2}\right) \right\} \qquad (8.2)$$

$$= e n_i^2 W (e^\eta - 1) \left\{ B^S + \frac{\exp\left[\tfrac{1}{2}(\eta_D + \eta)\right]}{(\eta_D - \eta)/2} \left[B_1 n_{po} + B_2 p_{no} \right] \sinh\left(\frac{\eta_D - \eta}{2}\right) \right\} \qquad (8.3)$$

In (8.2) the concentrations n_p and p_n have been used in the sense of Table 7.1. In (8.3) the corresponding equilibrium concentrations have been used:

$$n_{po} \equiv n_i \exp(\delta - \tfrac{1}{2}\eta_D), \quad P_{no} \equiv n_i \exp(-\delta - \tfrac{1}{2}\eta_D),$$

whence by (7.16,17)

$$n(-W/2) = n_{po}\exp \eta, \quad p(W/2) = P_{no}\exp \eta,$$

(8.3) was given as eq.(28) in reference 6 who have also fig. 8.1.
For $\qquad V \leqslant V_D - 3kT/e$, i.e. $\eta_D - \eta \geqslant 3$,

$$\sinh\left[\tfrac{1}{2}(\eta_D - \eta)\right] \sim \tfrac{1}{2}\exp\left[\tfrac{1}{2}(\eta_D - \eta)\right],$$

so that only a weak η - dependence is left inside the braces in (8.2)
and (8.3). Thus $J_{tr}(bb)$ depends mainly as $(\exp \eta - 1)$ on the voltage
across the junction. Equation (8.3) is illustrated in fig. 8.1.

For Si at room temperature one would expect

$$B^S \sim 10^{-14} \text{ cm}^3 \text{ s}^{-1}, \quad B_1 + B_2 \sim 10^{-30} \text{ cm}^6 \text{ s}^{-1}, \quad n_i \sim 10^{10} \text{ cm}^{-3}$$

$$B^S/(B_1 + B_2)n_i \sim 10^6.$$

For narrower band gaps the ratio will be smaller and a value of 100
has been adopted in fig. 8.1. Even so, the Auger effect increases
the recombination current only slightly. For large enough $\eta_D - \eta$
as assumed in fig. 8.1 ($\eta_D \sim 20$ at room temperature), the actual
value of η_D need not be specified since one then has

$$J_{tr}(bb) = en_i^2 W(e^\eta - 1)\left\{B^S + \frac{2 e^{\eta_D}}{\eta_D - \eta}\left[B_1 n_{po} + B_2 P_{no}\right]\right\} \qquad (8.3a)$$

It is convenient to write (8.3) as

$$J_{tr}(bb) = A_{tr}(bb)(\exp \eta - 1), \qquad A_{tr}(bb) \equiv \frac{en_i W}{\tau_{tr}(bb)}, \qquad (8.5)$$

where

$$\frac{1}{\tau_{tr}(bb)} \equiv \left\{ \qquad \right\} n_i .$$

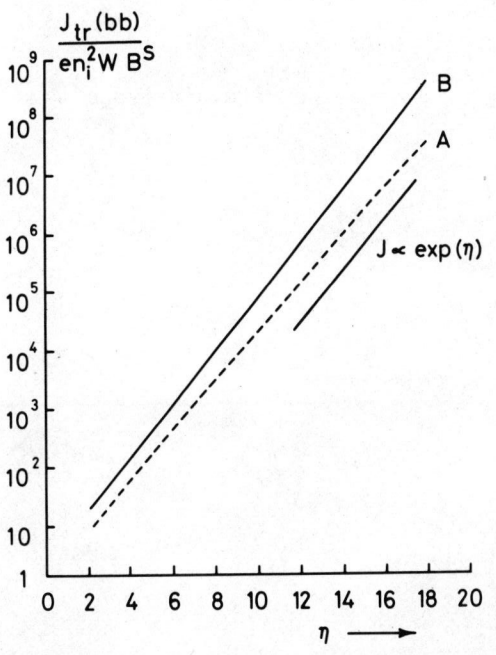

Fig. 8.1   The transition region band–band recombination current
density according to equation (8.3) with η_D = 20.

A: $B_1 = B_2 = o$ (no Auger effects)

B: $B^S\!/(B_1+B_2)n_i$ = 100

The recombination current via traps in the transition region will be considered next, and this requires the second term in (6.15). As there are now six recombination coefficients, let us suppose for simplicity that

$$T_1^S \sim T_2^S (\equiv T^S) \ , \ T_1 \sim T_2 \sim T_3 \sim T_4 (\equiv T) \tag{8.6}$$

whence, using (6.8)

$$G = H = T^S + T(n + p) \tag{8.7}$$

The steady-state recombination rate via traps is

$$U_t = n_i^2 (e^\eta - 1) \ N \int_{-W/2}^{W/2} \frac{T^S + T(n + p)}{n_1 + p_1 + n + p} \ dx$$

$$= n_i^2 (e^\eta - 1) \ NTWQ$$

where, using (7.19,20) and

$$\phi = \frac{n + p}{(n_1 + p_1)\cosh y} = 2n_i e^{\eta/2} / (n_1 + p_1) = e^{\eta/2} / \cosh[\eta_t(x) - \gamma_i(x)] \tag{8.8}$$

$$Q \equiv \frac{1}{W} \int \frac{\alpha + \phi \cosh y}{1 + \phi \cosh y} \ dx \qquad \left(\begin{array}{l} \alpha \equiv T^S / T(n_1 + p_1) \\ \\ y \equiv \delta + \theta x/W \end{array} \right) . \tag{8.9}$$

One finds that

$$Q = \frac{1}{\theta} \int_{\delta-\theta/2}^{\delta+\theta/2} \frac{\alpha + \phi \cosh y}{1 + \phi \cosh y} \ dy = 1 - \frac{1-\alpha}{\theta} \ I(\phi)$$

where

$$I(\phi) \equiv \int_{\delta-\theta/2}^{\delta+\theta/2} \frac{dy}{1 + \phi \cosh y} . \tag{8.10}$$

Thus this recombination current density is

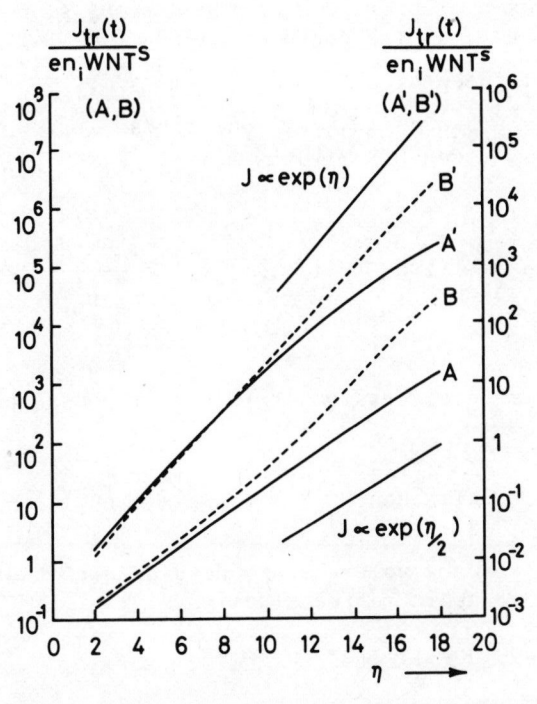

Fig. 8.2   The transition region recombination current density via
traps according to equation (8.11). Key to the curve
labels

	T = 0 (no Auger effects)	$T^S/T n_i =$ 1000
$\eta_t = \eta_i$	A	B
$\eta_t = \eta_i \pm 5.7$	A'	B'

It is also assumed that $\eta_D \sim 20$, $\delta = 0$ (symmetrical
doping). Note that the primed curves are separated
from the unprimed one by use of different scale.

$$J_{tr}(t) = en_i^2 \, NWT(e^{\eta} - 1) \left\{ 1 - \frac{1}{\eta_D - \eta} \left[1 - \frac{T^S}{T(n_1 + p_1)} \right] I(\phi) \right\} \quad (8.11)$$

The integral $I(\phi)$ can be evaluated (see Appendix), but numerical evaluation gives a clearer idea of the various slopes which can be exhibited by the $\ell n \, J_{tr}(t)$ dependence on η (see fig. 8.2). Assuming for a reasonably deep trap $T^S \sim 10^{-12} \, cm^3 \, s^{-1}$, $T \sim 10^{-25} \, cm^6 \, s^{-1}$, $n_i \sim 10^{10} \, cm^{-3}$ one finds $T^S/Tn_i \sim 10^3$ as a typical value. In analogy with (8.5), it is convenient to put (8.11) in the form

$$J_{tr}(t) = A_{tr}(t)(\exp \eta - 1) \,, \quad A_{tr}(t) \equiv \frac{en_i W}{\tau_{tr}(t)} \,, \quad \frac{1}{\tau_{tr}(t)} \equiv \left\{ \begin{array}{c} \\ \end{array} \right\} n_i NT.$$

$$(8.12)$$

An approximate discussion of (8.11) is possible. Normally one may expect

$$|E_t - E_i| \gtrsim 2kT \quad \text{whence} \quad \phi \sim 2 \exp\left[\frac{\eta}{2} - |\eta_t(x) - \gamma_i(x)| \right]$$

Thus if higher forward voltages are developed so that $\phi \gtrsim 2$ and if single-electron trapping effects dominate

$$J_{tr}(t) \propto W(\exp \eta - 1) \, I(\phi)/(\eta_D - \eta)$$

$$\sim W \exp \eta \, \exp(- \frac{\eta}{2})$$

$$\sim \exp(\eta/2) \,.$$

where weak η-dependences have been successively neglected[+] This last expression is the result pointed out in [1], see here curve A of fig. 8.2. The other extreme case is a variation as $(\exp \eta - 1)$, which is indistinguishable in form from the ideal diode characteristic which is expected if no recombination current arises from the transition region. This ideal characteristic is reproduced if Auger effects dominate:

$$J_{tr}(t) \propto \left[1 - I(\phi)/(\eta_D - \eta) \right] \left[\exp \eta - 1 \right] W$$

By (A.6) the first factor can go to zero for small ϕ but for $\phi \gtrsim 2$ it lies between 2/3 and unity, so that the dependence is as

---

[+] A voltage dependent ideality factor has in effect been derived here.

$(\exp \eta - 1)$. The same dependence is found if single-electron trapping occurs (and, possibly, trap Auger effects as well), but then only if the voltage developed is small enough

$$\eta < 2|n_t(x) - \gamma_i(x)| \quad .$$

In that case $I(\phi) \sim \theta = \eta_D - \eta$

$$J_{tr}(t) \propto W(\exp \eta - 1) \, T^S/T(n_1 + p_1)$$

$$\sim \exp \eta - 1 \quad .$$

A fuller study of transition region recombination in solar cells must take into account surface recombination, possible doping profiles, the absorption coefficient and the nature of the incident spectrum. This can complicate the theory very considerably, but has been done by various authors. Nearest in spirit of the present exposition is the paper on p-n junction solar cells by Mallinson and Landsberg[9]. See also Choo's paper on p-n junctions[10].

9. THE BULK-REGIONS RECOMBINATION CURRENT DENSITY

To the assumptions noted in section 7 we must now add an assumption (d) concerning the domination of majority carriers in the bulk regions:

$$p \simeq p_p \gg n, \, n_1, \, p_1 \quad (x \leqslant -W/2) \tag{9.1}$$

$$n \simeq n_n \gg p, \, n_1, \, p_1 \quad (x \geqslant W/2) \tag{9.2}$$

This means that the trap energy level must not be too far from the intrinsic level. For example $p/n_1 \gg 1$ implies that for $x < -W/2$, using (7.7), (7.8),

$$1 \ll \frac{N_v}{N_c} \exp(\eta_v + \eta_c - \gamma_p - \eta_t) = \exp(2\gamma_i - \gamma_p - \eta_t) \quad ,$$

i.e. $E_i - F_p > E_t - E_i$. \tag{9.3}

Under these conditions one finds in the p-type region

$$\frac{np - n_i^2}{n - n_p} \sim \frac{np_p - n_p p_p}{n - n_p} \sim p_p \quad . \tag{9.4}$$

It follows that there exists a minority life time τ_n defined by

$$U = (n-n_{po})/\tau_n \tag{9.5}$$

i.e. $\dfrac{1}{\tau_n} = \left\{ B^S + B_2 P_p + \dfrac{(T_1^S + T_2 P_p)(T_2^S + T_4 P_p)N}{(T_1^S + T_2 P_p)n_1 + (T_2^S + T_4 P_p)(P_p + P_1)} \right\} P_p$,

$$\tag{9.6}$$

where (9.4), (6.8) and (6.10) have been used. One finds approximately

$$\frac{1}{\tau_n} = B^S P_p + B_2 P_p^2 + [T_1^S + T_2 P_p]N \tag{9.7}$$

and similarly in the n-type region the minority carrier lifetime is given by

$$\frac{1}{\tau_p} = B^S n_n + B_1 n_n^2 + [T_2^S + T_4 n_n]N \tag{9.8}$$

If ε_p is the (small and approximately constant) electric field in the p-type region, the minority current density is

$$J_e(x) = eD_e\left(\frac{dn}{dx} + E_p n\right) \qquad \left(E_p \equiv \frac{e\varepsilon_p}{kT}\right) \tag{9.9}$$

From (5.1)

$$dJ_e/dx = eU = e(n-n_{po})/\tau_n$$

and (9.9) yields

$$\frac{d^2 n}{dx^2} + E_p \frac{dn}{dx} - \frac{n-n_{po}}{L_n^2} = 0 \qquad (L_n^2 \equiv D_e \tau_n) \tag{9.10}$$

In introducing E_p, and later E_n, we use them to illustrate some of the effects of a voltage drop across the bulk regions. However, as E_p, E_n will be assumed small we do not depart too much from our assumption that these drops shall be neglected.

The solution of (9.10) is conveniently written as

$$n(x) - n_{po} = A \exp(z/L_1) + C \exp(-z/L_2) \ , \ z \equiv x + W/2 \quad (9.11)$$

One finds that with

$$\Lambda_n \equiv L_n/(1 + E_p^2 L_n^2/4)^{\frac{1}{2}} \qquad\qquad\qquad\qquad (9.12)$$

$$\frac{1}{\Lambda_n} = \frac{1}{L_1} + \frac{E_p}{2} = \frac{1}{L_2} - \frac{E_p}{2} \qquad\qquad\qquad\qquad (9.13)$$

From (8.4) one can put as a first boundary condition

$$n = n_{po} \exp \eta \qquad \text{at} \qquad x = -W/2$$

Assuming surface recombination velocity to be fast enough to reduce n to n_{po} at $x = -a$, we can put as a second boundary condition

$$n = n_{po} \qquad \text{at} \qquad x = -a.$$

One then finds

$$A = \frac{\exp(W_p/\Lambda_n)}{2 \sinh(W_p/\Lambda_n)} \, n_{po}(\exp \eta - 1)$$

$$C = -\frac{\exp(-W_p/\Lambda_n)}{2 \sinh(W_p/\Lambda_n)} \, n_{po}(\exp \eta - 1)$$

where W_p (fig.5.1) is the width of the bulk region: $W_p = a - W/2$. This completes the identification of solution (9.11).

We now insert (9.5) and (9.11) into (5.5) to find the current density due to recombination in the bulk p-type region:

$$J_{bp} = \frac{en_{po}(\exp \eta - 1)L_n^2}{\tau_n \Lambda_n} \left\{ \coth\left(\frac{W_p}{\Lambda_n}\right) - \frac{\exp\left[\frac{W_p E_p}{2}\right]}{\sinh(W_p/\Lambda_n)} + \frac{E_p \Lambda_n}{2} \right\} \quad (9.14)$$

For small E_p one finds

$$J_{bp} = \frac{en_{po}(\exp \eta - 1)L_n^2}{\tau_n \Lambda_n} \, \frac{\cosh(W_p/\Lambda_n) - 1}{\sinh(W_p/\Lambda_n)} \qquad\qquad (9.15)$$

By a similar argument one also finds

$$J_{bn} = \frac{ep_{no}(\exp \eta - 1)L_p^2}{\tau_p \Lambda_p} \left\{ \coth \frac{W_n}{\Lambda_p} - \frac{\exp\left(\frac{W_n E_n}{2}\right)}{\sinh(W_n/\Lambda_p)} + \frac{E_n \Lambda_p}{2} \right\} \quad (9.16)$$

where, in analogy with (9.12),

$$\Lambda_p \equiv L_p / (1 + E_n^2 L_p^2 / 4)^{\frac{1}{2}} . \quad (9.17)$$

If $E_p \sim E_n \sim 0$ one sees that $\Lambda_p \sim L_p$, $\Lambda_n \sim L_n$. Thus one sees that the bulk region recombination current depends on the voltage developed across the device according to

$$J_{bp} + J_{bn} \propto (\exp \eta - 1) \quad (9.18)$$

It is convenient to put (9.14,16) in the form

$$J_{bp} = A_{bp}(\exp \eta - 1), \quad A_{bp} \equiv \frac{en_{po}L_n}{\tau_n'} \quad (9.19)$$

$$J_{bn} = A_{bn}(\exp \eta - 1), \quad A_{bn} \equiv \frac{ep_{no}L_p}{\tau_p'} \quad (9.20)$$

where the minority carrier life times τ_n , τ_p are now modified to τ_n', τ_p' given by

$$\frac{1}{\tau_n'} = \frac{L_n}{\tau_n \Lambda_n} \left\{ \quad \right\} , \qquad \frac{1}{\tau_p'} = \frac{L_p}{\tau_p \Lambda_p} \left\{ \quad \right\} \quad (9.21)$$

There the expressions in braces in (9.14,16) have not been written out again.

For a fuller treatment of bulk recombination in solar cells one needs to introduce the electron-hole pair generation rate due to the absorption of radiation in (9.10). As our main interest is here in recombination effects, however, this term was omitted for the sake of simplicity. The effect of its inclusion, together with the other items noted at the end of section 8, in a spirit closely related to that of the present exposition was described in [10a] Among other attacks on these problems we can note only a small selection from the very large literature:

Analytical approaches to p-n junction theory: Ritner[11].
Optimisation for concentrating cells: Crook and Yeargan[12].
Computer models: Dunbar and Hauser[13].
M.I.S. cells: D.C. Olsen[14], Klimpke and Landsberg[15], Pulfrey[16].
An additional entry to the literature can be obtained from the
proceedings of the recent Photovoltaic Specialist Conferences (No. 13
at Washington 1978 and No.14 at San Diego 1980) and review articles
by Bucher[17], Rothwarf and Böer[18], and Wilson et al[19].

10. SUMMARY OF P-N JUNCTION CURRENT DENSITIES FROM SECTIONS 8 AND 9

In the above treatment the light current has not been calculated
at the same level of microscopic detail as the other currents. In
fact, it has been justified merely by the argument leading to (3.9).
On the other hand, by assuming the potential distribution we have
been able to calculate the transition region and bulk recombination
current densities. The absorption of light should, however, be taken
into account. If one does so, the bulk region differential equation
(9.10) is altered as a carrier generation term

$$\frac{\eta \alpha \phi_o}{D_e} e^{-\alpha(x + a)}$$

appears on the left-hand side. Here α is the absorption coefficient,
η the quantum efficiency (number of electron-hole pairs produced per
absorbed quantum) and ϕ_o is the flux density of radiation falling
on the left-hand face of the cell. This leads to additional complications if the field E_p is neglected, and makes a treatment of the
transition region current rather tedious. This is actually often
neglected. However, our interest is here in recombination effects,
and so it seemed of interest to follow the path of sections 5 to 9.

Collecting together our results, the total current (3.4) or
(4.1) or (5.4) is by (8.5,12) and (9.19,20)

$$J(V) = J_e(-a,V)+J_h(c,V)+ \left[A_{tr}(bb)+A_{tr}(t)+A_{bp}+A_{bn}\right]\left[\exp \eta -1\right] -J_L(V)$$

$$(10.1)$$

Hence, as in section 3,

$$J_{sc} = -J_L(0) + J_e(-a,0) + J_h(c,0) \qquad (10.2)$$

Also assuming

$$J_L(V_{oc}) \sim J_L(0), \qquad\qquad\qquad (10.3)$$

$$\exp\frac{|e|V_{oc}}{kT} = \frac{J_e(-a,0)-J_e(-a,V_{oc})+J_h(c,0)-J_h(c,V_{oc})-J_{sc}}{A_{tr}(bb)+A_{tr}(t)+A_{bp}+A_{bn}} + 1 \quad (10.4)$$

Thus for given short-circuit currents each recombination path limits the open-circuit voltage, as has been discussed most recently by Godlewski et al[20].

.One can put (10.1) into a different form by using (10.3) and (10.4) to eliminate $J_L(V)$ in favour of J_{sc} to find

$$J(V) = J_{sc}+\left[A_{tr}(bb)+A_{tr}(t)+A_{bp}+A_{bn}\right]\left[\exp\eta -1\right] + B(V)$$

The last term gives the minority current at the contacts at potential V reduced by its value at short-circuit (V = 0) in the presence of light and it is reasonable to suppose that this quantity has the voltage dependence shown:

$$B(V) \equiv J_e(-a,V)-J_e(-a,0)+J_h(c,V)-J_h(c,0) = A_{cont}(\exp\eta -1).$$

One then recovers the more familiar expressions of the form (4.1) to (4.3):

$$J(V) = J_o(\exp\eta - 1) - \left[J_L(0) - J_e(-a,0) - J_h(c,0)\right] \qquad (10.5)$$

where

$$J_o = A_{tr}(bb) + A_{tr}(t) + A_{bp} + A_{bn} + A_{cont} \qquad\qquad (10.6)$$

Note that the light current and the short-circuit current differ numerically by the minority contact currents. Thus even though $J_L(V)$ may well be broadly independent of the voltage developed, the voltage dependence of the minority contact currents has the effect that the dark current density $J_o(\exp\eta - 1)$ with applied voltage V does not simply differ by a constant from $J(V)$:

$$J_{dark}(V) - J(V) = J_L(V) + \underbrace{\left[J_e(-a,V) + J_h(c,V)\right]}_{dark} - \underbrace{\left[\begin{matrix}same\\terms\end{matrix}\right]}_{light}$$

It is these additional V-dependent terms which can clearly cause a break-down of the "shift theorem" which says

$$J_{dark}(V) - J(V) = \text{independent of } V. \tag{10.8}$$

This result was already touched on in connection with equation (3.9).

PART III

11. CONFIGURATION AND ELECTROSTATICS OF THE SCHOTTKY BARRIER SOLAR CELL

As an alternative configuration in which to study recombination effects, consider a metal-semiconductor junction solar cell. For generality it will be supposed that an insulating layer is inserted as shown in Fig.11.1 and 11.2, making this an M.I.S. cell. Such cells can be of class n or of class p type, as shown. This corresponds to the use of an n-type or p-type semiconductor respectively. Light can fall onto the cell from the left, and the structure is best approached from the right, where the electric field $E_s(x)$ is

smallest and indeed zero in a first approximation. The plane at which it is first practically zero has been denoted by w. In an ideal junction (which extends to $x = \infty$) one would define w by

$$E_s(x) = 0 \quad \text{for} \quad x \geqslant w, \tag{11.1}$$

but the physics cannot sustain such rigour.

We now study the electrostatics of this contact. If $\phi_s(x)$, $\rho_s(x)$, ε_s be respectively the electrostatic potential, the space charge density and the dielectric constant, in the semiconductor, then Poisson's equation

$$(-\phi_s)'' = \frac{4\pi\rho_s}{\varepsilon_s} \tag{11.2}$$

yields, on integrating from x to w and treating ρ_s as a constant,

$$E_s(x) = (-\phi_s)' = -\frac{4\pi}{\varepsilon_s}\rho_s(w - x) \tag{11.3}$$

The potential across the semiconductor depletion region, marked $V_b - V_s$ in the figures, is

$$-\phi_s(w) - [-\phi_s(0)] = \int_0^w E_s(x)dx = -\frac{2\pi}{\varepsilon_s}\rho_s w^2. \tag{11.4}$$

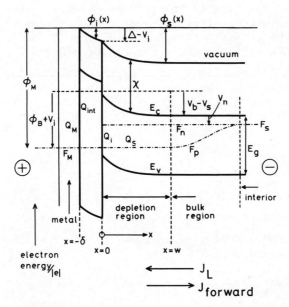

Fig. 11.1 A MIS, class n, Schottky barrier solar cell. The
           polarity shown corresponds to an externally applied
           forward potential.

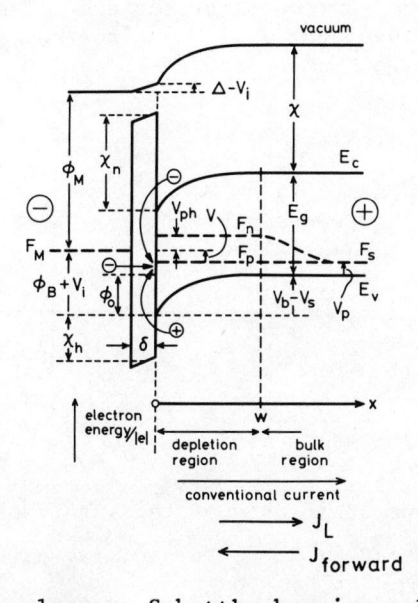

Fig. 11.2   A MIS, class p, Schottky barrier solar cell. The
            polarity shown corresponds to an externally applied
            forward potential.

The space charge per unit area in the depletion layer is

$$Q_s = w\rho_s = \sqrt{-\frac{\varepsilon_s\{-\phi_s(w) - [-\phi_s(0)]\}}{2\pi\rho_s}} \; \rho_s \tag{11.5}$$

One can check from Table 11.1 that no imaginary quantities occur. The sign of the quantities (11.3) and (11.4) confirm the band profile in Fig. 11.1 and 11.2. The upward bend in Fig. 11.1 is due to the positive space charge; the downward slope in Fig. 11.2 is due to the negative space charge.

Our next move is to calculate the charge density Q_i in the states at $x = 0$ on the semiconductor surface. The suffix i stands for "interface" or "insulator". In the interfacial layer of an assumed constant charge density

$$[-\phi_i(x)]'' = \frac{4\pi}{\varepsilon_i}\,\rho_i,$$

so that, on integrating from negative x to $x = 0$,

$$E_i(0) - E_i(x) = -\frac{4\pi}{\varepsilon_i}\,\rho_i x,$$

i.e.

$$\varepsilon_i E_i(0) = \varepsilon_i E_i(x) - 4\pi\rho_i x \; . \tag{11.6}$$

If Q_i is the charge per unit area of the interface, the boundary condition for the displacement vectors D^* is, using (11.3,5,6),

$$4\pi Q_i = D_s^* - D_i^* = \varepsilon_s E_s(0) - \varepsilon_i E_i(0) = -4\pi Q_s - \varepsilon_i E_i(x) + 4\pi\rho_i x \; .$$

Hence

$$E_i(x) = \frac{4\pi}{\varepsilon_i}\left[\rho_i x - Q_s - Q_i\right] \tag{11.7}$$

Lastly, integrating across the interfacial layer from $x = -\delta$ to $x = 0$,

$$-\phi_i(0) - \left[-\phi_i(-\delta)\right] = -\frac{4\pi\delta}{\varepsilon_i}\left[\frac{\rho_i\delta}{2} + Q_s + Q_i\right] \tag{11.8}$$

This drop has been labelled $\Delta - V_i$ in the figure. One finds [23]

$$Q_i = -\frac{\varepsilon_i}{4\pi\delta}(\Delta - V_i) - \frac{\rho_i\delta}{2} - Q_s \tag{11.9}$$

The analysis given applies whether or not there is an insulating layer, and can be used to study the effect of it on the electrostatic regime. Consider first the n-type case. Photogeneration of carriers causes the electron Fermi level to rise in the semiconductor, and more acceptor-type surface states are able to capture electrons from the semiconductor. The semiconductor space charge density $\rho_s(x)$, now no longer considered to be x-independent, is increased. (This assumes that the surface states communicate better with the semiconductor than with the metal.) By the appropriately generalised form of (11.4) this increases the band bending. For simplicity we shall assume that there are no donor-type surface states. Now from (11.4) and Table 11.1 the numerical value of the potential drop across the semiconductor increases, thus favouring a bigger open-circuit voltage V_{oc}.

For a p-type semiconductor we assume that the surface states are all donors. Photogeneration lowers the hole Fermi level and electrons are released into the semiconductor. This increases the negative space charge there and hence band-bending. (This effect was not considered in Part ii where the effect of surface states in the junction was neglected). The positive drop (11.4) increases, favouring a bigger V_{oc}.

The summary of the signs in Table 11.1 shows that the fields E_s and E_i and the potential drops are in the same sense for insulating layer and semiconductor for any one cell. One can infer from these considerations:

Table 11.1 The signs of various quantities.

Column number	1	2	3	4	5	6	7	8
	ρ_s	Q_i	Q_s (11.5)	E_s (11.3)	E_i (11.7)	$V_b - V_s$ (11.4)	$\Delta - V_i$ (11.8)	Quantity under √ in (11.5)
Class n cell	$\sim e N_D$ +	Acceptors −	+	−	−	−	−	+
Class p cell	$\sim e N_A$ −	Donors +	−	+	+	+	+	+

The equation number used is given in brackets.

Band-bending and hence open-circuit voltage in Schottky barrier
solar cells are favoured by an insulating layer with acceptor/
donor surface states which are in better communication with the
n-type/p-type semiconductor than they are with the metal.

The symbols V_s and V_i used in columns 6 and 7 will be defined
to be zero in thermal equilibrium (no incident radiation). The
equilibrium potential drops are then Δ for the insulating layer,
for which V_b is the band-bending parameter. The sign conventions
can be read off Figs. 11.1 and 11.2. They include, for example,
that the voltage V developed has the value

$$(-\phi)_{metal} - (-\phi)_{\substack{semiconductor \\ bulk}} = F_s - F_M$$

One sees that

$$\phi_B + V_i = V_b - V_s + V + \begin{cases} V_n \\ V_p \end{cases} \quad \text{[general V]} \qquad (11.10)$$

and $\qquad \phi_B = V_b + \begin{cases} V_n \\ V_p \end{cases} \qquad \left[\begin{array}{c} \text{thermal equilibrium} \\ V = 0 \end{array}\right] \quad (11.11)$

$$V = V_i + V_s \qquad \left[\text{taking } (11.10)-(11.11)\right] (11.12)$$

The voltage developed by the cell thus consists by (11.12) of an
insulator and a semiconductor component, as is clear also from
the figures.

12. THE PLACE OF RECOMBINATION EFFECTS IN (P-TYPE) SCHOTTKY BARRIER
SOLAR CELLS

We must revert to our theme of recombination process in solar
cells, but now with special reference to Schottky barrier solar
cells. For simplicity attention will be confined to a p-type semiconductor. We know from Fig. 3.3(a) that the electrostatic field
acts to the right, and from Fig. 3.1 that the light-induced
current flows to the right. The photogenerated holes move therefore
into the semiconductor, while the photogenerated electrons tend to
accumulate near the plane x = 0. This causes a rise of the electron
quasi-Fermi level F_h above the equilibrium Fermi level, which is
essentially the Fermi level F_M in the metal; this is practically
the same in equilibrium and under radiation. This rise has been
called the photovoltage[15]

$$V_{ph} = F_n(0) - F_M,$$

and Fig. 13.2 shows that the usual implicit assumption that it
vanishes is not correct. (It was neglected in Part II since smooth
variation of potentials across the p-n junction was there assumed).
This accumulation of photogenerated electrons can be dissipated by
the processes described in Table 12.1 which involve also holes in
the valence band near the plane x = 0.   In addition one must consider
the   recombination in the depletion layer.   This may be done by
using (6.15), together with expressions for the spatial dependences
of hole and electron concentrations, corresponding to (7.3,16,17)([22]):

$$n(x) = N_c \exp \left\{ \left[\frac{eN_A}{2\epsilon_s} (w - x)^2 + V_p - E_g + V_{ph} + V \right] \frac{e}{kT} \right\} \quad (12.1)$$

$$p(x) = N_v \exp \left\{ -\left[\frac{eN_A}{2\epsilon_s} (w - x)^2 + V_p \right] \frac{e}{kT} \right\} \quad\quad (12.2)$$

where $0 \leqslant x \leqslant w$.   This looks after processes 1 of Table 12.1 in the
depletion layer. Analogous processes in the bulk semiconductor are
less important as they are driven by the difference in the quasi-
Fermi levels and this difference goes rapidly to zero for $x > w$ in
the model of Figs. 11.1 and 11.2.

The processes   2 to 4 can be incorporated into an electron
and a hole interfacial recombination term U_e^{ifr}, U_h^{ifr} of the type
(6.15), but with the band-band terms F equal to zero.   It is of the
generalised Shockley-Read form (6.10).   However N is replaced by
the number of interfacial surface states per unit area in an energy
range E_t, $E_t + dE_t$, to be denoted by $D_s dE_t$, where $D_s \sim 10^{15} m^{-2} eV^{-1}$.
This value of D_s should be compared with a (1,1,1) single crystal
cleavage plane of lattice constant a in a diamond structure.   The
number of lattice sites is $4/\sqrt{3}a^2$.   The maximum density of interface
sites is with a = 0.542nm(Si) distributed over the energy gap is

$$4/\sqrt{3}a^2 E_g \sim 7.1 \times 10^{18} m^{-2} eV^{-1} \quad .$$

There is also an additional generalisation arising from the
tunnelling in processes 3 and 4.   These can be specified by a
tunnelling time constant τ, and a metal Fermi occupation probability
of a quantum state

Table 12.1 Six mechanisms for the dissipation of photogenerated
carriers accumulated near x = 0 in the conduction band of a p-type
semiconductor.

Number of process	
1	Band-band recombination in the depletion layer (bad) (a) direct ($B^S \sim 10^{-17} m^3 s^{-1}$) (b) via deep traps ($T_1^S \sim 4.4 \times 10^{-16} m^3 s^{-1}$; $T_2^S \sim 3 \times 10^{-13} m^3 s^{-1}$) Dissipation via interfacial states $(T_1^S \sim 4.4 \times 10^{-15} m^3 s^{-1}$, $T_2^S \sim 3 \times 10^{-14} m^3 s^{-1})$
2	(a) Through band-band recombination (bad)
3	(b) electrons from the conduction band pass into the metal (good) via interfacial states
4	(c) metal electrons pass into the semiconductor valence band (bad) via interfacial states Thermionic emission
5	(a) of electrons from semiconductor conduction band into the metal (good)
6	(b) of metal electrons into the semiconductor valence band (there are not enough metal electrons opposite the semiconductor conduction band to enter it) (bad).

$$f_M \equiv \frac{1}{\exp\left[\dfrac{E_t - F_M}{kT}\right] + 1}$$

(12.3)

One finds

$$U_e^{ifr} = I \times \left[(np - n_i^2)H - (n + n_1) \frac{f_M}{\tau} - \frac{n}{\tau} \right] G$$

$$U_h^{ifr} = I \times \left[(np - n_i^2)G + (p + p_1) \frac{f_M}{\tau} - \frac{p_1}{\tau} \right] H$$

(12.4)

where

$$I = \int_{F_M - \phi_B}^{F_M - \phi_B + E_g} \frac{D_s \, dE_t}{G(n+n_1) + H(p+p_1) + \frac{1}{\tau}}$$

Here n, p, G, H, n_1, p_1, are all evaluated at $x = 0$. In the limit $\tau \to \infty$ one recovers (6.10) provided $D_s dE_t \to dN$.

The processes 5 and 6 may be dealt with by using the results of Lundstrom and Svensson[21], as we did[15,22]. The thermionic emission is facilitated by passing via a defect in the insulating layer, and this has been investigated[23a].

The above is a description of the theory of the Schottky barrier solar cell, rather than a detailed exposition of it. Before we discuss typical results, we note the approximations still involved. They are:

(i) A uniform surface state density D_s (i.e. D_s is independent of energy E_t).

(ii) The depletion approximation, with the assumed Fermi level shown in Figs. 11.1 and 11.2.

(iii) The slope of the quasi-Fermi levels is neglected for the majority carriers in the bulk semiconductor and for both types of carriers in the depletion layer.

All of these can be removed. By dropping (i) one has the complication of integrals over functions of E_t as in (12.4) except that D_s would also be a function of E_t. One can develop computational procedures which enable one to jettison (ii) and (iii), but these are fairly complicated[7,23]

Table 13.1 Numerical values

$$D_s \sim 10^{11} cm^{-2} eV^{-1} = 10^{15} m^{-2} eV^{-1}$$

$$N_A = 10^{16} cm^{-3}$$

$$V_p = 0.18 eV$$

$$T = 300 K$$

$$E_g = 1.1 eV$$

For B^s, T_1^s, T_2^s see Table 12.1

13. RECOMBINATION CURRENTS AND VOLTAGE DROPS IN (P-TYPE) SCHOTTKY BARRIER SOLAR CELLS

It is desirable to consider now some of the results which are obtained from the model outlined in sections 11 and 12 using the numerical values given in Table 13.1.

Consider first the substrate and interfacial hole recombination current densities as a function of output voltage. For low D_s processes 2,3,4 of Table 12.1 are relatively unimportant. For a reasonably thick interfacial layer ($\delta \sim 25 A^{\circ} \sim 2.5 nm$) processes 5 and 6 are also unimportant because of the large tunnelling barrier. Therefore the electron quasi-Fermi level F_n rises and $V_{ph} > 0$. This encourages considerable recombination in the substrate (bulk and depletion layer) by process 1. Because of the low value of D_s and as the hole concentration near $x = 0$ is small, process 2 is rather weak.

A strong internal field, causing holes to be swept from the region $x = 0$, while electrons are bunched up there, is expected at short-circuit. This leads to minimal recombination mainly from process 1 (Fig. 13.1), maximal V_{ph} (Fig. 13.2), and most of the photogenerated electrons are carried by process 5. Minimal recombination can also be recognised from the maximal output current J in Fig. 13.3.

Now let us consider the variation of these recombination processes with output voltage. As the output voltage is increased, the internal field becomes weaker, holes can increasingly be found near $x = 0$, and the recombination rate via process 1 increases. In fact, it can rise to a maximum (Fig. 13.1) and later, as more holes become available near $x = 0$, process 2 (via interfacial states) is liable to come on reasonably strongly.

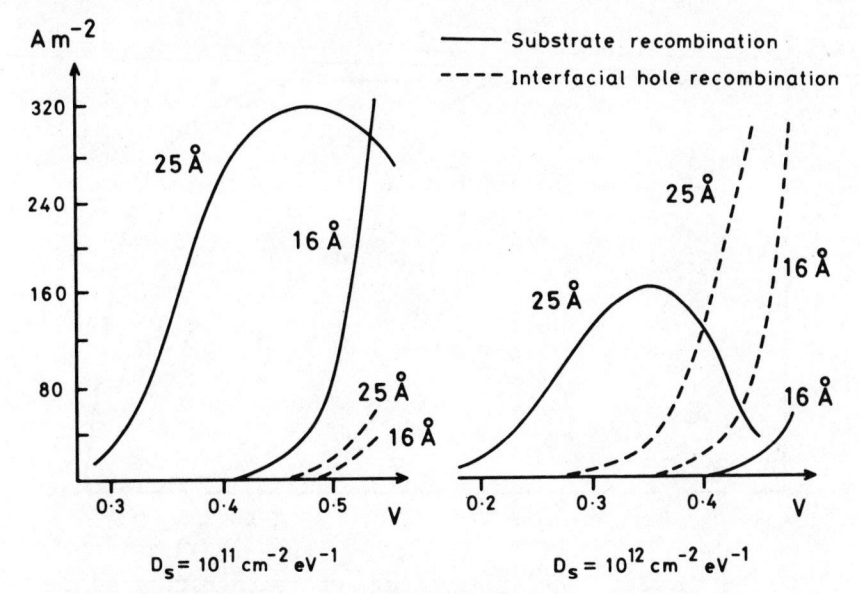

Fig. 13.1   The substrate (full lines) and interfacial hole
            (dashed lines) recombination current densities as a
            function of output voltage V for two interfacial
            layer thicknesses δ and two interfacial surface
            state densities D_S . The parameters of Table 13.1
            have been used.

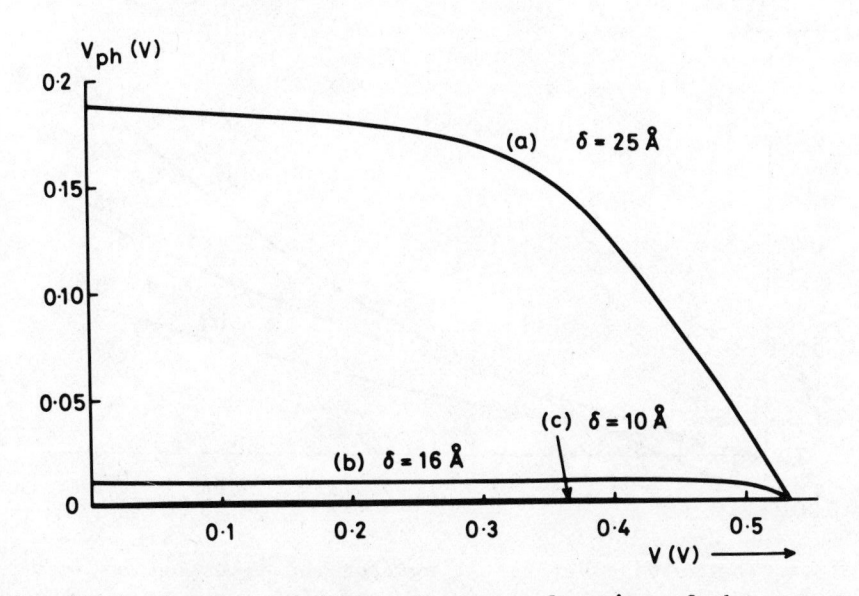

Fig. 13.2   The photo-voltage V_{ph} as a function of the output
            voltage V for various interfacial layer thicknesses δ.
            The parameters of Table 13.1 have been used.

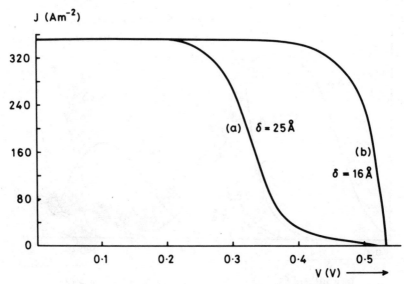

Fig. 13.3   The current density - voltage characteristics of the p-
type Schottky barrier solar cell for various interfacial layer
thicknesses δ. The parameters of Table 13.1 have been used.

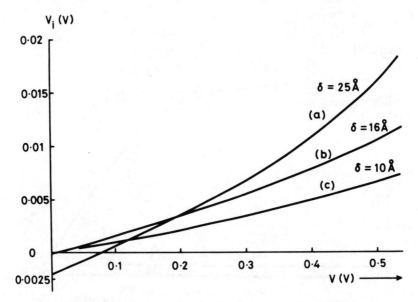

Fig. 13.4   The voltage V_i, resulting from the illumination, developed across the insulating interfacial layer as a function
of the output voltage V, for various interfacial layer thicknesses δ. The parameters of Table 13.1 have been used.

It is clear that as D_s is increased, interfacial state recombination (process 2) will push substrate recombination (process 1) into second place. For large thicknesses $\delta(\sim 25\text{A}^{\text{o}})$ of the insulating layer, thermionic emission into the metal (process 5) will also play a part. However, for thinner insulator layers the electron hopping between the metal and the semiconductor via the interfacial states (processes 3 and 4) must also play a significant part.

Note that if process 1 recombination is radiative, then photons are produced which can ionise traps again. The non-radiative recombination is therefore more damaging in permanently withdrawing carriers from the light-induced current. The loss of efficiency due to non-radiative processes is even more obvious in light-emitting diodes.

One can summarise this part of the discussion as follows:- In a p-type cell low D_S and large δ lead to the accumulation of photoelectrons near $x = 0$, and trap non-radiative recombinations in the bulk semiconductor is a dominant recombination mechanism as a result. From this extreme case other situations with larger D_S or smaller δ can also be understood.

We shall now discuss the voltage drop across the interfacial layer.

Consider again a p-type Schottky barrier solar cell with reasonably low density of interfacial surface states $(D_s \sim 10^{11}\text{cm}^{-2}\text{eV}^{-1})$ and a reasonably thick insulating layer $(\delta \sim 25\text{A}^{\text{o}})$. Starting in the short-circuit condition consider what happens to the potential drops $\Delta - V_i$ across the insulating layer, and to the band bending potential $V_b - V_s$. We note equation (11.2), $V = V_i + V_s$, and that in thermodynamic equilibrium $V_i = V_s = 0$.

Under illumination and at short-circuit V is again zero so that V_i and V_s must be of opposite sign. It will be shown that at short-circuit $V_s > 0$ and hence $V_i < 0$. The photogeneration of carriers raises the electron Fermi level at $x = 0$ above its equilibrium value, so that the photo voltage V_{ph} is positive. Some donor surface states previously positively charged are now neutralised. The electrons come from the semiconductor, whose space charge and band-bending are therefore reduced relative to equilibrium. Thus V_s is positive,

V_i is negative (Fig. 13.4). At short-circuit the potential drop across the insulator is thus increased by the illumination, a little noted interesting effect[15].

Reducing the load current allows a voltage V to be developed. The hole quasi-Fermi level drops by V below the metal Fermi level and the interfacial Fermi levels are pulled downwards. This now increases the space charge in the semiconductor and also band bending. So V_s is reduced and becomes negative for large enough V, while V_i becomes positive.

For a thinner insulating layer the communication between metal and semiconductor improves and there is less accumulation of photo-generated electrons near x = 0. The photovoltage at short-circuit is less and all the effects just discussed are reduced. The (V_i,V)-curve therefore becomes shallower. The V_i values are also reduced if the surface state density is reduced.

We next make a remark about the optimal thickness of the insulating layer. For large values of $\delta(\sim 30A^{\circ})$ we noted a reasonably large photovoltage. But there is little chance of the desirable processes 3 and 5 occurring which require a somewhat thinner layer. However, on reducing δ to as little as $10A^{\circ}$ the photovoltage is reduced practically to zero because of the better communication between semiconductor and metal. The Fermi level of some of the interfacial states will now be below F_M. The desirable process 3 is therefore replaced by the deleterious process 4 which reduces the efficiency. One can therefore expect an optimum thickness δ.

Indeed the degradation of cells by corrosion due to water vapour has been attributed to the growth of the insulating layer or the development of an additional layer[24]. Fig. 13.5 gives experi- mental points marked by crosses on current-voltage curves for p- type Schottky barrier solar cells[25]. They are compared with the theory described in these lectures, using a series resistance of 3.2 Ω(in an equivalent circuit) for a cell area of $1cm^2$, and adopting $D_s = 6 \times 10^{15}m^{-2}eV^{-1}$. One sees the severe depression in the current density due to loss of photogenerated carriers by recomb- ination in the cell.

To summarise: An optimally thick insulating layer is thick enough to allow only a negligible electron current from the metal to the valence band of the semiconductor; but it is not so thick as to produce a wide separation between the two carrier quasi-Fermi levels and hence deleterious substrate recombination.

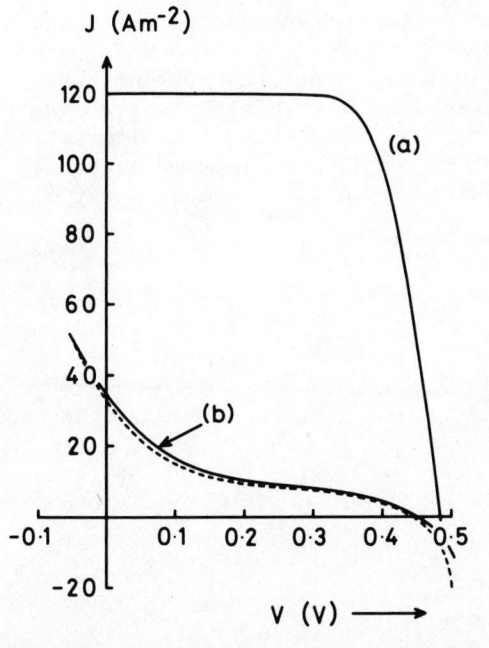

Fig. 13.5 P-type Schottky barrier solar cell (a) before
(b) after degradations. --; Experimental curve.
Curves are theoretical:

(a) δ = 13A; $\chi_n = \chi_h$ (electron and hole tunnelling
barrier heights) $\sim$ 0.5 eV; (b) δ = 29A, χ_e = 0.7 eV,
χ_h = 0.5 eV.

14. CONCLUSION

The recombination coefficients are affected by band structure, by type of defect etc., and the theory should therefore be able to investigate the effect on solar cell performance of a wide range of values for each of the coefficients occurring in (6.8). Although this has been done, there is no space to discuss these matters here. Similarly one wants to know if the notion of a region within a solar cell of constant and parallel quasi-Fermi levels is a good enough approximation. Again this matter has been looked at[7,23], but is beyond the present scope and we must refer to the literature. One can learn also from investigations of devices other than solar cells. One finds for example that for high current densities, such as occur in power rectifiers, the Auger effect is a dominant recombination mechanism, since it grows as n^2p or p^2n with carrier concentration[26]. Its importance, though undoubted in some contexts, has in some aspects of solar cells been the subject of controversy[27].

A FEW MORE GENERAL TOPICS

(I) Thermodynamic efficiency

The Carnot efficiency for ambient temperature T and pump temperature T_p is at $1-3000/600 \sim 95 \%$ too high to be realistic. A somewhat lower efficiency can be estimated as follows [28].

A system receives energy and entropy from a pump and emits energy and entropy into surroundings at steady rates $\dot{E}_p$, $\dot{S}_p$, $\dot{E}_s$, $\dot{S}_s$. It also emits heat and performs work at steady rates $\dot{Q}$, $\dot{W}$. If the rate of irreversible entropy generation is $\dot{S}_g$ then the rate of charge of energy and entropy of the system at temperature T is

$$\dot{E} = \dot{E}_p - \dot{E}_s - \dot{Q} - \dot{W} = 0$$

$$T\dot{S} = T\dot{S}_p - T\dot{S}_s - \dot{Q} + T\dot{S}_g = 0$$

Eliminating $\dot{Q}$, the efficiency of working is

$$\eta \equiv \frac{\dot{W}}{\dot{E}_p} = \eta^* - \frac{T\dot{S}_g}{\dot{E}_p} \leqslant \eta^*$$

where

$$\eta^* = 1 - T\frac{\dot{S}_p}{\dot{E}_p} - \left(1 - \frac{T\dot{S}_s}{\dot{E}_s}\right)\frac{\dot{E}_s}{\dot{E}_p}$$

We first check that this formula for a thermodynamically maximal temperature yields the Carnot efficiency. As heat transfer is included explicitly in the formulae, we simply remove the sink ($\dot{E}_s = \dot{S}_s = 0$) and associate a temperature $T_p \equiv \dot{E}_p/\dot{S}_p$ with the pump. Then, indeed

$$\eta^* = 1 - T/T_p$$

Next we take black-body radiation to be the pump, and regard black-body radiation as emitted. Then we may put

$$\dot{S}_i = \frac{4}{3} AT_i^3 , \quad \dot{E}_i = AT_i^4 , \quad \dot{S}_i/\dot{E}_i = 4/3 \, T_i$$

with i = s or p. We find

$$\eta^* = 1 - \frac{4}{3} \frac{T}{T_p} - (1 - \frac{4}{3} \frac{T}{T_s}) (\frac{T_s}{T_p})^4 .$$

As system temperature T and ambient temperature T_s are in this case nearly the same, we find

$$\eta^* = 1 - \frac{4}{3} \frac{T}{T_p} + \frac{1}{3} (\frac{T}{T_p})^4 .$$

In equilibrium (T_p = T) η^* = 0. If T = 0, η^* = 1. Both limits are shared by the Carnot efficiency, but (between these limits) the new efficiency lies lower. It is still unrealistically high, however.

A further reduction may be achieved by assuming radiation to pass in sequence through n ideal solar cells of decreasing band gaps so that longer wavelengths are absorbed later. At the same time each cell emits radiation to keep in a steady state and this is absorbed by the two neighbouring cells. This radiation becomes important as n → ∞, since for many cells the photons absorbed by the radiation passing through the cells becomes smaller in amount. The optimised gaps have been calculated [29], and the overall efficiency has been calculated for various values of n and for concentration C = 1 of one sun and maximum concentration to reproduce the solar temperature C_m.   The results are given below. It is seen that the theoretical maximum efficiency

n	1	2	3	4	∞
$\eta^*(C=1)(\%)$	30	42	49	53	68.2
$\eta^*(C=C_m)(\%)$	40	55	63	68	86.8

reduced further and thus becomes more realistic. Considerations leading to the same result exist (see J. Loferski's lectures).

This is hopeful as showing the possible benefit of "tandem cells".

(II) Simple theory to see that an optimum energy gap exists

Suppose a two-level device is illuminated by a distribution of $n_\nu(E)$ dE photons in energy range $(E, E + \Delta E)$. Expressing n_ν as a function of $y \equiv E/kT_\nu$ for simplicity,

$$\eta = \frac{\text{energy used}}{\text{incident energy}} = \frac{E_G \int_x^\infty n_\nu(y)\,dy}{kT_\nu \int_0^\infty y n_\nu(y)\,dy} = \text{const} \cdot x \int_x^\infty n_\nu(y)\,dy$$

where $x \equiv E_G/kT_\nu$. There is a maximum since $\eta = 0$ at both $x = 0$ and $x = \infty$. To determine it, put $dy/dx = 0$ (with T_ν fixed) whence

$$\int_{x_{opt}}^\infty n_\nu(y)\,dy = x_{opt}\, n_\nu(x_{opt})$$

For each distribution of radiation this yields the energy gap, assuming all photons are absorbed and E_G of their energy is used whenever $h\nu \geqslant E_G$. For black body radiation in particular, $x_{opt} = 2.17$. This particular case yields

$$E_G = 2.17 \times \frac{6000}{300} \times \frac{1}{40} \text{ eV} \sim 1.1 \text{ eV}$$

and $\eta = 44\ \%$ as optimal results. Other losses and a more accurate theory yield a larger energy gap (~ 1.5 eV, corresponding to GaAS) and a lower maximum efficiency ($\sim 25\ \%$).

(III) Is dollars per peak watt a good unit ?

Let solar energy be incident at intensity ϕ (power per unit area). Then a cell of area A and efficiency η will yield power

$$p = \eta\phi A \ .$$

Suppose the cell costs K dollars or k = K/A dollars per unit area.
Then the price fo the cell in dollars per peak watt is

$$x = \frac{K\phi}{p} = \frac{K}{A\eta} = \frac{k}{\eta}$$

where ϕ is taken as kWm^{-2}. However, if price and power output
are linear in ϕ, then ϕ actually cancels out of the expression.
We see that the price

dollars per peak watt = cost per unit area/η .

The right-hand side has clearly the unit "dollars per m^2" which
is unambiguous [30]. The left-hand side is not manifestly in these
units, and this is a drawback, as it can be very misleading.
The "dollar per peak watt" is best consigned to the history books,
where it will find appropriate partners such as the British Thermal Unit.

(IV) Energy unit for global use

Very large powers of 10 occur in the wide-spread discussions of
global energy schemes. An example is the Exa Joule (10^{18} J). As
this number is larger than the number of seconds since the big
bang ($\sim 5 \times 10^{17}$), its size is difficult to appreciate, and a
more intuitive unit is desirable.
Annual energy consumption is dimensionally power, and a good unit
of power is a 60 W continuously burning light bulb. (One could
make it a 100 W bulb, but that is a slightly less modest unit).
Furthermore it makes sense to express the power per capita of the
world population, and one then arrives, typically at Table 1.
Food consumption by an individual could be a starting point at
300 kcal day^{-1} or 3 kWh day^{-1} or 120 W, which is two light bulbs.
This has to be multiplied by 16.5 to allow for the individual's
share (on average) in roads, factories, cars, etc. Table 1 shows
at once the low average efficiency of photosynthesis, so that
there is scope for the development of high energy-yielding crops.

To see the proposed unit in action, we ask to what extent the deserts of the world – a somewhat fuzzy concept, but see Table 2a – could supply world energy requirements by direct conversion of solar radiation (to electricity, say). This is not a practical idea at present, but it is nonetheless important to have answers to such questions. Assuming rather poor devices of 1 % efficiency and an insolation at one tenth of the solar constant, one finds that the deserts would give an adequate yield for some decades to come (Table 2b). Although there is in fact a great deal in hand at present, this advantage is being eroded (by the increasing population coupled by the increasing living standards) as time goes on. The numbers for the energy consumption are reasonably small and, it is hoped that they carry an intuitive meaning.

Table 1. Important energy flows (earth averages)

	60 W bulbs p.c.	$J \ yr^{-1}$
1975 (World population 4×10^9)		
Food consumption	2	1.5×10^{19}
Energy consumption	33	2.5×10^{20}
Photosynthesis yield	410	3.1×10^{21}
Solar energy incident on earth	396,000	$3 \ \times 10^{24}$
2000 A.D.(World population 6×10^9)		
Energy consumption	51	5.8×10^{20}
2020 A.D.(World population 7×10^9)		
Energy consumption	76	1.0×10^{21}

Table 2. Power from the deserts ?

(a) Areas on the Earth (10^7 km^2)

Antarctica (in the south) 1.4		Arctica (in the north) 0.95 - 1.23 depending on whether adjoining seas are included
Total land area including Antarctica	(29 % of total)	15
Total sea area	(71 % of total)	36
Total surface area		51
Desert areas	(17 % of land, 5 % of total surface)	2.5

(b) Possible power yield from deserts p.c.

Assumed : Mean insolation 135 W m^{-2}; Conversion efficiency 1 %

Year	Population	Yield (60 W bulbs p.c.)	Requirements from Table 1 (60 W bulbs p.c.)
1975	4×10^9	141	33
2000	6×10^9	94	51
2020	7×10^9	81	76

(V) <u>When will solar conversion be economically viable</u> ?

Economic projections will perhaps be made by other lectures.
Here we note that fuel price inflation over and above a general
inflation rate is an argument for investment in solar devices [31]
which is sometimes neglected.
Turning to history the time gap from scientific verification of
an effect to its economic exploitation is of interest, and details
are given below.

Electricity	: Faraday (1831) to Siemens (1856) is		25 years
Steam engine	: Newcomen (1712) to 1785	is	73 years
	Watt (1765)		20
Fission	: 1942 to 1965	is	23 years
Solar cells	: 1954 to 1990	is	36 years

On this basis we can perhaps expert a wide use of economic solar
cells by the year 1990.

Appendix

An integral occurring in §8

The transition region recombination current density depends on the voltage developed across the junction through a factor $(\exp \eta - 1)$, but the integral

$$I(\phi,\delta,\theta) \equiv \int_{\delta-\theta/2}^{\delta+\theta/2} \frac{dy}{1 + \phi\cosh y} \qquad (A.1)$$

also enters with

$$\delta \equiv \ln\sqrt{n_n/p_p} \,, \quad \theta = \eta_D - \eta \,, \quad \phi^{-1}\exp(\tfrac{1}{2}\eta) = \cosh\left[\eta_t(x)-\gamma_i(x)\right] \,.$$

To discuss this integral, let $z = \exp y$ whence $dy = dz/z$ and

$$I = \frac{2}{\phi} \int_{\ell}^{u} \frac{dz}{(z + \phi^{-1})^2 + 1 - \phi^{-2}}$$

$$\left.\begin{array}{l} \ell \equiv \exp(\delta - \theta/2) = \left(\dfrac{n_n}{p_p}\right)^{\frac{1}{2}} \exp\left(\dfrac{\eta - \eta_D}{2}\right) \\[3mm] u \equiv \exp(\delta + \theta/2) = \left(\dfrac{n_n}{p_p}\right)^{\frac{1}{2}} \exp\left(\dfrac{\eta_D - \eta}{2}\right) \end{array}\right\} \qquad (A.2)$$

A number of cases arise:

Case 1. $\phi = 1$.

$$I = \tfrac{1}{2}\sqrt{\frac{p_p}{n_n}} \frac{\sinh\left[\tfrac{1}{2}(\eta_D - \eta)\right]}{1 + \cosh\left[\tfrac{1}{2}(\eta_D - \eta)\right]} \qquad (A.3)$$

Case 2. $\phi < 1$.

$$I = \frac{1}{\sqrt{1-\phi^2}} \int_{\ell}^{u} \left[\frac{1}{z + \phi^{-1} - \sqrt{\phi^{-2} - 1}} - \frac{1}{z + \phi^{-1} + \sqrt{\phi^{-2} - 1}}\right] dz$$

$$= \frac{1}{\sqrt{1 - \phi^2}} \ln \frac{\phi z + 1 - \sqrt{1 - \phi^2}}{\phi z + 1 + \sqrt{1 - \phi^2}} \Bigg|_{\ell}^{u}$$

$$= \frac{1}{\sqrt{1 - \phi^2}} \ln \left\{ \frac{u\phi + 1 - \sqrt{1 - \phi^2}}{u\phi + 1 + \sqrt{1 - \phi^2}} \cdot \frac{\ell\phi + 1 + \sqrt{1 - \phi^2}}{\ell\phi + 1 - \sqrt{1 - \phi^2}} \right\} \quad (A.4)$$

The variation of I with η is small under all conditions of cases 1 and 2, so that the voltage dependence of the recombination current is given by $(\exp \eta - 1)$.

Case 3. $\phi > 1.$ (This is the most important case).

$$I = \frac{1}{\sqrt{\phi^2 - 1}} \cos^{-1} \left\{ \frac{\phi + \cosh y}{1 + \phi \cosh y} \right\} \Bigg|_{y = \delta - \theta/2}^{y = \delta + \theta/2} \sim \phi^{-1} \quad (A.5)$$

An approximation

Since $\cosh y \geqslant 1$,

$$I < \int_{\delta-\theta/2}^{\delta+\theta/2} \frac{dy}{1 + \phi} = \frac{\theta}{1 + \phi}$$

whence for $\theta = \eta_D - \eta > 0$

$$1 > 1 - \frac{I}{\theta} > \frac{\phi}{1 + \phi} \, . \quad (A.6)$$

References

[1] Sah C.T., Noyce R.N. and Shockley W. Carrier generation and recombination in p-n junctions and p-n junction characteristics. Proc. I.R.E. 45, 9 (1957).
[2] Lindholm, F.A. Fossum J.G. and Burgess E.L., Basic corrections to predictions of solar cell performance required by non-linearities, 12th I.E.E.E. Photovoltaic Specialist Conference, Baton Rouge, p.33 (1976).
[3] Landsberg, P.T. An introduction to the theory of photovoltaic cells, Solid-State Electronics 18 1043 (1975).

[4] J. Lindmayer. Theoretical and practical fill factors in
        solar cells. Comsat Tech. Rev. 2 105 (1972)

[5] P.T. Landsberg. Principles of solar cell operation, U.K. –
        I.S.E.S. one day technical conference C21 on
        Photovoltaics Solar Energy Conversion held at the
        Royal Society in September 1979 by the U.K. section
        of I.S.E.S.

[6] D.A. Evans and P.T. Landsberg. Recombination statistics for
        Auger effects with applications to p-n junctions,
        Solid-State Electronics 6 169 (1963).

[7] C.M.H. Klimpke and P.T. Landsberg. An improved analysis of
        n-type Schottky barrier solar cells, 2nd European
        Photovoltaic Solar Energy Conference, Berlin, April,
        1979 (Dordrecht : Reidel 1979), p.678.

[8] A. de Mari. An accurate numerical steady-state one-dimensional
        solution of the p-n junction Solid-St. Electr.
        11 33 (1968).

[8a] P.T. Landsberg. Proc. Inst. Elect. Engrs. A review of
        recombination mechanisms in semiconductors B106,
        908 (1959).

[9] J.R. Mallinson and P.T. Landsberg Transition region recom-
        bination in solar cells (1st European) Photo-
        voltaic Solar Energy Conference, Luxemburg, Sept.,
        1977 (Dordrecht: D. Reidel, 1978), p.1027.

[10] S.C. Choo. Carrier generation-recombination in the space
        charge region of an asymmetrical p-n junction
        Solid-St.. Electr. 11 1069 (1968).

[10a] J.R. Mallinson and P.T. Landsberg. Meteorological effects on
        solar cells Proc. Roy . Soc. A355, 115 (1977).

[11] E.S. Rittner. Improved theory of silicon p-n junction solar
        cell J. of Energy 1 9 (1977).

[12] D.L. Crook and J.R. Yeargan. Optimisation of silicon solar
        cell design for use under concentrated sunlight
        I.E.E.E. Trans. E.D. 24 330 (1977)

[13] P.M. Dunbar and J.R. Hauser. Theoretical effects of surface
        diffused region lifetime models on silicon solar
        cells. Solid-St.. Electr. 20 697 (1977).

[14] L.C. Olsen. Model calculations for MIS solar cells Solid-
        St. Electr. 20 74 (1977).

[15] C.M.H. Klimpke and P.T. Landsberg. Recombination effects
        in p-type silicon S.B.S.C. Solid-St. Electr. 21,
        1539 (1978).

([16]) D.L. Pulfrey. MIS Solar cells : A. review I.E.E.E. Trans. ED 25, 1308 (1978).

([17]) E. Bucher. Solar cell materials and their basic parameters Appl. Phys. 17, 1 (1978).

([18]) A. Rothwarf and K.W. Böer. Direct conversion of solar energy through photovoltaic cells, Progr. in Solid-state Chemistry 4, 1 (1975).

([19]) J.I.B. Wilson, J. Mc Gill and D. Weaire. Recent Progress in thin-film solar cells Adv. in Physics 27 , 365 (1978).

([20]) M.P. Godlewski, T.M. Klucher, G.A. Mazaris and V.G. Weizer. Open-circuit voltage improvements in low resistivity solar cells 14th I.E.E.E. Photovoltaic Specialist Conference, San Diego, January 1980.

([21]) I. Lundstrom and C. Svensson. Tunneling to traps in insulators, J. App. Phys. 43, 5045 (1972).

([22]) P.T. Landsberg and C.M. Klimpke. Surface recombination effects in an improved theory of a p-type MIS solar cell Solid-State Electronics (to be published 1980).

([23]) C.M.H. Klimpke and P.T. Landsberg. An improved analysis of Schottky Barrier solar cells, submitted to Solid-State Electronics.

([23a]) A.H.M. Kipperman, S.C.M. Backerra, H.J. Maaskamp and R.J.C. van Zollingen. On the trapping-assisted-tunneling through the insulating layer of MIS-solar cells, E.E.C. Photovoltaic Solar Energy Conference, Luxemburg, Sept. 1977, p. 961.

([24]) C.M.H. Klimpke and P.T. Lansberg. Illumination and temperature dependence of p-type solar cells 14th I.E.E.E. Photovoltaic Specialist Conference, San Diego, January 1980.

([25]) D.R. Lillington and W.G. Townsend, 1978. Private Communication.

([26]) N.G. Nilsson. The influence of Auger recombination on the forward characteristic of semiconductor power rectifiers at high current densities, Solid-State Electronics 16, 681 (1973).

([27]) M.A. Shibib, F.A. Lindholm and J.G. Fossum. Auger recombination in heavily doped shallow-emitter silicon p-n-junction solar cells, diodes and transistors I.E.E.E. Trans. ED 26, 1104 (1979).

(^{28}) P. Landsberg.   Photochemistry and Photobiology 26, 313 (1977)

(^{29}) A. de Vos.   J. Phys. D13, 839 (1980).

(^{30}) D.J. Toop, P.T. Landsberg.   Solar energy 22, 571 (1979).

(^{31}) P.T. Lansberg and D.J. Toop.   IEE Conference Publication
          N° 171, London, Januari 1979, pp. 433.

SCHOTTKY BARRIER SOLAR CELLS

W.G. Townsend

Royal Military College of Science
Shrivenham
Swindon Wilts UK

1.   INTRODUCTION

Photovoltaic solar cells based on homojunctions in silicon and heterojunctions in GaAs and CdS have shown that useful amounts of power can be generated on earth using incident solar energy. Efficiencies approaching 20% have been demonstrated for silicon single crystal homojunction cells and up to 23% using heterojunctions in GaAs and concentrated sunlight. However, if current goals are to be met for the reduction in the cost per peak watt of generated power, alternative technologies have to be considered that necessitate using both cheaper substrates and cheaper and less energy demanding production methods.

One of the significant possibilities for large scale cost reduction lies in getting away from cells having p-n junctions and replacing them by simply deposited metal-semiconductor junctions. This form of contact known as a Schottky Barrier has been widely used for many years to make other types of semiconductor device. Its attraction for photovoltaic solar cells lies in the fact that it should lead to cheaper processing since thin film metal deposition techniques are highly developed and have already shown themselves capable of fast high yield processing. Since the process usually involves depositing the metal onto a low temperature substrate, this requires a lower energy input than forming a p-n junction and also avoids the degradation of semiconductor minority carrier properties that occurs with high temperature processing. This in turn suggests that barriers formed on thin film and non-single crystal substrates should lead to useful efficiencies; this in fact has been shown to be the case for both silicon and GaAs. High temperature processes of junction formation on the other hand

suffer from the diffusion of impurities down grain boundaries
leading to uncertainties regarding device performance.

In practice it has been found that intimate contacts formed
between metal films and crystalline semiconductors exhibit poor
photovoltaic response. This is caused by the fact that the
thermionic emission dark current at the Schottky barrier leads to
significantly higher dark currents than is normally encountered in
a homojunction or heterojunction structure. This problem, however,
can be got round while still preserving the potential advantages
of the Schottky barrier by allowing a very thin oxide or insulating
layer to be formed between the semiconductor and the metal contact.
The introduction of this layer leads to the common form of Metal-
Insulator-Semiconductor Schottky Barrier Solar Cell (MIS SBSC) with
which we are primarily concerned here.

In what follows the operation of two types of MIS cell will be
studied - those depending primarily on thermonically emitted Majority
carriers and Minority carrier tunnelling through the thin insulating
layer.   The theory of operation of these cells will be outlined
and the necessary parameters for efficient operation studied.

The experimental work done primarily on silicon will then be
reviewed from the point of view of both single and semicrystalline
substrates. Finally an assessment will be given of the long term
prospects for MIS cells in both Si and GaAs paying some attention
to the problem of the stability of the cell characteristics and
their likely lifetime.

2.   THE SCHOTTKY BARRIER CELL PRINCIPLE

The electrical characteristics of the Schottky barrier are
well known and have been considerably studied and reviewed[1-4].
Their application to photovoltaic conversion is a more recent study
but there is now an extensive literature that is rapidly expanding.
See references[5-9] for example. A review by Pulfrey[10] in 1978 is
well referenced and gives a good starting point for further study.

In the Schottky barrier solar cell light energy is transmitted
into a semiconductor substrate through an extremely thin semitransparent metal layer*. The metal and semiconductor are chosen
so that the difference in their work functions causes a depletion
region and hence an electric field to be produced below the semiconductor surface. Those photons absorbed within the semiconductor
produce electron holes pairs which are separated under the action
of the electric field and flow to opposite sides of the junction.
The separated charge tends to forward bias the junction causing a

----

*This is the usual configuration although in some cells back
illumination can take place through the semiconductor.

photo-electric current to flow in an external load connected between the metal and semiconductor.

SBSCs may be divided into two categories: those in which the semiconductor is n-type and those in which it is p-type. Both types are considered here although the discussion is limited to the semiconductor silicon.

The usual physical configuration of an SBSC is shown in Fig 1. Although the SBSC is strictly a metal-semiconductor contact it is nearly always fabricated with a thin insulating interfacial layer typically 2 nm thick situated at the metal-semiconductor interface to give an MIS SBSC. The properties of this interfacial layer are such that although it is essentially transparent to electrons it is able to withstand potential across it. It will be shown later how this may be used to advantage in the MIS SBSC to increase the open circuit voltage and hence the efficiency of the device. Since the interfacial layer is often produced by the thermal oxidation of the semiconductor during the fabrication process, the terms interfacial layer and interfacial oxide will be used synonymously here.

The satisfactory operation of the SBSC relies largely upon efficient light transmission into the semiconductor substrate.

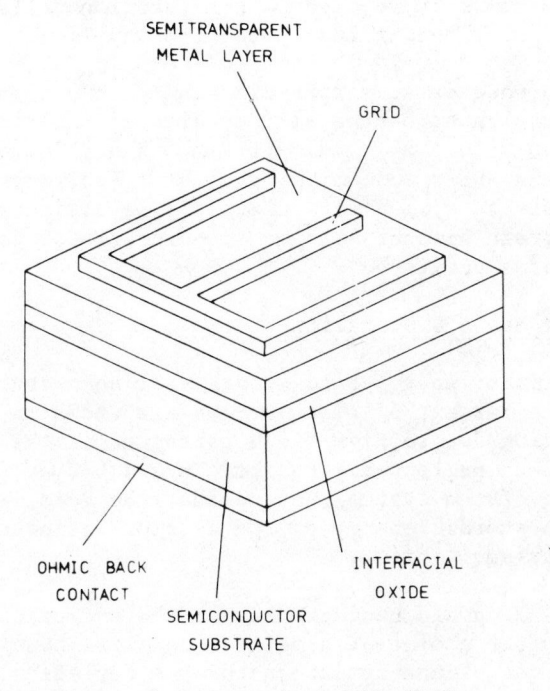

Figure 1. A typical MIS SBSC

Since the absorption of light in a thin metal film is an exponential
function of its thickness, it is important that the film should be
made as <u>thin</u> as possible. The solar cell is also a low impedance
device and as such should ideally have a series resistance of
less than $\sim 10^{-4}$ Ω m^2. In this respect the metal layer should be
made as <u>thick</u> as possible so that its sheet resistance does not
contribute significantly to the cells series resistance. These
requirements are conflicting and a compromise must be made between
the two. Generally it is found that a metal film thickness of
between 6 and 10 nm is a satisfactory compromise although it is of
course a function of the particular metal used. The maximum
permissible sheet resistance is also dependent upon the current
collection grid spacing, but generally lies within the range 30 to
200 Ω per ☐. Closer grid spacings enable higher film sheet
resistivities and therefore thinner films to be used. The limitation
on the grid spacing is that the total grid surface area should be
made as small as possible. For commercial p-n junction cells in
which the grid lines are defined photolithographically a minimum
grid spacing of 1 mm and line width of $\sim$ 75 μm are normally employed
to give a grid coverage area of $\sim$ 7% of the total cell area.

Because most metals are highly reflective in the near u.v to
near i.r. region of the electromagnetic spectrum it is standard
practice to apply a quarter wavelength anti-reflection coating to
SBSCs as the final step in the fabrication process. The dielectrics
used are those commonly applied to p-n junction cells for anti-
reflection purposes and include, ZnS, SiO and TiO$_2$.

The semiconductor substrate used in a conventional silicon SBSC
is single crystal n or p type silicon with an impurity concentration
usually between $\sim 10^{21}$ m^{-3} and $\sim 10^{22}$ m^{-3} and thickness $\geqslant 100$ μm.
In order to make SBSCs economically viable a polycrystalline or
amorphous substrate will almost certainly be ultimately used. Whatever the substrate however, the basic principle of solar cell
operation remains unaltered.

2.1 Principle of SBSC Operation

The simplified energy band diagrams of an n and p-type silicon
MIS SBSC at the instant of illumination are shown in Figures 2 and 3.
Schottky lowering due to high field strengths at the semiconductor
surface have been neglected. For both n and p-type SBSCs the
thermodynamics of the system require that the Fermi level, which
represents the average energy of the carriers be constant throughout
the junction at equilibrium.

In Figure 2 upward band bending at the n-type silicon surface
is produced by the choice of a metal whose work function is much
greater than that of the semiconductor. A depletion region and also
a strong electric field are produced below the silicon surface.

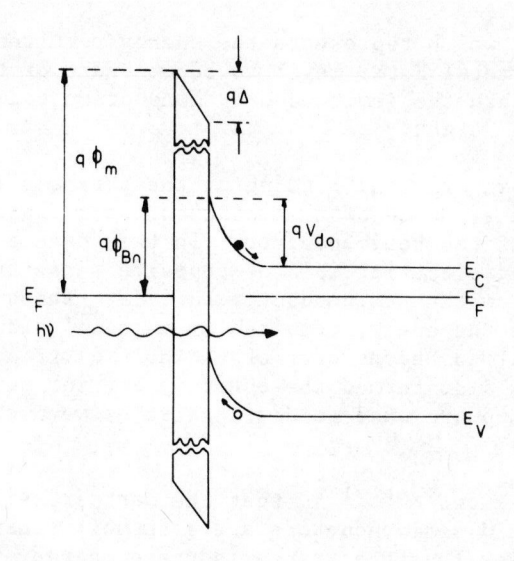

Figure 2.   Simplified energy band diagram of an n-type
            MIS SBSC at instant of illumination

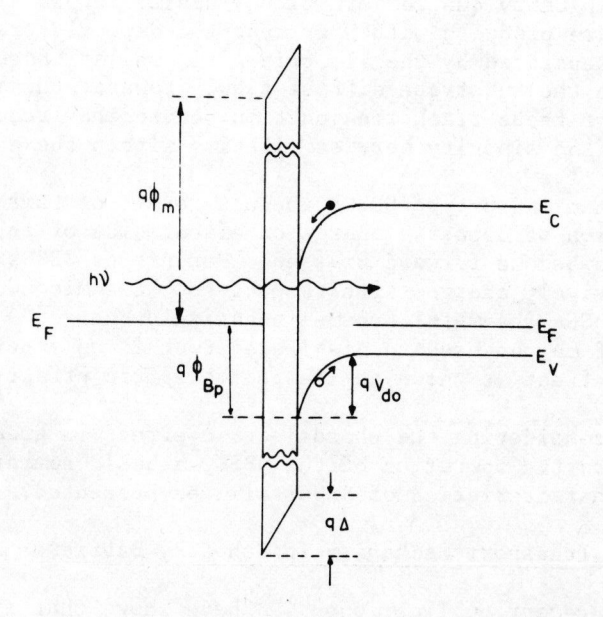

Figure 3.   Simplified band diagram of a p-type
            MIS SBSC at instant of illumination

The quantity ϕ_{Bn} which represents the energy required by an electron
situated at the metal Fermi level to become part of the electron
distribution within the semiconductor conduction band is termed the
Schottky barrier height.

In Fig 3 downward band bending at the p-type silicon surface
is produced by the choice of a metal whose work function is much
less than that of the semiconductor. In this case a depletion region
and also a strong electric field of opposite sense to that of Fig 2
are produced below the semiconductor surface. The quantity ϕ_{Bp}
which represents the energy required by a 'hole' situated at the
metal Fermi level to become part of the hole distribution in the
semiconductor is also termed the Schottky barrier height. The
subscripts n and p are used to distinguish between the semiconductor
material 'type'.

It is well known[4],[11-14]   that the barrier heights existing
between metals and semiconductors are extremely sensitive to the
surface properties (surface states, surface charge etc) of the
semiconductor and often bear little relation to the difference in
work functions between the two. For this reason barrier heights
are best determined experimentally for the particular metal and
semiconductor surface properties.

Considering an incident photon of energy $h\nu$ it can be seen that
if $h\nu > E_g$ an electron hole pair will be produced within the silicon
substrate on a unity quantum efficiency basis. Those electrons and
holes which are produced within or near the depletion region are
immediately separated by the electric field whilst those produced
deeper within the substrate diffuse slowly towards the junction.
Whether or not these reach the junction before they recombine
depends upon the minority carrier lifetime within the substrate.

For both n and p-type SBSCs the net result of this action is
an accumulation of opposite charge on either side of the junction
causing it to become forward biassed. For n-type SBSCs the metal
becomes positively charged with respect to the semiconductor and
for p-type SBSCs the metal becomes negatively charged. In both
cases the emf may be used to dissipate power in an external electrical
load. This effect is known as the photoelectric effect.

Before considering the photoelectric effect in greater detail
in relation to the operation of the SBSC, a basic summary of the
electrical characteristics of SBSCs will be presented.

2.2 Current Transport Mechanism in Schottky Barriers

Several recent publications [10]   have shown that single crystal
SBSCs may exhibit AMI conversion efficiencies comparable to p-n
junction cells when a thin interfacial layer is present at the metal

semiconductor interface. Although the effect of interfacial oxides on the electrical characteristics of silicon SB diodes has been recognised by Card and Rhoderick[1], the application to solar cell technology is relatively recent and has stimulated renewed interest in the electrical characteristics of these devices.

The principal current transport mechanism in SBs in which little or no interfacial layer is present is by thermionic emission of majority carriers over the energy barrier. Under small forward bias and no illumination the current voltage relationship is exponential and is given by

$$J_F = J_0 \exp \left(\frac{qV_F}{kT} - 1 \right) \qquad\qquad 1$$

$$\text{where } J_0 = A^{**} \, T^2 \exp - \frac{q\phi_B}{kT} \qquad\qquad 2$$

A^{**} is the effective Richardson constant taking into account Schottky lowering of the barrier

V_F is the forward bias voltage

J_F and J_0 are forward bias and reverse saturation current densities

Equations 1 and 2 are equally valid for n or p-type SBs where ϕ_B is replaced by ϕ_{Bn} or ϕ_{Bp} respectively and the appropriate value of A^{**} is used. It should be remembered however that for n-type SBs the metal is biassed positively with respect to the semi-conductor and for p-type SBs the semiconductor is biassed positively with respect to the metal. Andrews and Lepselter[15] have determined A^{**} at room temperature for n and p-type silicon in which the impurity concentration was 10^{22} m^{-3} and have shown that it remains essentially constant at 1.1×10^6 A m^{-2} K^{-2} and 3.2×10^5 A m^{-2} K^{-2} respectively for field strengths within the range 5×10^5 V m^{-1} and 5×10^7 V m^{-1}. Equations 1 and 2 predict that for $V_F > \frac{3kT}{q}$ the semilogarithmic plot of $\ln J_F$ vs V_F is linear with slope $\frac{q}{kT}$ and intercept on the J axis of J_0.

This behaviour is indeed experimentally observed in 'ideal' silicon Schottky barriers where the barrier has been achieved by cleavage of the silicon crystal under ultra high vacuum in a stream of the evaporated metal vapour[14] However, in devices in which less stringent precautions have been taken to ensure that there is no interfacial layer at the metal-semiconductor interface or where an interfacial layer has been purposely introduced, the slope of the $\ln J_F$ vs V_F forward characteristic is considerably smaller than

$\frac{q}{kT}$ and may even approach $\frac{q}{2kT}$ or less. This is inexplicable on the model of equations 1 and 2.

The role of the interfacial layer in metal-insulator-semiconductor Schottky barriers (MIS SBs) has been investigated by Card and Rhoderick[1] and Fonash[16]. Considering figures 4 and 5 which represent the simplified energy band diagrams for n and p-type silicon MIS SBs respectively, under forward bias V_F, it can be seen that with no illumination (J_{ph}^{e-} and $J_{ph}^{h} = 0$) a voltage V_s is developed across the semiconductor and V_i across the interfacial layer such that

$$V_F = V_s + V_i \qquad\qquad\qquad 3$$

V_F, V_s and V_i are defined such that they are positive for both n and p-type MIS SBSCs.

Neglecting the effects of minority carriers within the semiconductor, the forward bias current density for the case of an n-type silicon MIS SB is given by

$$J_{Fn} = J_{sm}^{e-} - J_{ms}^{e-} \qquad\qquad\qquad 4$$

where J_{Fn} is the current density flowing from the metal into the semiconductor under forward bias V_F.

J_{sm}^{e-} is the rate of flow of negative charge per unit area due to electrons flowing from the semiconductor into the metal

J_{ms}^{e-} is the rate of flow of negative charge per unit area due to electrons flowing from the metal into the semiconductor.

For p-type silicon MIS SB the current density flowing from the semiconductor into the metal is given by

$$J_{Fp} = J_{sm}^{h} - J_{ms}^{h} \qquad\qquad\qquad 5$$

where J_{sm}^{h} is the rate of flow of positive charge per unit area due to holes flowing from the semiconductor into the metal.

J_{ms}^{h} is the rate of flow of positive charge per unit area due to holes flowing from the metal into the semiconductor.

Following the method of Fonash[16]; if Schottky lowering is

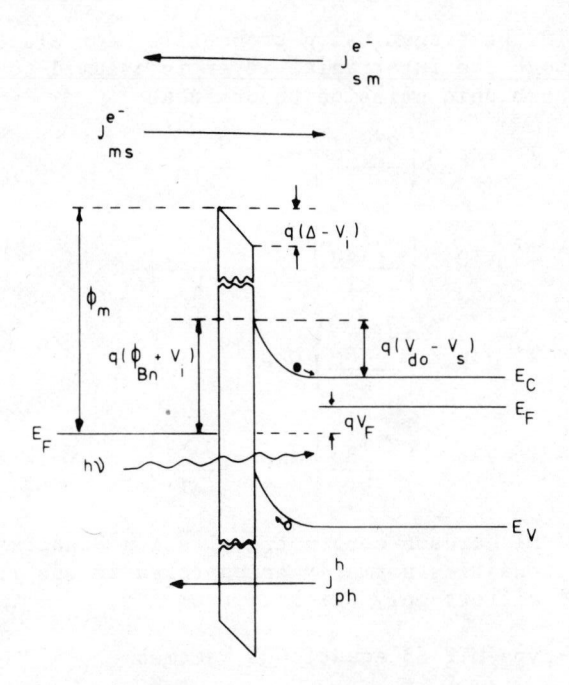

Figure 4. Simplified energy band diagram of an n-type
MIS SBSC under forward bias V_f

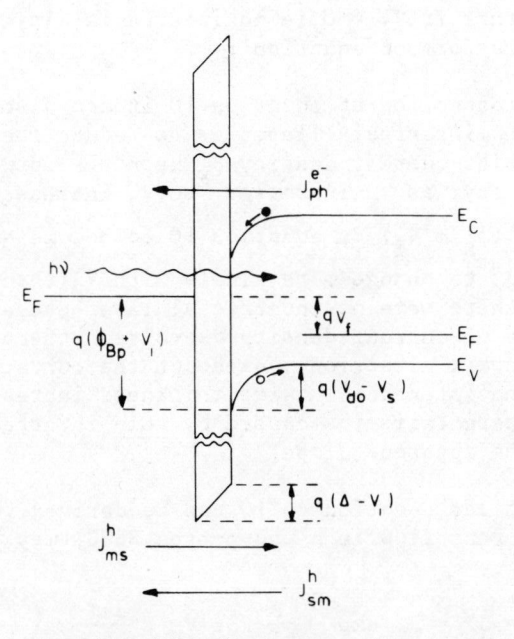

Figure 5. Simplified band diagram of a p-type MIS SBSC
under forward bias V_f

neglected and if the transmission probability for electrons
tunnelling through the interfacial layer is assumed to be unity it
follows from thermionic emission theory that

$$J_{sm}^{e-} = A^* \, T^2 \, \exp - \left(\frac{q \, \phi_{Bn}}{kT} \right) \exp \left(\frac{q \, V_s}{kT} \right) \qquad\qquad 6$$

$$J_{ms}^{e-} = A^* \, T^2 \, \exp - \left(\frac{q \, \phi_{Bn}}{kT} \right) \exp - \left(\frac{q \, V_i}{kT} \right) \qquad\qquad 7$$

$$J_{sm}^{h} = A^* \, T^2 \, \exp - \left(\frac{q \, \phi_{Bp}}{kT} \right) \exp \left(\frac{q \, V_s}{kT} \right) \qquad\qquad 8$$

$$J_{ms}^{h} = A^* \, T^2 \, \exp - \left(\frac{q \, \phi_{Bp}}{kT} \right) \exp - \left(\frac{q \, V_i}{kT} \right) \qquad\qquad 9$$

where A* is the Richardson constant neglecting Schottky lowering.
For the doping densities normally encountered in SBs it has been
shown[17] that A* differs only slightly from A**.

For the p-type MIS SB equation 5 becomes

$$J_{Fp} = A^* \, T^2 \, \exp - \frac{q}{kT} \, (\phi_{Bp} + V_i) \left[\exp \frac{q \, V_F}{kT} - 1 \right] \qquad\qquad 10$$

It can be seen that if $V_i = 0$, ie no interfacial layer, then equation
10 reduces to the form of equation 1.

On closer inspection of equation 10 it can also be seen that
the effect of the interfacial layer is to reduce the dependence
of the forward bias current density J_F upon the forward bias voltage
V_F. That is to say; as V_F increases so V_i increases causing the
factor $\exp - \frac{q}{kT} \, (\phi_B + V_i)$ in equation 10 to become smaller. This
in turn causes J_F to change more slowly with V_F than would otherwise
be expected if there were no interfacial layer present. The semilogarithmic plot of current density vs voltage therefore becomes
curved in a downward direction. Although the curvature becomes more
pronounced as the interfacial oxide thickness increases the only
change in the characteristics caused by thin interfacial oxides is
a decrease in the apparent slope.

Since a similar equation to 10 can be derived for an n-type
MIS SBSC the current flow in n and p-type SBSCs may be written in
the form

$$J_{Fn} = J_{Fp} = J_F = J_0 \, \exp \frac{q \, V_F}{nkT} \quad \text{for } V_F > \frac{3kT}{q} \qquad\qquad 11$$

where $J_0 = A^* \, T^2 \, \exp - \dfrac{q \, \phi_B}{kT}$ 　　　　　　　　　　　　　　12

　　　$n = \dfrac{V_F}{V_F - V_i}$ 　　　　　　　　　　　　　　　　　　　13

　　The diode quality factor n, is in fact a function of the applied bias voltage V_F, for all interfacial oxide thicknesses. However, for many devices in which the oxide is <u>thin</u> the change of n with V_F over the bias range of interest is <u>small</u> and n may be treated as a constant. From equation 13 it is clear that as the interfacial oxide thickness increases and the potential V_i, which appears across the oxide, increases, so n increases. A quantitative analysis of the relationship between n and oxide thickness δ is given in Section 2.3.

　　Some care must be exercised when using the reverse saturation current J_0 obtained from the semilogarithmic current voltage plot and equation 12 to determine the metal-semiconductor barrier height ϕ_B. Card and Rhoderick[1] have shown that if the interfacial oxide is sufficiently thick so that the electron tunnelling transmission coefficient is no longer unity then the reverse saturation current is reduced to a value equal to the product of the reverse saturation current when no interfacial layer is present and the transmission coefficient of the interfacial oxide, that is

　　　　$J_{0oxide} = T \, (\delta) \, J_0$ 　　　　　　　　　　　　　14

where J_{0oxide} is the reverse saturation current when an interfacial oxide of thickness δ is present having a transmission coefficient $T(\delta)$.

　　If the forward bias characteristic is appreciably linear over most of the bias range then J_{0oxide} may be used together with equation 12 to determine an effective barrier height for the metal semiconductor contact. From equations 12 and 14

　　　　$T \, (\delta) \, J_0 = A^* \, T^2 \, \exp - \dfrac{q \, \phi_{Beff}}{kT}$ 　　　　　　　15

where ϕ_{Beff} is the effective barrier height when an interfacial oxide of thickness δ is present.

$\therefore$ 　　$\phi_{Beff} = - \dfrac{kT}{q} \, \ell n \left(\dfrac{T \, (\delta) \, J_0}{A^* \, T^2} \right)$ 　　　　　　　16

ie

　　　　$\phi_{Beff} = \phi_B - \dfrac{kT}{q} \, \ell n \, T \, (\delta)$ 　　　　　　　　17

   Since T (δ) is less than unity, the effective barrier height
is higher than the barrier height when no interfacial oxide is
present. Card and Rhoderick[1] have attempted to quantify T (δ) in
terms of the thickness and work function of the interfacial oxide
in gold n-type silicon MIS SBs. However, calculations based on
their experimental data showed that for oxide thicknesses between
0.8 nm and 2.6 nm the work function properties did not necessarily
resemble those of bulk silicon dioxide and made theoretical calculation of T (δ) impossible. Because of this, ϕ_{Beff} is best
thought of as simply that value of metal-semiconductor barrier
height which when used in conjunction with equations 11 and 12
provides an accurate prediction of the forward bias current
voltage relationship.

2.3 Effect of the MIS Potential Distribution upon the Diode Quality Factor n

   The effect of the interfacial layer upon the diode quality
factor n has been investigated by Card and Rhoderick[1] on the
assumptions that:

1)   The forward current voltage characteristic was truly
exponential, ie n, as described by equations 11 and 13 is constant
within the range of bias considered.

2)   The density of interface states at the semiconductor surface
could be subdivided into two groups. One of these groups
communicated most readily with the metal, the other with the semiconductor. This was a result of the distribution of the interface
states in space as well as in energy over the semiconductor
forbidden band gap. The density of these interface states in each
group was dependent upon the interfacial oxide thickness, and the
communication of the states at the semiconductor surface with
those at the metal surface was assumed to decrease with increasing
oxide thickness.

   With reference to Figures 4 and 5 and under the assumption
(1) above equation 13 may be differentiated wrt V_F to give

$$\frac{1}{n} = 1 - \frac{d\,V_i}{d\,V_F} = -\frac{d\,V_D}{d\,V_F} \qquad\qquad 18$$

where $\dfrac{d\,V_i}{d\,V_F}$ is +ve and $\dfrac{d\,V_D}{d\,V_F}$ is negative for both n and p-type MIS SBs.

   For the case of the n-type device evaluation of $\dfrac{d\,V_i}{d\,V_F}$ in terms

of the density of interface states which equilibrate with the metal

D_{sa}, and with the semiconductor D_{sb}, gives a value of n

$$n = 1 + \frac{\delta/\varepsilon_i \ (\varepsilon_s/W + qD_{sb})}{1 + \delta q \ D_{sa}/\varepsilon_i} \qquad\qquad 19$$

where ε_i and ε_s are permittivities of the oxide and semiconductor,
W is the depletion width and δ oxide thickness.

For gold n-type silicon MIS SBs in which the interfacial oxide
was between 1 nm and 2.6 nm thick it has been shown[1] that the slope
of the semilogarithmic current voltage characteristic may be
described with reasonable accuracy by equation 19 with $D_{sa} = 0$,
ie all the interface states are assumed to equilibriate with the
semiconductor.

Assuming this is also true for p-type MIS SBs in which the
interfacial oxides are of comparable thickness, equation 19 then
reduces to

$$n = 1 + \frac{\delta}{\varepsilon_i} \ \left(\frac{\varepsilon_s}{W} + q \ D_{sb}\right) \qquad\qquad 20$$

It is also readily verified that for a surface potential of > 0.1 V
and for an interfacial oxide < 2.5 nm the term $\frac{\varepsilon_s}{W}$ in equation 20
is negligibly small. Equation 30 then further reduces to

$$n = 1 + \frac{\delta}{\varepsilon_i} \ q \ D_{sb} \qquad\qquad 21$$

Since the interfacial oxide in SBSCs is generally less than 2.5 nm
and surface potentials are high (usually > .5 V) equation 21
together with equations 11 and 12 should describe the MIS SB
under forward bias in the region where the semilogarithmic current
voltage characteristic is approximately linear.

For many MIS SBSCs this region was found to cover almost the
entire forward bias range of interest ($\sim$ 0.1 V to $\sim$ 0.5 V).

2.4 The MIS SBSC under Illumination

The electrical characteristics of the MIS SBSC under forward
bias have been presented in sections 2.2 and 2.3. In this and the
following sections of the chapter the performance of the cell under
illumination and the factors limiting its conversion efficiency
will be discussed.

Considering Figure 4 which represents an n-type MIS SBSC under forward bias V_F it can be seen that under illumination the output current density flowing from the cell under load is given by

$$J_{out} = J_{ph}^{\ h} + J_{ms}^{\ e-} - J_{sm}^{\ e-} \qquad\qquad 22$$

where $J_{ph}^{\ h}$ is the current density due to holes photogenerated within the semiconductor bulk and space charge regions which cross the barrier into the metal.

J_{out} is the current density flowing out of the metal of the MIS SBSC.

$J_{ms}^{\ e-}$ and $J_{sm}^{\ e-}$ are the rates of flow of negative charge per unit area due to electrons flowing from the metal to semiconductor and from the semiconductor to metal respectively.

But from equations 4 and 11

$$J_{sm}^{\ e-} - J_{ms}^{\ e-} = J_0 \exp\frac{q\,V_F}{n\,kT} \text{ for } V_F > \frac{3kT}{q} \qquad\qquad 23$$

Therefore from equations 22 and 23 the output current density is given by

$$J_{out} = J_{ph}^{\ h} - J_0 \exp\frac{q\,V_F}{n\,kT} \text{ for } V_F > \frac{3kT}{q} \qquad\qquad 24$$

Using a similar analysis it may also be shown that for a p-type silicon MIS SBSC the output current density is given by

$$J_{out} = J_{ph}^{\ e-} - J_0 \exp\frac{q\,V_F}{n\,kT} \text{ for } V_F > \frac{3kT}{q} \qquad\qquad 25$$

where $J_{ph}^{\ e-}$ is the rate of flow of negative charge due to photo-generated electrons which cross the barrier into the metal.

J_{out} is the output current density flowing out of the semiconductor.

It is convenient to determine $J_{ph}^{\ h}$ and $J_{ph}^{\ e-}$ in terms of the short circuit current density so that one equation may be used to describe both n and p-type MIS SBSC behaviour.

Under short circuit conditions ($V_F = 0$) equations 24 and 25 require that $J_{out} = J_{sc}$, where J_{sc} is the short circuit current density.

Therefore $J_{ph}^{e-} = J_{ph}^{h} = J_{sc} - J_0 = J_{sc}$ for $J_{sc} \gg J_0$ 26

The output characteristic for both n and p-type silicon MIS SBSCs under illumination may then be written in the form

$$J_{out} = J_{sc} - J_0 \exp \frac{q\, V_F}{n\, kT}$$ 27

The factors determining the magnitude of J_{sc} are examined later.

Under open circuit conditions J_{out} in equation 27 is equal to zero, and V_F is equal to the open circuit voltage V_{oc}.

Equation 27 then becomes

$$0 = J_{sc} - J_0 \exp \frac{q\, V_{oc}}{n\, kT}$$

Hence $V_{oc} = \dfrac{n\, kT}{q} \ln \dfrac{J_{sc}}{J_0}$ 28

The maximum power output density delivered by the cell is given by the maximised product of the output current and voltage

ie $P_{max} = V_{mp}\, J_{mp}$ 29

where V_{mp} and J_{mp} are the voltage and current respectively at the maximum power point.

P_{max} may also be conveniently expressed in terms of V_{oc} and J_{sc} by a fill factor F which describes the shape of the V-I output characteristic.

If F is defined by the ratio

$$F = \frac{V_{mp}\, J_{mp}}{V_{oc}\, J_{sc}}$$ 30

then equation 29 becomes

$$P_{max} = F\, V_{oc}\, J_{sc}$$ 31

The conversion efficiency of which, is defined as the ratio of the power output density and the input power density, is then given by

$$\eta = \frac{F\ V_{oc}\ J_{sc}}{E} \ \times \ 100\% \qquad\qquad 32$$

where E is the incident power density of the illumination.

Before considering the MIS SBSC parameters affecting the conversion efficiency η the mechanism of the minority carrier MIS cell (min MIS) will be outlined.

2.5 The Minority Carrier MIS SB Cell

The so-called min MIS cell has recently established itself as probably the most important type of silicon MIS cell since it has been demonstrated to have a markedly better V-I characteristic and efficiency than cells working on the thermionic emission of Majority carriers already described[10].

The theory of this cell has been fully described by Green and his co workers[18-20]. An outline of the principles involved will now be given.

The forward bias dark current of a Schottky barrier diode has already been demonstrated to be caused by the thermionic emission of majority carriers from the semiconductor to the metal. It can be represented by the equation

$$J_F = A^*\ T^2\ \exp\left(\frac{-q\phi_b}{kT}\right)\left[\exp\frac{qV_j}{nkT}\right] \qquad\qquad 33$$

where ϕ_b is determined of course both by the difference between the metal work function and the semiconductor electron affinity and by the interface states at the metal semiconductor interface. As considered by Pulfrey[21] the presence of an interfacial insulating layer affects ϕ_b directly and equation 33 still gives a value for J_F implying that current control is through the thermionic emission of carriers from the semiconductor to the metal and is not limited by transport through the interfacial layer. For this case calculations for silicon show that barrier height can either decrease or increase with oxide thickness depending on the value of the work function ϕ_m. Alternatively the presence of the insulating layer can serve to control the diode current. If the majority carrier current, which forms the dominant component of the dark current could be reduced without any corresponding reduction in the photo-generated minority carrier current then an increase in photovoltage could result. Card and Yang [22] have shown this is possible and that it can lead to a reduction in the usual Schottky dark current by a tunnelling probability factor of the form

$$\exp - ((4\pi/h)(2m^*\chi)^{\frac{1}{2}}\delta)$$

where m* is the carrier effective mass, δ the oxide thickness
and χ the tunnelling barrier height. Fonash[16] has shown that
increases in V_{oc} can be achieved in this case if a particular
set of localised states can be introduced at the semiconductor-
insulator interface having a majority carrier cross-section much
greater than that for minority carriers.

A further way of controlling current transport in MIS barriers
occurs when the metal work function is chosen such that the semiconductor becomes inverted at the semiconductor-insulator interface
over the range of forward bias of interest. For an n-type semiconductor this requires a high value of ϕ_m, for a p-type a low
value. Under these circumstances the system Fermi level at the
interface is nearer in energy to the valence band edge than to the
conduction band edge as shown in Figure 6 for an n-type semiconductor. Here we see that the electron concentration is
significantly less than the hole concentration at the interface
and therefore the majority carrier current is greatly reduced.
Correspondingly there is no reduction in minority carrier current
and in practice under low forward bias conditions that correspond
to solar cell operation large minority tunnel currents can flow
leading to a pinning of the minority carrier quasi Fermi level
in the semiconductor to the metal Fermi level[18,20]. Here then

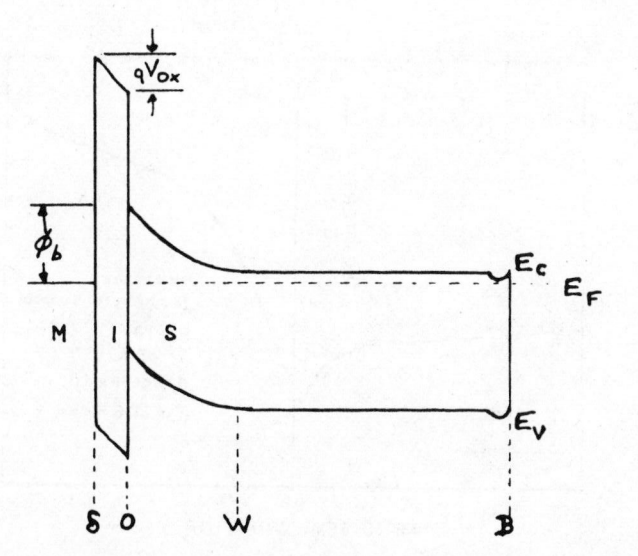

Figure 6.   Energy band diagram for metal/insulator/n-type
            semiconductor solar cell.

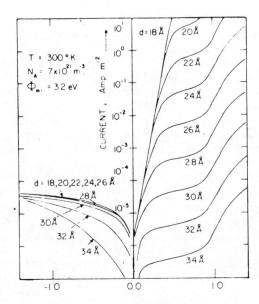

BIAS TO SEMICONDUCTOR, volts ⟶

Fig. 7a.   Computed I–V characteristics of the MIS tunnel diode demonstrating "nonequilibrium" effects for minority-carrier devices. The substrate is 2-Ω p-type ⟨100⟩ silicon, ϕ_{mi} is equal to 3.2 eV, and the variable parameter is d, the thickness of the oxide layer. (Ref 23)

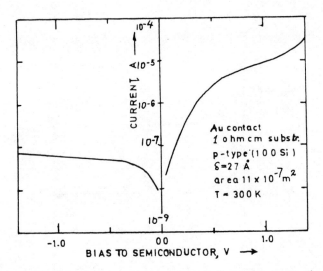

BIAS TO SEMICONDUCTOR, V ⟶

Figure 7b.   Experimental measurement of V–I characteristic of a majority carrier MIS device. Gold barrier on 27 Å SiO$_2$ insulating layer on p-type (100) Si. (Ref 24).

the dark current becomes controlled by processes within the semiconductor consisting of injection-diffusion and recombination
components appropriate to a p-n junction within the semiconductor
brought about by the presence of the inversion layer. Under these
conditions it can be shown that the dark current will be given
primarily by the sum of the injection-diffusion current J_{diff} and
the depletion layer recombination generation current J_{rec}.
Thus we have $J_F \simeq J_{diff} + J_{rec}$ or

$$J_F \simeq J_{0_{diff}} \left[\exp\left(\frac{q V_j}{kT}\right) - 1 \right] + J_{0_{rec}} \exp\left(\frac{q V_j}{2kT}\right) \qquad\qquad 34$$

This value of J_F may be significantly lower than that found
from equation 33 for the thermionic emission of electrons.

Shewchun et al[23] have computed I-V characteristics demonstrating
minority carrier behaviour for an Al-SiO$_2$ (p-type) Si diode with a
2 Ω cm substrate. The calculations are for the ideal surface-state
and oxide-charge-free cases and Figure 7a shows for different
oxide thicknesses that below a critical thickness $\sim$ 30 Å non
equilibrium tunnel MIS diodes are formed. Here over a limited bias
range the diode current is semiconductor limited due to generation-
recombination current in the bulk. If the metal work function is
increased to 4.2 eV, replacing the aluminium by gold for example, the
calculated I-V curves are as shown in Figure 7b which corresponds
to majority carrier behaviour. In this case of course there will
be no inversion layer. The transition between a minority to a
majority carrier diode occurs about 3.6 eV and in this region
tunnelling via surface states dominate.

If the thickness of the insulator is reduced below about 10 Å
the concept of a tunnel MIS diode apparently becomes invalid, based
at least on experimental evidence, and these thin structures
perform as basic Schottky barriers. Above 28-30 Å the diodes
behave as equilibrium tunnel diodes. From Fig 7a it can be
observed that even in the minority carrier regime under forward
bias the region over which ideal p-n junction diode behaviour is
predicted is insulator thickness dependent. Since in the case of
p-n junctions in silicon under normal AM1 illumination about
0.5 - 0.7 V is developed across the junction this means that for
significant conversion efficiencies in these min MIS devices
insulator thickness should not exceed about 20 Å. At greater
thickness there will be some suppression of the photo-current due
to the shape of the I-V characteristic rather similar to that
observed in p-n junction solar cells with large series resistance.

Figure 8 shows calculated variations of three parameters,
V_{oc}, I_{sc} and Fill factor as a function of insulator thickness for

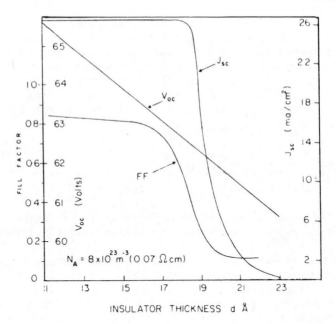

Fig. 8.   The variation of short-circuit current J_{sc}, open-circuit
voltage V_{oc}, and fill factor FF as a function of insulator thickness for an Al–SiO$_2$–Si (p-type) diode with
carrier concentration of 8 x 10^{23}/m^3. (Ref 23)

an Al on p-type cell; the dramatic effect on I_{sc} should be noted.

    The role of surface states and oxide charge in conventional
MIS devices has been well established and there is an extensive
literature. Similar effects are observed in tunnelling structures
where both effects give rise to a change in V_{oc} caused by a change
in the barrier height which is modified to be

$$\phi'_{mi} = \phi_{mi} - q\left[\frac{Q_{ss} + Q_i}{\varepsilon_i}\right]d \qquad\qquad 35$$

where Q_{ss} is surface state charge and Q_i is effective charge on the
insulator.

    The result of this modified barrier beight is that lowering
of V_{oc} occurs with increase in $Q_{ss} + Q_i$. Since the surface states
also act as recombination-generation centres and provide additional
tunnelling paths between the metal and semiconductor there is also
an effect on J_{sc}.   Figure 9 gives the effect of three different
values of surface state density and insulator charge on conversion
efficiency. Note from this a theoretical maximum efficiency of
∿ 21% for this non optimised MIS solar cell under AM2 illumination.

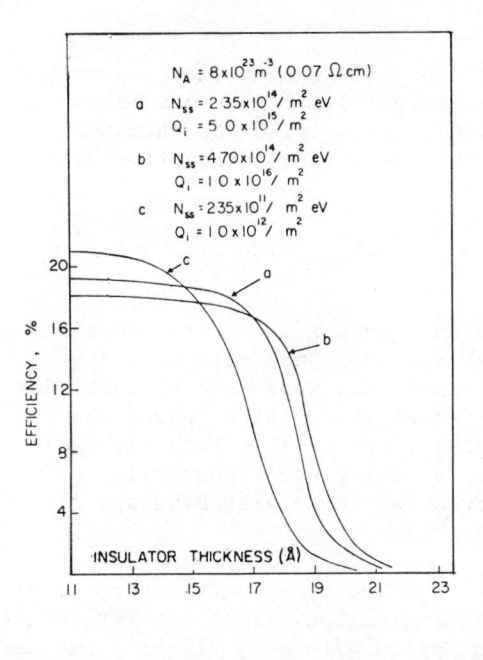

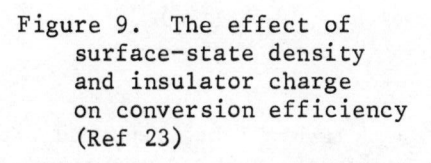

Figure 9. The effect of surface-state density and insulator charge on conversion efficiency (Ref 23)

3. SOLAR CELL PARAMETERS AND DESIGN CONSIDERATIONS

3.1. Metal-Semiconductor Barrier Height

The maximum power output from an SBSC is directly proportional to the open circuit voltage of the cell under illumination (eqn 31) assuming F and J_{sc} to be constant. From equations 2 and 28 it may be shown that

$$V_{oc} = \frac{n\ kT}{q} \ln \frac{J_{sc}}{A^* \ T^2 \ \exp\left(\frac{-\phi}{kT}\right)}$$

$$= \frac{n\ kT}{q} \left[\ln \frac{J_{sc}}{A^* \ T^2} + \frac{\phi}{kT} \right]$$

36

where ϕ is the metal semiconductor barrier height for an n or
p type SBSC and it has been assumed that $J_{sc} \gg J_0$.

For MIS SBSCs in which a thin interfacial oxide layer is present ϕ is replaced by ϕ_{eff}. From equation 35 it may be seen that in order to achieve maximum open circuit voltage and hence maximum power output from the cell the metal semiconductor barrier height should be made as large as possible by suitable choice of the metal and semiconductor surface properties.

3.2 Diode Quality Factor n

The open circuit voltage of an SBSC is also directly proportional to this quantity (see equation 28). It has been shown in Section 2.3 that n increases with interfacial oxide thickness and the density of interface states D_{sb} according to equation 21

$$n = 1 + \frac{q \, \delta \, D_{sb}}{\varepsilon_i}$$ 37

Fabre [9] has reported titanium p-type silicon MIS SBSCs in which the value of n was as high as 3.8. Such high values of n could not be explained by equation 37 and were most probably due to the presence of a thicker interfacial oxide which caused the forward current to become tunnel limited[18,24]. Such high values of n do not necessarily contribute to a greater conversion efficiency since the thickness of the interfacial layer also causes a reduction in the short circuit current density.

Figure 10 shows the effect of n and ϕ_B upon the open circuit voltage of MIS SBSCs under AMO illumination [25]. A similar result would be expected for AMI illumination with only slightly smaller values of V_{oc}. The importance of a high value of n is apparent.

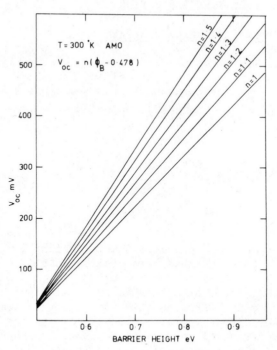

Fig. 10. Theoretical relationship between V_{oc}, ϕ_B, and n (22)

3.3 Interfacial Oxide Thickness

The effect of the presence of a thin interfacial oxide on the
dark current voltage characteristics of an SBSC has been examined
in sections 2.2 and for the min MIS cell in 2.5. The exact role
of the interfacial oxide is not fully understood at present,
although it is known that both minority and majority carrier flow
over the barrier decrease with increasing oxide thickness. If the
oxide is sufficiently thick so that its transmission coefficient
is no longer unity, the reverse saturation current density obtained
from the forward J_F vs V_F characteristics is reduced causing an
increase in the effective barrier height ϕ_{eff}. This effect is
further enhanced in p-type material by positive charge trapped within
the oxide causing an increase in the band bending (and hence
barrier height) at the semiconductor surface. For n-type silicon
the effect of positive charge within the oxide is of course detrimental to the metal-semiconductor barrier height.

The use of interfacial oxides to increase the open circuit
voltage and conversion efficiency of SBSCs has been extensively
reported[5-9,20,26]. The optimum oxide thickness reported,
varies widely between 0.5 and 8 nm[25] ; this is in part due to the
difficulty in measuring such thin films usually with an ellipsometer.

To summarise, the interfacial oxide thickness should be such
that the maximum values of n and ϕ_{eff} are achieved without significantly reducing J_{sc}. These requirements are conflicting and a
compromise must be made. What has been done in practice to
optimise the interfacial oxide layer or to replace it with an
alternative insulating film? Consider how this has been done and
the methods used for insulator formation.

A number of techniques have been used for providing the I
layer in the MIS solar cell; these have been summarised by Pulfrey[10]
and are repeated in Table I below. The standard procedure is to
thermally grow a layer of SiO_2 on the parent substrate. The problem
however is that for the thin oxide layers required, ~ 20 Å thick,
the thermally grown oxides are silicon rich and are not likely to
be homogeneous SiO_2 until a thickness of 50 Å is reached.
In this case there is a likelihood of further oxygen being
incorporated into the oxide during subsequent processing stages or
during actual operation of the cell. In the case of gold on n-type
SBSCs oxygen migration through the thin gold Schottky barrier is
known to cause significant changes to occur to the device characteristics as the cells age [6, 27] . The whole question of cell stability
will be dealt with in the final section of this chapter. As in the
case of the thicker SiO_2 layers incorporated into MOS transistors
control of the conditions under which the oxide is grown is vital
if a low value of surface state charge Q_{ss} is to be achieved. In
practice this requires a clean oxidation system using dry oxygen.

TABLE 1[10]

INSULATING LAYER GROWTH METHODS USED IN MIS CELLS

Semiconductor	Method	Time	Reference
Si	hot HNO_3	8 min	37, 38
"	boiling H_2O		37
"	10-percent HF in H_2O	1 min	39
"	SiO_2 in 40-percent HF	1.5-2 min	30
"	anodization		37, 40
"	evaporation of SiO		38
"	150°C in air	5-70 h	26
"	400°C in air	5-60 min	6
"	620°C in air	15-30 min	41
"	420°C in dry oxygen	10 min	28
"	420°C in wet oxygen	5-40 min	42
"	514°C in oxygen	12 min	40
"	450°C in wet nitrogen	15-30 min	43
"	450°C in N_2/O_2	∿ 10 min	44
"	$480-500^{\circ}$C in N_2/O_2	∿ 30 min	45
"	$700-900^{\circ}$C in steam	< 5 min	19
GaAs	evaporation of Sb_2O_3		46
"	oxygen plasma	10-20 s	47
"	anodization in buffered 3-percent H_3PO_4	∿ 10 min	47
"	anodization in acetone/$KMnO_4$	< 1 min	47
"	120°C in ozone	20 min	47
"	25°C in H_2O/O_2	50-100 h	47
"	105°C in air	70 h	8

A typical process used by Lillington & Townsend uses dry O_2 with a substrate held at 420°C for 10 minutes [28].

Alternative forms of oxide formation such as evaporation of SiO and growth in strong oxidizing agents should lead to complete oxidation and to homogeneous layers of oxide. However cells made in this way have not had either better characteristics or shown better stability. A promising technique introduced by Kipperman et al[29,30] oxidises the silicon in a solution of $(H_3O)_2SiF_6$. This gives a self limited oxide of reproducible thickness. It has the potential advantage that because of the presence of fluorine in the oxide in the form $SiOF_2$ there should be a reduced interface state density.

Work with non native oxides as insulators has been confirmed almsot exclusively to the use of Si_3N_4 [10]. This can be deposited by evaporation and has the advantage that its formation may remove the one high temperature stage in the cell processing,as well as simplify the overall cell production. This advantage is of course lost if a post oxidation annealing stage is used (∿ 400°C) in an attempt to lower the surface state charge. This stage is often combined with the formation of the rear contact.

Whatever the process used to form the insulating film, there is a necessity to systematically clean the substrate initially and to remove any existing native oxide film so that each wafer will have a common starting point. This cleaning procedure has been

described by various workers and is usually followed by a brief
etching stage [6, 10].

If MIS SBSCs are to be formed on GaAs of course there is again
a wide range of possibilities some of which are included in table 1.

3.4   Transmission Properties of the Metal

All incident photons which generate electron hole-pairs within
the semiconductor bulk and space charge regions must first pass
through the semitransparent metal layer. Photons which are
reflected at the air-metal interface or are absorbed within the
metal do not contribute to the photo current. It is therefore of
utmost importance that the transmittance of light into the semiconductor is made as high as possible.

The optical constants of all metals and semiconductors which
absorb light in some regions of the spectrum may be characterised
by a complex refractive index n-jk. The transmittance of light at
the metal-semiconductor interface, defined as the fraction of
light energy incident at the air metal interface (assuming no anti
reflection coating) which passed into the semi-conductor is
dependent upon the optical constants of the metal and semiconductor.
Although it is not usually possible to choose the Schottky barrier
metal from the optical constants point of view, since other factors
(ie metal-semiconductor barrier height and sheet resistance of the
metal in thin film form) are of overriding importance, it is of
interest to know these so that theoretical estimates of optical
film transmittance may be made.

It is well known [31] that all metals reflect light energy
extremely efficiently, particularly in the infra red region of the
spectrum. Although the light energy is only partially reflected
from a thin metal film on a semiconductor substrate, the energy
losses in SBSCs due to reflection from the surface may be several
times greater than that for p-n junction cells. It is therefore
standard practice to apply a quarter wavelength anti reflection
coating as the final step in the fabrication of an SBSC. The
dielectrics most commonly used are similar to those used for the
anti-reflection coating of p-n junction cells, TiO_2, SiO, Ta_2O_5
and ZnS.

The theory of the multi layer optical problem of the SBSC
structure has been described in the literature [31]. For single
layer dielectric coatings on silicon the optimum refractive index
is $\sim 2.1 - 2.4$. Coatings with refractive indices within this
region lead to practically identical results with theoretical
reflection losses as low at 6% for both Si and GaAs substrates.

3.5 Spectral Response

The spectral response of any solar cell is important since it
dictates how much of the solar spectrum may be utilised in the
generation of photocurrent.

The absorption coefficient of silicon[32] is shown in Figure 11.
Considering this curve it can be seen that the short and long wavelength cut off points of the cells spectral response are determined
by two factors.

1)   In the long wavelength region of the spectrum the absorption
coefficient of silicon falls off rapidly until at wavelengths of
$\sim$ 1100 nm and above there is no appreciable absorption within the
silicon.   There is therefore no cell response beyond this wavelength.

2)   In the short wavelength region of the spectrum the absorption
coefficient of silicon is extremely high and photons are absorbed
within a very small distance of the semiconductor surface.   For
p-n junction cells where the surface layer is heavily doped and
minority carrier lifetimes are short, recombination is such that
very few minority carriers reach the junction.   Response in this
part of the spectrum is small.   For SBSCs there is no such layer
and a greater response within this region is to be expected.

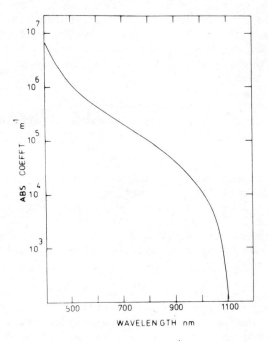

Figure 11.   Absorption coefficient of silicon (32)

However, Li et al [33] have shown that the short wavelength response
of SBSCs may be reduced by the presence of an inversion region
below the semiconductor surface, particularly if the metal semiconductor barrier height is large and strong inversion occurs.
Qualitatively this fall off occurs for the following reason. The
current resulting from short wavelength excitation consists primarily
of carriers generated near the semiconductor surface within the
inversion region. Most of the electrons and holes generated there
however ultimately flow towards the surface whereupon they recombine.
The minority carriers are directed towards the surface under the
action of the electric field within the inversion and depletion
regions whilst majority carriers flow to the surface by diffusion
which dominates the drift term in this region.

   Figure 12 shows the calculated dependence of the quantum yield
of a gold n-type silicon Schottky barrier photodiode upon the
impurity concentration within the substrate. An increase in the
impurity concentration results in an enhanced blue response at the
expense of red response. This consideration can be important for
thin film silicon solar cells where good short wavelength response
is vital. A similar result to that of Figure 12 would be expected
for a p-type silicon SBSC with a similar barrier height to that of
the gold n-type silicon contact although some difference would
occur due to the differences in lifetime within the substrate.

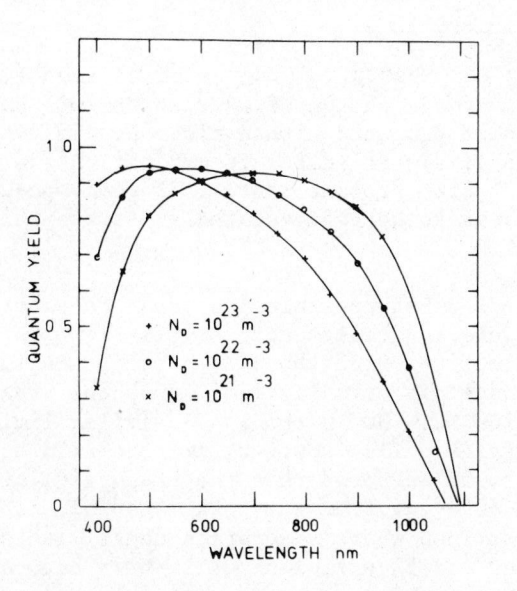

Figure 12.   Calculated quantum yield of a gold n-type silicon
             photodiode (33)

3.6 Substrate Resistivity

The spectral response of SBSCs within the useful region of
the spectrum ($\sim$ 400 nm to 1100 nm) is to some extent determined
by the substrate resistivity. The relationship between minority
carrier lifetime and bulk impurity concentration has been investigated by Ross and Madigan [34]. For both n and p-type silicon the
minority carrier lifetime decreases with increasing impurity
concentration (decreasing substrate resistivity). Since good red
response of all solar cells relies almost entirely on long minority
carrier lifetimes within the substrate, SBSCs fabricated on low
resistivity substrates show little response at long wavelengths.

It has been shown [33] that the presence of an inversion region at
the semiconductor surface of an SBSC causes a reduction in the blue
response of the cell. The thickness of this region however
decreases with increasing doping concentration so the net effect
of lowering the substrate resistivity is to shift the spectral
response towards the blue end of the spectrum as is seen in Figure
12.

For most SBSCs the use of silicon substrates with resistivities
within the range 1 - 10 Ω cm (impurity concentration between
$\sim 10^{21}$ m^{-3} and $\sim 10^{22}$ m^{-3}) provides a satisfactory match between
spectral response and the solar spectrum.

3.7 Substrate Thickness

Considering the absorption curve for silicon (Figure 11) the
penetration depth for a photon of wavelength 1000 nm is $\sim$ 100 μm.
Electron hole pairs produced within close proximity to the rear
contact are also likely to suffer recombination due to the high
recombination velocity present there. If good spectral response at
long wavelengths is to be achieved the substrate thickness should
therefore be > 100 μ.

The effect of substrate thickness upon the short circuit
current of p-n junction cells under AM1 is shown in figure 13. In
practice the extent to which the thickness of single crystal solar
cells may be reduced is normally limited by the fragility of the
substrate - typically $\sim$ 300 μ thick. A similar limitation exists
for Schottky Barrier cells, however, if thin film substrates are
to be used it should be noted that the properties of the films
will be widely different from the bulk material. This is particularly
the case for α-silicon where absorption coefficients are more than
10^3 times greater and 5% efficient cells have been achieved using
active regions $\sim$ 1 μ thick [35].

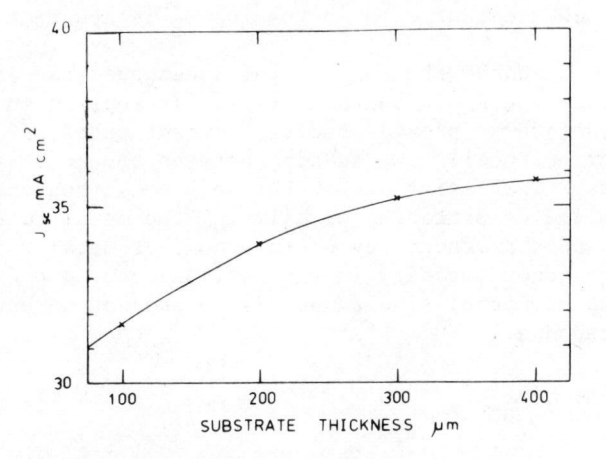

Figure 13. Effect of substrate thickness upon J_{sc} under AMO (70)
            (British Crown Copyright)

3.8 Series Resistance

The SBSC in common with other solar cells is a low impedance
device and as such should possess a low series resistance. The
three main factors which contribute to increased series resistance
are contact resistance, sheet resistance of the semitransparent
metal layer forming the Schottky barrier and bulk resistance of
the semiconductor substrate.

Customarily the series resistance of the solar cell has been
considered to be a parasitic element. However Dubey[36] has pointed
out that the optimum series resistance of an SBSC is non-zero owing
to contribution to the series resistance from the bulk resistivity
of the substrate. For a 1 cm^2 gold n-type gallium arsenide SBSC
this was calculated to be 0.2 Ω. This is also true of other SBSCs
and a series resistance of 0.15 Ω may be calculated for a silicon
SBSC fabricated on a 5 Ω cm substrate 300 μm thick. In many cases
however this contribution is small compared to the other contributing
factors already mentioned and may be neglected.

When a series resistance component is present in an SBSC the
V-I output characteristic under illumination is no longer given by
equation 27 but by the following equation.

$$J_{out} = J_{sc} - J_0 \left(\exp \left(\frac{q \, (V_F + J_{out} \, R_s)}{n \, kT} \right) - 1 \right) \qquad 38$$

where R_s is the series resistance component.

Figure 14 shows the output characteristics of a typical 1 cm^2
gold n-type silicon SBSC (ϕ_{Bn} = 0.81 eV, n = 1, J_{sc} = 25 mA cm^{-2}
T = 300°K) for several different values of R_s. The large decrease
in available power output with increasing R_s is apparent.

In order to reduce the contact and sheet resistances to an
acceptable value the front contact is usually applied in the form
of a grid structure to provide shorter current paths for the minority
carriers. For most cells the spacing between the grid lines is
∿ 2 mm or less. The resistance of the back ohmic contact is
usually minimised by sintering or alloying the metal to the
semiconductor back surface. By using a contact metal which is either
an acceptor or donor impurity in the semiconductor a back surface
field may also be formed simultaneously to produce an enhanced long
wavelength response.

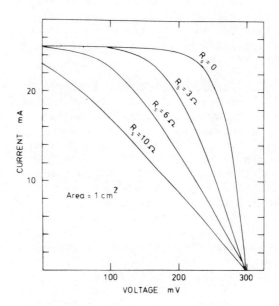

Figure 14.   Effect of R_s on V-I characteristics of gold n-type
             silicon SBSC s

4.   RESULTS AND DISCUSSION OF TYPICAL SILICON MIS CELLS

It is convenient to consider the mass of data available for silicon and GaAs MIS cells in terms of the four factors, open circuit voltage V_{oc}, short circuit current density J_{sc}, fill factor FF, and efficiency η. Pulfrey [10] has compared values of these parameters obtained by numerous workers who have made cells based on both majority and minority carrier tunnelling. Before considering the results for the min MIS and layer cells ayer cells in more detail let us summarise the current position for the four parameters.

4.1 Open Circuit Voltage

Owing to the fact that the commonly used thermal oxidation process leads to silicon rich layers at the silicon - insulator boundary and a consequent increased number of donor states in this region, cells made using n-type silicon suffer from an unfavourable decrease in band bending. This results in n-type substrates leading to cells having reduced values of V_{oc} usually < 450 mV as shown in table 2.

TABLE 2[10]

V_{oc} VALUES FOR Si MIS CELLS

Type	Substrate Orientation	Resistivity (Ω . cm)	Barrier Metal	Oxidation Method[1]	V_{oc} (mV)	Reference
p	100	1	Al	Th	615	44
p	100	0.1	Al	Th	610	44
p	100	0.2	Al	Th	600	48
p	100	2	Cr	Th	600	49
p		1	Al	Th	560	28
p		35.1	(SiO_2)	Th	554	50
p	100	1	Ti	Chem	550	37
p	100	2-8	Al	Th	470	26
p		3-15	Al	Nat	470	5
n	111	1-10	Au	Dep	536	25
n	100	0.15-0.25	Au	Th	444	40
n	100	8-19	Au	Anod	427	40
n	111	5	Au	Th	420	6
n	111	5-10	Pt	Th	410	42
n	111	1-10	Au	Chem	400	25

[1]Th = thermal, Chem = Chemical, Nat = natural oxide (spontaneous), Dep = evaporated Anod = anodized.

The exception for n-type substrates occurs when deposited SiO is used [25] - in this case the silicon rich layer is not present giving rise to a more favourable distribution of surface states and a value of $n \sim 1.34$.

In order to profit from the properties of the silicon rich oxide layer, p-type substrates are used. In this case significantly higher values of $V_{oc} \sim 500 - 600$ mV have been reported using Al,

Cr and Ti as barrier metals (see Table 2). Consider one of these
in more detail, the Cromium barrier cell produced by Anderson,
Vernon   et al.[41] The cell construction is shown in Figure 15; it has
a barrier consisting of an initial 40-50 Å cromium film covered by
a similar thickness of either copper or silver to enhance film
conductivity. The contact grid is 1μ thick aluminium and the
insulator a thermally grown oxide $\sim$ 15 Å. A typical V-I curve for
a 1.75 cm^2 cell is shown in Figure 16.

    Considering the data in table 2 the first three entries with
the highest values of V_{oc} all claim to be min MIS cells where low
work function metals are used together with low resistivity substrates
to create an inversion layer at the semiconductor surface leading
to minority carrier tunnelling. In order to distinguish the mode
of operation of these p-type cells a measurement of dark forward
conduction characteristic may be plotted. For low bias (< 0.5V)
minority carrier diodes have an inverse slope $\sim$ 2 to their log I
vs Vcurve and an inverse slope $\sim$ 1 at higher bias. For the case of
a dominant thermionic emission current on the other hand a single
exponential curve gives an inverse slope or n value between 1 and 2.

    In order to optimise V_{oc} two methods used in Si homojunction
cells may be considered. These are the use of low resistivity base
material and the use of a back surface field with high resistivity
base material. Both of these modifications have been made for
p-type MIS cells and they are discussed later in para 4.5. The
other parameter that can be optimised is ϕ_B and to do this various
barrier metals have been used as indicated in table 2.

4.2 Short Circuit Current Density

    The major factor limiting the values of J_{sc} for MIS cells is
the existence of highly reflecting metal barrier layers. This makes
the provision of a well matched AR coating even more essential for
barrier cells than for homojunction cells. This fact has been dealt
with in para 3.4 where the different types of AR coating used have
been discussed. The second factor that limits J_{sc} is the presence
of the interfacial insulating layer between the semiconductor and
and the barrier. As we have seen this layer substantially increases
V_{oc} but it has to be sufficiently thin in order to minimise any
effect on the tunnel transmission coefficient or the effective
series resistance of the cell. This means for silicon using silicon
oxides < 25 Å thick and in the case of GaAs using oxides < 35 Å.

    The other factor of importance in determining J_{sc} is the degree
of coupling of the light into the semiconductor and the recombination
of carriers that occurs both in the bulk and at the surface. Here
MIS cells are at an advantage compared with homojunction cells since
due to the requirement of comparatively low processing temperatures
< 500 C, there is significantly less degradation to the carrier

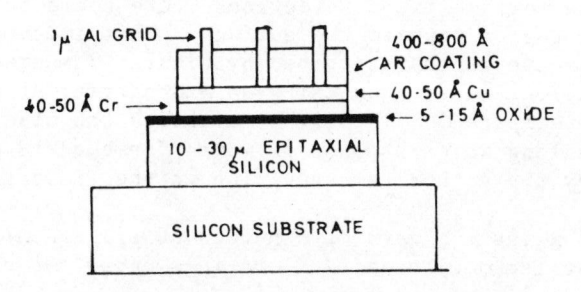

Figure 15.   Construction of Cr barrier MIS cell on an epitaxial
                                                substrate (Ref 41).

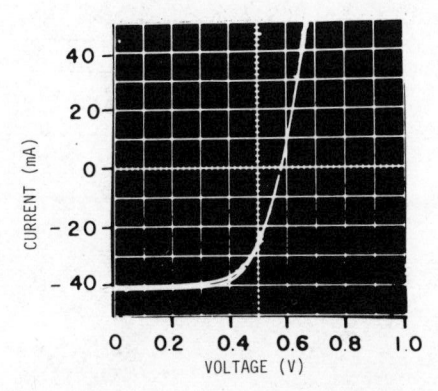

Figure 16.   AM1 V-I curve for a typical Cr barrier MIS cell (Ref 41)

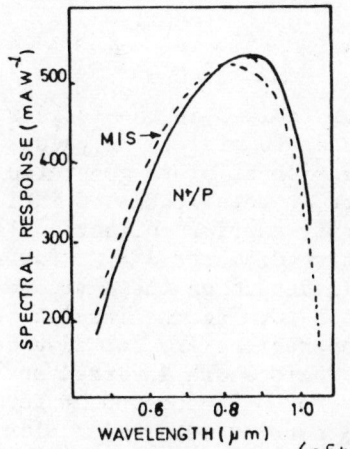

(after Fabre see Ref 10)

Figure 17.   Spectral response of MIS cell and diffused junction n+
             p cell made from similar Si starting material.

minority lifetime in the bulk material. This preservation of
lifetime should provide a better long wavelength response for MIS
cells. The shortwave or blue response should be enhanced by the
action of the surface field in devices where there is a depletion
or inversion region next to the surface. This improvement in
spectral response is characterised by Figure 17 where two cells
made on identical substrates using an MIS barrier at a n+ p
junction are compared[37]. Note that although the blue response is
improved the long wave response is worse for the MIS cell, a fact
attributed by the author to absorption in the Ti barrier used.

Table 3 gives I_{sc} data of a number of silicon MIS cells with
differing barrier metals and A-R coatings under AM1 conditions.

TABLE 3[10]

J_{sc} VALUES FOR Si MIS CELLS

Type	AR Coating		Barrier Metal		Total Area (cm^2)	Peripheral Area[1] %	J_{sc} (mA.cm^{-2})		Reference
	Material	Thickness (nm)	Material	Thickness (nm)			Quoted	Corrected	
p	TiO$_2$		Ti	5	3.0	1	28.6	28.3	37
p			SiO$_2$	100	1.1	0	30.3	30.3	50
p	SiO	67	Cr/Cu/Cr	4/6/0.5	1.0	3	22.3	21.6	41
n			Au	10	0.33	13	24.0	21.2	6
p			Al		0.3	7	23.0	21.5	44
n			Au		0.196	19	22.1	18.6	40
n			Au	10	0.18	20	23.3	19.4	25
n	In$_2$O$_3$	70	Pt	8	0.12	23	29.2	23.7	42
p	SiO/ZnS	60/45	Al	10	0.026	25	26.5	21.2	5
p			Au	7	0.019	74[2]	43.7	25.1	39
n			Au		0.0078	96	42.0	21.4	51

[1] Computed as in [52] taking Δ = 100 µm for p-type and Δ = 200 µm for n-type.

[2] Δ taken as 200 µm in this case because of likely inverted surface due to "dry" oxide[39].

One point to note from Table 3 is the large range of cell areas
reported on. This leads to misleading values of J_{sc} for the smaller
cells since the collection of photo generated carriers from regions
peripheral to the barrier metal tends to inflate the values of J_{sc}[52]
If an effective minority carrier collection length Δ is defined
this yields an effective diameter d + 2Δ for a circular metal barrier
of diameter d. Δ will depend on the semiconductor surface treatment
and the method used for forming the insulator; in practice Δ will be
smaller when depletion regions are formed under the bare oxide than
is the case when the regions are inverted or accumulated[52]. Using
a value of Δ = 100 µm for p-Si and 200 µm for n-Si and basing J_{sc}
calculations on total barrier metal area plus the extra peripheral
collection area the corrected values of J_{sc} shown in Table 3 were
obtained. This corrected set of figures shows less spread than the
uncorrected values and brings out the good performance of the Ti

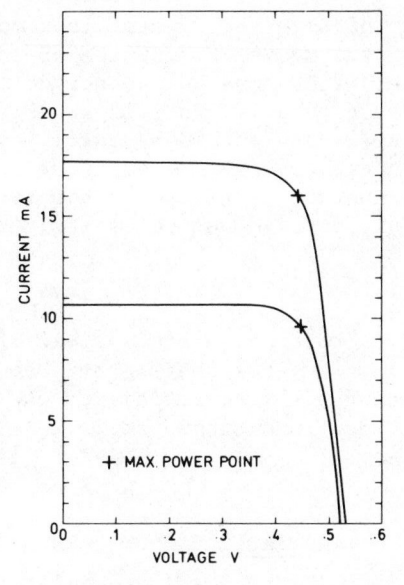

Fig. 18. V-1 characteristics for the SBSC of Fig. 1 under AMI illumination before and after application of antireflection coating. (Ref. 28)

barrier cell, the spectral response of which was given in Figure 17. The thin ~ 50 A Ti layer used in this cell together with the long minority carrier diffusion length ~ 100 μm are important factors making this cell outperform most of the others.[37] The cell with the highest value of J_{sc} is different from the remainder in that it has no continuous barrier metal layer. It is in fact an inversion layer cell of a type considered later in para 4.6. The fact this cell has a high value of I_{sc} highlights the importance of reflection from the barrier metal in determining I_{sc}. The development of cells using conducting metal oxide films such as indium tin oxide (ITO) may lead to an improvement in I_{sc} values since conducting oxides deposited on top of very thin barrier metals may allow combination of the features required of the AR coatings to be achieved at the same time as a reduction in the series resistance in a single deposition stage [53].

The fall off in J_{sc} found when semicrystalline substrates are used is not very large as is shown in Figure 18 which shows an I-V curve for an Al on p-type MIS cell based on Wacker Silso cast silicon. Cells up to 4 cm^2 active area have shown values of J_{sc} ~ 18 mA/cm^2.[28]

4.3 Fill Factor

For Si MIS cells the best fill factors (FF) obtained have been $\sim 0.73 - 0.75$[6, 25-6, 28, 48] and even higher values have been reported

for GaAs. The problem of high fill factor is almost entirely
determined in terms of the series resistance of the cell. In some
cases where high values of FF have been observed they have been
accompanied by low values of J_{sc}. This is usually caused by too
thick barrier metal layers and should be easily corrected. In order
to reduce series resistance however more effort has to be put into
the design of the top contact particularly for larger area cells.
This is a more difficult problem but not a limiting one.

4.4 Efficiency

Here the problem is to combine high values of V_{oc}, J_{sc} and FF
in the one cell. Table 4[10] gives the highest recorded values of
these quantities together with the efficiency one would get if
these values were combined in a single cell.

TABLE 4[10]

| Semiconductor | Maximum Reported Values | | | | Present Technology η_{max} |
	V_{oc} (mV)	J_{sc}[1] (mA.cm^{-2})	FF	η[1] (%)	
GaAs	830	27	0.82	17.0	18.34
Si	615	28.3	0.75	10.2	13.1

[1] Based on total collector area.

Also included are values of actual efficiencies found in
practice. From this it can be seen that although the maximum
attainable efficiency is close to what should be available from
junction cells, $\sim$ 20% for Si and $\sim$ 24% for GaAs there is a considerable margin to make up before these values are reached in practice.
For GaAs cells the main progress needs to be made in improving V_{oc},
however in silicon it is J_{sc} that is presently limiting overall
cell performance. To improve J_{sc} one needs better I-R coatings,
better contact grids with finer fingers or perhaps transparent
conducting oxides, and probably thinner metal barrier layers.

4.5 The min MIS Cell

As already discussed in chapter three it has been discovered by Green, King and Shewchun[18] that if the metal work function and substrate doping of an MIS solar cell are chosen to ensure strong inversion of the semiconductor surface at equilibrium, then for small forward bias the majority carrier thermionic emission dark current can be reduced to a negligible value. Under these conditions the current is dominated by the injection of minority carriers from the metal into the semiconductor and the subsequent recombination of these carriers in either the depletion region or the quasineutral base. In this device the dark current-voltage characteristic is therefore similar to that of a diffused junction p-n diode formed on an identical substrate for small forward bias and reverse bias conditions. This being the case, under illuminated conditions both structures should give the same value of V_{oc} when they are illuminated to give the same value of J_{sc} [10,20]. It should also be possible to apply the same techniques for improving V_{oc} to MIS structures that have been applied to diffused junction cells, namely the employment of either low resistivity base material or the use of a high resistivity base together with a back surface field (BSF).

Considering the first of these techniques for improving V_{oc}. In silicon homojunctions the dark current decreases with base doping until the back injection current component of holes diffusing from the base to the surface layer of an n+ p cell becomes dominant. In this type of cell however due to the very heavily doped surface layer it has been shown that band gap shrinkage and an increased intrinsic carrier concentration impose a lower bound on the back injection current[10]. For the min MIS cell on the other hand these high doping density effects should not be present owing to the induced nature of the equivalent surface layer. In this case, in principle at least, it should be possible to reduce the diffusion current in the minority carrier MIS cell as the base level doping is increased according to the simple diffusion theory[54]. This should lead to an increase in V_{oc} as shown in Figure 19 where some experimental points are also plotted for MIS and p-n junction cells[10,54]. The data for the MIS cell is not very encouraging and shows little improvement over the junction cell at low base resistivities. This could be due to there being a requirement for a metal with a lower work function than aluminium to give inversion in such heavily doped substrates.

The second technique for improving V_{oc} by introducing a back surface field has been demonstrated by Tarr and Pulfrey[55,56]. Al-SiO$_x$-p-Si solar cells were fabricated on 10 Ω cm substrates that were identical to those used to form diffused junction BSF cells. A necessary requirement for a BSF action in the cell to occur is that the minority carrier diffusion length in the base must be greater than the base width. In this work the base width was

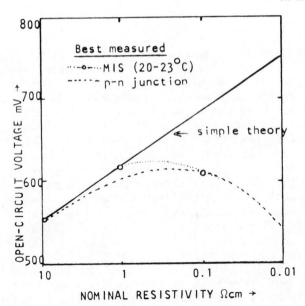

Figure 19.   Comparison of best measured open circuit voltages
            for MIS and p-n junction devices. Note the
            departure from simple theory in both cases.

typically ∿ 300 μm and the electron diffusion length > 800 μm. The
BSF region was formed in the usual manner for junction cells by
first screen printing an aluminium paste into the back surface of
the wafers and then alloying the aluminium at 840 C for 400 secs.
This formed a p+ region ∿ 1 μm thick at the back of the cell.
Following a chemical cleaning process ∿ 0.3 μm of aluminium was
evaporated onto the back surface and the wafer then heated in
dry oxygen at 500 C for 20 min. This process served the purpose
of forming a good ohmic contact to the back of the wafer and also
forming a thin oxide layer on the front onto which ∿ 80 Å of
Aluminium was evaporated to form the MIS barrier. Following the
evaporation of a further 0.3 μm of aluminium to form a front contact
grid the BSF MIS cells were ready for test under an ELH tungsten-
halogen lamp the intensity of which was set to give a short circuit
current density in the cell of 300 Am^{-2}. Cells were always fabricated
in pairs, one with and one without the BSF. The results showed that
in every case the value of V_{oc} was higher for BSF cells than for
those without this feature. Mean values were

    V_{oc} for BSF cells   577 ± 4 mV

    V_{oc} for non BSF cells   539 ± 8 mV

The fill factors for the BSF cells was ∿ 0.6 – 0.7 and thought to
be limited by the non optimised contact grid pattern. Certainly
the significant increase in V_{oc} observed does provide support for

the idea that it is possible to make Al-SiO$_x$-p Si diodes in which majority carrier thermionic emission current is small compared with the minority carrier injection diffusion current. This belief has been further confirmed by Tarr and Pulfrey [57] who have made measurements of J_{sc} and V_{oc} over a range of illumination levels at various temperatures for this type of MIS cell without a BSF. In this work they found that at high illumination levels the data they obtained satisfied the relation that

$$J_{sc} = J_o \exp \left(\frac{q\,V_{oc}}{kT} \right) \qquad\qquad 39$$

when J_o is a temperature dependent constant. Examination of the variation of J_o with temperature showed again that the dark current in the diodes was dominated by minority carrier flow.

How may the value of V_{oc} be further increased for the BSF cells? Some clue may be gained by looking at the dark J-V curve shown in Figure 20. Here it is seen that at low forward biases < 400 mV, the n factor is significantly greater than one would expect for an MIS cell ($\sim$ 2). Comparison with the curve for a similar device lacking the BSF showed that whereas the dark current above 500 mV was less for the BSF device, at lower forward biases it was very much larger. This suggests that to achieve further increase in V_{oc} in the BSF-MIS device it will be necessary to identify the mechanisms giving rise to these dark currents and to find ways of reducing them. Tarr and Pulfrey [55] suggest one possible explan-

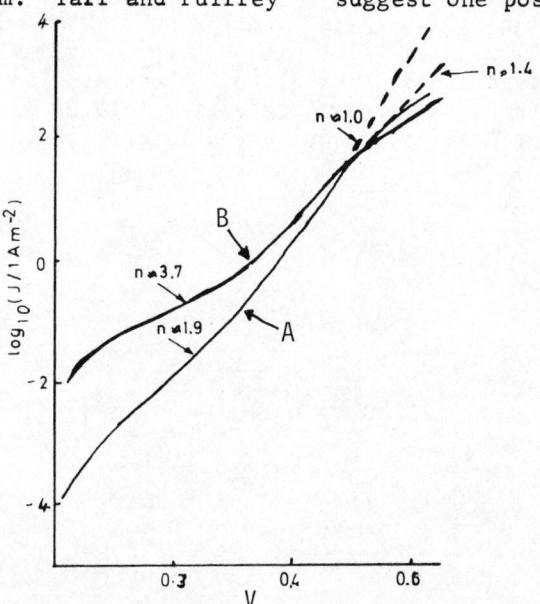

(Ref 55)

Figure 20.   Dark J-V and J_{sc}-V_{oc} (shown dashed) characteristics for representative non BSF cell (curve A) BSF cell (curve B)

ation in their case that contamination present in the aluminium
paste used for the BSF junction may have diffused into the silicon
causing the shunt leakage current to be increased together with
the depletion region recombination current. More work needs to
be done on these interesting cells to clarify the true position
and to enable them to achieve their true potential.

4.6   The MIS Inversion Layer Cell

The inversion layer cell has been studied by a number of
workers [50,58,72]. This type of cell utilises the fact that inherent
to a thermally oxidised silicon surface is a sheet of positive charge
located in the oxide near the oxide-silicon interface. This charge
is fixed, stable, process dependent and highly predictable[59].
If a lightly doped - 10 Ω cm p-type substrate is used the inherent
interface charge will induce an n-type surface inversion layer to
which contact may be made by means of a fine grid to form a solar
cell structure simular to that shown in Figure 21.

The advantage of this type of cell over the conventional MIS
technique is that it removes the necessity for the very thin semitransparent barrier metal layer. It is advantageous compared with
the diffused junction cell in that it gives an ultra thin junction
leading to an enhanced UV and blue response and requires no high
temperature processing.

Green and his co-workers [72,73,74] have produced inversion
layer cells using SiO as the AR coating on .1 Ω cm substrates having
active area efficiencies > 18% with V_{oc} up to 642 mV. Cells using
coarser contact gratings, similar to those used on space cells
have demonstrated total area efficiencies $\sim$ 13.5% with reduced fill
factors $\sim$ 0.68 caused by the increased lateral resistance.

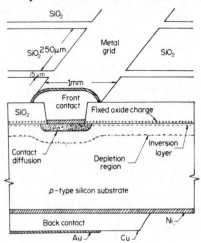

Figure 21.   Inversion layer solar cell structure. (Ref. 58)

Two types of cell have been described by Van Overstraeten and his co-workers [50]. One is based on titanium oxide TiO_x as the insulator and the other on SiO_2. Figure 22 a and b shows the structure of the two cells. For the TiO_x cell one masking stage is involved and there are no high temperature stages, other than that required to grow the thin silicon dioxide layer during the sintering of the backside metallisation in wet oxygen. The aluminium grid is formed from a 1 μm evaporated layer subsequently defined photolithographically. The 1000 Å Titanium oxide layer is spun on to bare silicon, the thin oxide having been previously removed between the contact fingers by an HF dip. Typical I-V curves for a cell with a 2.95 cm^2 active area is shown in Figure 22c. The cell efficiency was found to increase monotonically with substrate resistivity with a maximum obtained value of 8%. I_{sc} is a strong function of grid spacing decreasing as the spacing becomes too large.

Recently the performance of the TiO_x cell has been improved by the addition of a p+ BSF region on the wafer. Using 10 Ω cm p-type silicon wafers a 2% increase in efficiency has been observed compared with identical cells made without the BSF. This increased AM1 efficiency results from an increase in V_{oc} of 60 mV and an increase in $I_{sc} \sim 4.5$ mA/cm^2. This change in I_{sc} is markedly more than is found normally in conventional BSF cells.[60]

The SiO_2 MIS cell was more difficult to produce since two masking stages were required, one to define the thin oxide region and the second for the top contact metalisation. Cell efficiencies ~ 12-13% were obtained with values of $I_{sc} \sim 30$ mA/cm^2, $V_{oc} \sim 520$-550 mV and fill factors ~ 0.75. Cells of both types have been made on semicrystalline cast silicon substrates with very encouraging results [50].

In another recent paper by Thomas et al [61] using spun on Ta_2O_5 to create the inversion layer on p-type silicon, 1 cm^2 cells with total area efficiencies ~ 17% have been made which is the highest claimed for any MIS solar cell. Values of V_{oc} were ~ 55 mV, $I_{sc} \sim 40$ mA/cm^2 and fill factor ~ 0.8.

A significant advantage of all three types of inversion layer cell of course is that there is no additional stage necessary to deposit an A-R coating. The major snag is probably the need for an expensive photolithographic stage or stages.

4.7. Stability of MIS solar Cells

Before large scale production of MIS solar cells can be contemplated for terrestrial applications it is essential that the long term performance of the cells should be assessed and any mechanisms present that cause the cell efficiency to degrade identified. Because of the thin oxide and barrier metal films

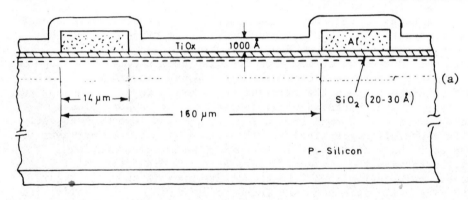

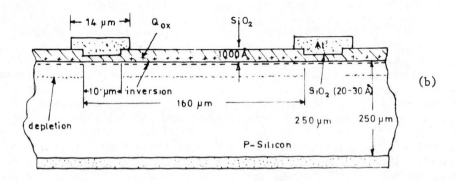

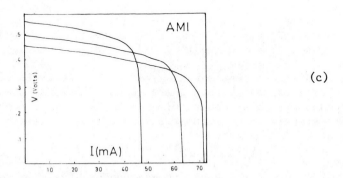

Figure 22. Cross sections through TiO$_x$ (a) and SiO$_2$ (b) MIS inversion cells and typical I-V characteristic for TiO$_x$ cells of active area 2.95 cm^2 (c) (Reference 50).

required for MIS cells there is currently some doubt regarding cell
stability over both long and short time intervals and a number of
papers have drawn attention to the difficulty [28], [62-64].

Consider the published data in a little more detail. It is
well known for gold n-type MIS cells that diffusion of gold through
the oxide gives rise to drastic reductions in V_{oc} as cells age [6], [65].
This is of course due to the effective elimination of the
insulating layer. Fill factor is also degraded due to an increase
in cell series resistance. Turning to the more promising p-type
cells, again there are doubts about long term stability owing to
the critical role played by the thin oxide layer.

A systematic study of the reliability of their cromium-MIS
p-type cells has been carried out by Anderson and Kim[62]. Over an
eighteen month period cells encapsulated in either polystyrene
or Sylgard showed reductions of $V_{oc} \sim$ 6-8%, $I_{sc} \sim$ 8-25%
(the higher figure caused by discolouration of the Sylgard)and FF
$\sim$ 5-10%. Investigation of the cells after prolonged use using
Auger techniques showed that there was no significant diffusion
of copper into the chromium but that there was some penetration
of the chromium into the oxide. The most serious degradation
in FF highlights the need for cells to be hermetically
encapsulated since only in this way will the cell series resistance
increase be avoided by preventing a decrease in the conductivity
of the thin barrier film. The encapsulant used should not discolour
with age or pick up dirt during exposure to the elements.

Earlier work by Townsend and his co-workers[28],[63] has cast
doubt on the stability of Aluminium p-type MIS cells and this has
been recently supported by Pulfrey[64]. The conclusion reached is
that the conditions under which the SiO_2 is grown and the barrier
deposited are critical. In particular it has been shown that the
presence of any water vapour during processing caused significant
decreases in V_{oc} and FF with ageing. This was thought to be
caused by the formation of an aluminium oxide interface in addition
to the grown silicon oxide layer. Cells annealed in hydrogen and
oxidised in dry oxygen at 400 C showed significantly less degradation.
Typically V_{oc} showed an initial 10% decrease during the first week
after production but this was compensated for by a similar increase
in I_{sc} so that the efficiency was preserved. Thereafter some cells
remained stable over a period of twelve months but others slowly
degraded.

A study of this long term degradation has shown that it is
enhanced by illuminating the cell with concentrated sunlight but
that the original performance can be restored by a simple heat
treatment of the cell at 85 C for 15 minutes. Figure 23[63] shows
how the V-I characteristic of one cell behaved when subjected to
alternate illumination and annealing at 85 C. Note even resting

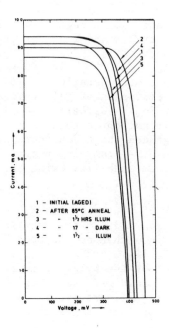

Figure 23. Degradation and
recovery in I/V characteristics for 1 cm^2 Al-oxide
single crystal silicon
SBSC under AMl illumination
(Ref 63).

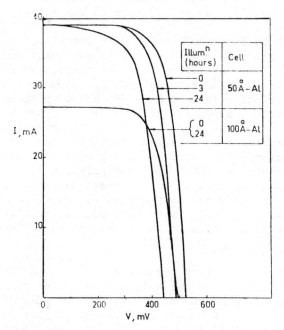

Fig. 24.   Effect of prolonged 1 Sun AMl illumination on coated
Al/SiO/pSi solar cells. The cells were on open circuit
during illumination. (Ref. 64)

the cell for a period in the dark is sufficient to partially restore V_{oc}. A subsequent thermal anneal of this cell fully restored V_{oc} to its original value.

Results very similar to these have recently been reported by Kleta and Pulfrey [64] for their min MIS p-type cells that used SiO as an AR coating and a 50 Å Al barrier layer. Figure 24 [64] compares results obtained for two cells, with 50 and 100 Å barrier layers after 24 hours exposure to AM1 illumination. It is seen that the cells with the thicker barrier did not degrade though of course it has a significantly lower value of I_{sc}. It has been further observed by studying the performance of cells illuminated through narrow band optical filters that the changes in V_{oc} are only observed when photons of relatively short wavelength < 4800 Å are incident on the cell. This observation supports the view that for the thicker metal barrier photons are prevented from reaching the semiconductor by increased absorption in, or reflection from, the Al layer.

How may the degradation process be explained? Grimshaw and Townsend have attributed it to photo-stimulated migration of oxygen from the oxide to the cell exterior, coupled with some migration of aluminium into the thin oxide layer and changes to the chemical bonding. This result it supported by Auger analysis and is not inconsistent with the observation of the wavelength dependence of the degradation process reported by Kleta and Pulfrey. Pulfrey has put forward an alternative explanation that the degradation is caused by a process of photo-neutralization of the effective positive charges occurring at the oxide/semiconductor interface. Such a mechanism would clearly be dependent on the photon wavelength and also, because of the change in effective barrier height induced, would lead to an instability primarily in V_{oc} rather than induce changes in I_{sc}. The temperature dependent restoration process can be seen in terms of thermal ionization leading to reactivation of the unspecified positive charge centres.

Whatever the cause of the degradation there is obvious cause for concern for the whole future of MIS cells if their characteristics cannot be preserved over a useful working life > 10 years for a finished cell. More work is necessary on this problem since it will obviously be fatal if the only way in which the cells can be made stable is to shield the cells with thicker metal barriers, so degrading their short wavelength response until it is worse than the corresponding more stable junction cell.

The problem of cell stability is likely to be an even greater problem in the case of amorphous silicon MIS cells where one has the additional problems caused by the properties of the thin α-Si layer in addition to the thin barrier and insulating films.

4.8   The Future for MIS cells - Cheaper Substrates?

On the assumption that it will be possible to get around the
long term stability problem discussed in the previous section the
future for Schottky Barrier solar cells must depend on our ability
to develop cells on cheaper substrates while maintaining cell
conversion efficiencies $\sim$ 10% or better. There are a number of
alternative substrates that are currently under investigation among
which the following are the most promising.

> Cast grain oriented polycrystalline silicon (SILSO
> type)
>
> Ribbon silicon either EFG or web dentritic
> Silicon coated graphite
> Amorphous silicon

Results so far obtained on poly silicon substrates show MIS cells
having efficiencies $\sim$ 10% less than on the corresponding single
crystal substrates [28]. Similarly encouraging results have been
reported for ribbon silicon[49]. Here the current emphasis is to
improve the growth processes to give ribbons wider than the 50 mm
ones presently available having smaller numbers of crystal defects
and inclusions. Silicon coated graphite and other processes using
metallurgical grade silicon to produce silicon in sheet form should
also show useful efficiencies with MIS cells when the substrate
materials have been better developed. Current work has tended to
emphasise the formation of junction cells.

Amorphous silicon is the long term hope for a cheap truly thin
film silicon solar cell. Numerous groups throughout the world are
working on the materials problem to produce α-silicon films having
low enough interband trap densities that can give solar cells of
useable efficiencies. MIS cells using n-type α-Si have achieved
efficiencies $\sim$ 5%[35] and theoretical predictions give hope that more
than double this value should be possible. This subject is beyond
the scope of the present paper as it is very much more dependent
on making good α-Si than on the process used to form the MIS
barrier. At the present time two methods are used for forming the
hydrogenated α-Si layers required. These are the pyrolosis of
silane in an h-f flow discharge, work pioneered by Spear at
Dundee[66-7] and r-f sputtering of silicon in an hydrogen atmosphere[68].
The Schottky barriers have usually been made using a platinum or
paladium barrier with TiO_2 deposited to form the insulating layer.

Work on GaAs MIS cells has likewise been very much constrained
by need to develop cheaper methods for growing GaAs substrates.
The advent of improved VPE processes based on metal organics for
growing GaAs films should lead in time to an efficient thin film MIS
solar cell.   A typical recent result [69] claims a 9 cm^2 cell having
an AM1 efficiency of 5% for a cell with no A-R coating. The

inclusion of an A-R coating would probably boost the efficiency to
7.5%. Work up to the present time, however, has been largely
concerned with the development of really high efficiency single
crystal cells that can be used with concentrated sunlight. The
state of this work has been reviewed by Pulfrey [10] and reported in
table 4.

Acknowledgement

The author is extremely grateful to his former colleague
Dr D R Lillington for his permission to quote freely from his PhD
thesis [71] during the first half of this paper. Many of the figures
in sections 2 and 3 are taken directly from this work.

References

1. H.C. Card and E.H. Rhoderick, J. Phys.D. Appl.Phys. 4:1589 (1971).
2. B.L. Smith and E.H. Rhoderick, Solid State Electron, 14:71
     (1971).
3. H.K. Henisch, 'Rectifying Semiconductor Contacts', Oxford
     University Press, Oxford, (1957).
4. E.H. Rhoderick, Metal-Semiconductor Contacts, Oxford University
     Press, Oxford (1978).
5. E.J. Charlson and J.C. Lien, J. Appl. Phys, 46:3982 (1975).
6. D.R. Lillington and W.G.Townsend, Appl. Phys, Lett 28:97 (1976).
7. W.A. Anderson A.E. Delahoy and R.A. Milano J. Appl. Phys.45:3913
     (1974).
8. R.J. Stirn and Y.C.M. Yeh, Appl. Phys. Lett. 27:95 (1975).
9. E. Fabre Appl. Phys. Lett. 29:607 (1976).
10. D.L. Pulfrey, IEEE Trans. on Elec. Dev ED25:1308 (1978).
11. A.M. Cowley and S.M. Sze J. Appl. Phys. 36:3212 (1965).
12. J.L. Saltish and L.C. Terry Proc. IEEE 57: 493 (1970).
13. F.G. Allen and G.W. Gobeli Phys. Rev. 127:150 (1962).
14. R.J. Archer and M.M. Attalla Proc. Conf. on Clean Surfaces,
     New York 1962, Amn Acad of Sci. 101:697 (1963).
15. J.M. Andrews and M.P. Lepselter, IEEE Solid State Devices Conf.
     Washington, Oct (1968).
16. S.J. Fonash, J. Appl. Phys. 46:1286 (1975).
17. C.Y. Chang, S.M. Sze, Solid State Electron 13:727 (1970).
18. M.A. Green, J. Shewchun and F.D. King, Solid State Electron,
     17:551 (1974).
19. J. Shewchun, M.A. Green and F.D. King, Solid State Electron,
     17:563, (1974).
20. M.A. Green and R.B. Godfrey, Appl. Phys. Lett. 29:610 (1976).
21. D.L. Pulfrey 'Photovoltaic Power Generation' Van Nostrand
     Reinhold, New York (1978).
22. H.C. Card and E.S. Yang Appl. Phys. Lett 29:51 (1976).
23. J. Shewchun, R. Singh and M.A. Green J.of Appl. Phys 48:765 (1977).

24. J. Shewchun and M.A. Green J. Appl. Phys. 46:5179 (1975).

25. J.P. Ponpon and P. Siffert J. Appl. Phys. 47:3248 (1976).

26. D.L. Pulfrey, Solid State Electron 20:455 (1977).

27. J.P. Ponpon, J.J. Grop, A. Grob, R. Stuck and P. Siffert,
    in Proc. 3rd Int. Conf. Ion Beam Analysis, (1977).

28. W.G.Townsend and D.R. Lillington   Proc. 1st CEC Photovoltaic
    Solar Energy Conf. 207 (1977). (D. REIDEL, Dordrecht)

29. A.H.M. Kipperman, S.C.M. Backerra, H.J. Maaskamp and R.J.C.
    van Zolingen, Proc. Ist CEC Photovoltaic Solar Energy
    Conf. 961 (1977).

30. A.H.M. Kipperman, S.C. Backerra and H.J. Maaskamp, Ibid, 956
    (1977).

31. O.S. Heavens 'Optical Properties of Thin Solid Films',
    Butterworth, London (1955).

32. W.C. Dash and R. Newman, Phys. Rev. 99:1151 (1955).

33. S.S. Li, F.A. Lindholm and C.T. Wang J. Appl. Phys. 43:4123 (1972)

34. B. Ross and R. Madigan, Phys. Rev. 108:1428 (1957).

35. D.E. Carlson and C.R. Wronski, Appl. Phys. Lett. 28:671 (1976).

36. P.K. Dubey, Appl. Phys. Lett. 29:435 (1976).

37. E. Fabre, J. Michel and Y. Bawdet, Proc. 12th IEEE Photo. Spec.
    Conf. 904 (1976).

38. J.P. Ponpon, R. Stuck, P. Siffert, Ibid 900 (1976).

39. A.H.M. Kipperman and M.H. Omar, Appl. Phys. Lett 28:620 (1976).

40. R. Childs, J. Fortuna, J. Geneczko and S.J. Fonash Proc. 12th
    IEEE Photo Spec Conf 862 (1976).

41. W.A. Anderson, S.M. Vernon, A.E. Delahoy, J.K. Kim and P. Mathe,
    J. Vac. Sci. Tech. 13:1158 (1976).

42. H. Matsumani, S. Matsumoto and T. Tanako, Proc. 12th IEEE
    Photo. Spec. Conf. 917 (1976).

43. P. van Halen, R.E. Thomas and R. van Overstraeten, Ibid 907 (1976)

44. M.A. Green, R.B. Godfrey and L.W. Davies Ibid 896 (1976).

45. E.J. Charlston, Final Rep. NSF Contract ENG 74-02918 (1976).

46. R.J. Stirn, Y.C.M. Yeh, E.Y. Wang, F.P. Ernest, C.J. Wu, Tech.
    Dig IEEE Int. Electron Devices Meeting, 48 (1977).

47. R.J. Stirn and Y.C.M. Yeh Proc. 12th IEEE Photo. Spec. Conf.
    883 (1976).

48. J.A. St. Pierre, R. Singh, J. Shewchun and J.J. Loferski, Ibid
    847 (1976).

49. A.E. Delahoy, W.A. Anderson and J.K. Kim Proc. 1st CEC Photo .
    Solar Energy Conf. 308 (1977). (D. REIDEL, Dordrecht)

50. P. van Halen, R. Mertens, R. van Overstraeten, R.E. Thomas and
    J. van Meerbergen, Ibid 280 (1977).

51. S, Shevenock, S.J. Fonashand, J. Geneczko, Tech. Dig. IEEE Int.
    Electron Devices Meet. 211 (1975).

52. R.B. Godfrey and M.A. Green, Appl. Phys. Lett 31:705 (1977).

53. J.J. Loferski, M. Spitzer and D. Burk Proc. 14th IEEE Photo
    Spec. Conf , 1376 (1980).

54. F.A. Lindholm, A. Neugroshel, C.T. Sah, M.P. Godlewski and
    H.W. Brandhorst, Proc. 12th IEEE Photo. Spec. Conf. 1 (1976).

55. N.G. Tarr and D.L. Pulfrey, Appl. Phys. Lett. 35, 258 (1979).

56. N.G. Tarr, D.L. Pulfrey and P.A. Iles, Proc. 14th IEEE Photo.
          Spec. Conf. 1345 (1980).
57. N.G. Tarr and D.L. Pulfrey Appl. Phys. Lett. 295:34 (1979).
58. G.C. Salter and R.E. Thomas, Solid State Electron 20:95 (1977).
59. B.E. Deal, M. Sklar, A.S. Grove and E.H. Snow, J. Electro. Chem.
          Soc. 114:266 (1967).
60. P. van Halen, K. Petit, R. Mertens, R. van Overstraeten and
          R. Girisch, Proc. 14th IEEE Photo. Spec. Conf.1366 (1980).
61. R.E. Thomas, C.E. Norman and R.B. North Ibid 1350 (1980).
62. W.A. Anderson and J.K. Kim, J. Appl. Phys. 17:401 (1978).
63. J.A. Grimshaw and W.G. Townsend Proc. 2nd CEC Photo. Solar
          Energy Conf. 197 (1979). (D. REIDEL, Dordrecht)
64. J.K. Kleta and D.L. Pulfrey, IEEE Elec. Dev. Lett. EDLI 107
          (1980).
65. G. Pananakakis, P. Viktorovitch and J.P. Ponpon Rev. Phys
          Appl.
66. W.E. Spear and P.G. LeComber Solid State Commun. 17:1193 (1976).
67. W.E. Spear and P.G. LeComber Phil. Mag. 33:935 (1976).
68. M.J. Thompson, J. Allison, M.M. Alkaisi and S.J. Barber
          Proc. 1st CEC Photo. Solar Energy Conf 231 (1977).
69. S.S. Chu, T.L. Chu and Y.T. Lee Proc. 14th IEEE Photo. Spec.
          Conf. 1306 (1980).
70. R.L. Crabb, F.C. Treble, RAE Tech. Report 67305 (1967).
71. D.R. Lillington An Investigation of the electrical and optical
          properties of silicon Schottky barrier photovoltaic cells
          PhD thesis (1978).
72. R.B. Godfrey and M.A. Green Appl. Phys. Lett. 34:790 (1979).
73. R.B. Godfrey and M.A. Green Appl. Phys. Lett. 33:637 (1978).
74. M.A. Green, R.B. Godfrey, M.R. Wilson and A.W. Blakers Proc.
          14th IEEE Photo. Spec. Conf. 684 (1980).

CdS-Cu$_x$S THIN FILM SOLAR CELLS

W. H. Bloss and H.-W. Schock

Institut fuer Physikalische Elektronik
Universitaet Stuttgart
D 7000 Stuttgart 1, F.R.G.

1. INTRODUCTION

Thin film photovoltaic generators based on a heterojunction have been considered for many years as interesting alternatives to single crystal solar cells. Thin film solar cells promise an economic operation and large scale fabrication due to character-istic features inherent to the systems. One of the most promising photovoltaic thin film generator which has been widely investi-gated and where much progress in research and technology has been obtained during the last years is represented by the CdS-Cu$_x$S-system /1,2/. It is expected that this type of photovoltaic gene-rator could reach the price goal which has been defined for eco-nomic operation competitive to conventional power plants.

Thin film solar cells in general require a minimum of mate-rial and energy input. This is an important fact in view of availability of materials and pay-back time of solar generators. Thin film technology provides means for large scale production and the feasibility to realize integrated photovoltaic generators by the same production process.

. The minimum thickness of the generator is primarily defined by the absorbing film which must absorb the major fraction of the incoming radiation and convert it in potential electrical energy. The Cu$_x$S film in the system to be discussed overtakes this function. Generally materials with high absorption constant α i.e. semiconducting materials with direct band transitions are required.

As can be derived from figure 1 the absorption constant α
of Cu_2S and of amorphous Si is by one to two orders of magnitude
higher than for crystalline Si.

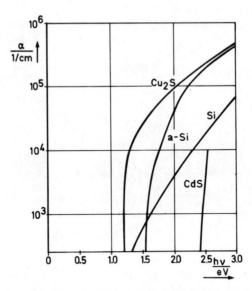

Fig. 1. Absorption coefficients of different materials.

This means that the thickness of crystalline Si solar cells
must be higher by the factor 10 to 100. Thin film solar cells
require absorbing films in the order of 1 μm or less. Crystalline
Si solar cells correspondingly exhibit thicknesses in the order
of 100 μm.

Only few materials used for thin films can be made p-type
and n-type equally. A p-n-junction therefore in many cases can
only be realized by using two different materials which leads
to a heterojunction which imposes additionally the problem of
compatibility of different chemical and crystallic structures.
For the heterojunction system CdS-Cu_xS conversion efficiencies
as high as 9 % have been demonstrated, a practical method of
fabrication technology has been established and operational tests
have been performed /3,4/. The results and experience obtained
prove the CdS-Cu_xS thin film solar cell as an alternative to
crystalline cells such as Si. The basic structure of the CdS-Cu_xS
thin film solar cell is illustrated in figure 2.

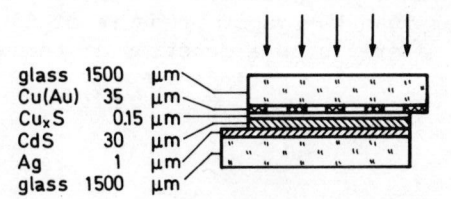

glass	1500	μm
Cu(Au)	35	μm
Cu$_x$S	0.15	μm
CdS	30	μm
Ag	1	μm
glass	1500	μm

Fig. 2.   Schematic structure of a CdS-Cu$_x$S thin film solar cell.

The glass substrate is covered by a metallic film (Ag) which serves as a back contact. The heterojunction system is based on a CdS film (between 10 and 30 μm thickness) which due to sulfur vacancies is n-type. On the top of the CdS film a Cu$_x$S film (x ≈ 2) is prepared which due to small deviations from stoichiometry is p-type. A highly transparent grid represents the front contact. The front glass plate through which the incoming radiation penetrates to the junction provides a hermetically sealed encapsulation which is required for long time stability.

In the following chapters the technological procedures, material properties and performance of photovoltaic generators based on the CdS-Cu$_x$S heterojunction will be discussed and future prospects will be evaluated.

2. CdS THIN FILM TECHNOLOGY

2.1. VACUUM VAPOR DEPOSITION OF CdS FILMS

The highest efficiencies of polycrystalline CdS-Cu$_x$S solar cells were achieved using vapor deposited CdS films /3,4/. However the reproducibility of the film properties necessitates special arrangements. For vapor deposition the properties of CdS powders or pellets have to be taken into account. CdS powders contain a certain amount of absorbed gases and volatile compounds, so that careful degassing is required. Furthermore, CdS sublimes and dissociates to Cd and S. From the subliming powder solid particles are ejected and are sticking to the film causing hereby defects such as growth anomalies and pinholes.

A similiar behaviour can be observed for ZnS which is a material of interest for the formation of $Cd_{1-x}Zn_xS$ compounds in order to get better matching of the electron affinity and lattice constants to the Cu_xS. The vapor pressure of ZnS is about one order of magnitude lower than the vapor pressure of CdS. The vapor pressure of both materials as a function of temperature is plotted in figure 3.

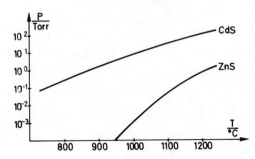

Fig. 3.   Vapor pressure of CdS and ZnS /5/.

It can be deduced from these curves that the composition of the vapor from mixed compound sources depends strongly on the temperature. Since CdS evaporates at higher rates films from mixed compound sources exhibit gradients of composition. Better reproducibility can be obtained by two source evaporation, yet a lateral gradient of the film composition makes this method inapplicable for large area devices.

The evaporation of both compounds from one coaxial source with nozzles having a hole size adjusted to the evaporation rates of both compounds proved to be the best deposition method. A review on the deposition of mixed compound films is given by Burton /6/.

The evaporation of CdS from open boats requires low source temperatures and a wide distance between source and substrate in order to prevent solid particles reaching the substrate ("spattering").

Due to the corrosiveness of CdS vapor the boat has to consist of non reactive material such as quartz, graphite or alumina. Tantalum gets brittle after a short time of operation due to the formation of Ta_2S_5. At high evaporation rates, a baffle has to prevent spattering of the powder. The design of a well-suited graphite source is shown in figure 4.

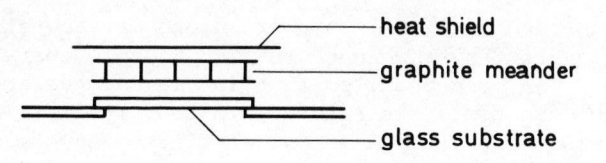

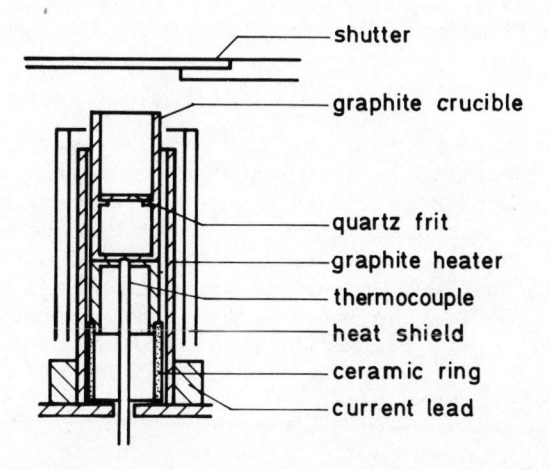

Fig. 4. Evaporation source for deposition of sublimating
         materials at high rates /7/.

Coaxial indirect heating provides a homogeneous temperature
distribution in the zone where the CdS load and the quartz frit
are located. Uniform layers of an area of 7 x 7 cm^2 and a thickness of more than 100 μm at a rate up to 2 μm/min can be deposited with this source /7/.

An economic production of thin film solar cells however can
only be achieved with large area evaporation systems with a material yield of more than 50 %. Such a system cannot be realized
by one "point source" but requires several sources distributed
within short distances below the substrate area in order to get
small material losses at the edges and uniform thickness of the
film /4/.

The growth conditions in this type of system differ somewhat
from those operating with a single point source due to the difference of solid angles. Nevertheless solar cells with good performance have been manufactured with those films /4/.

Other methods such as flash evaporation and coevaporation of Cd and S have been tried by some authors /1/. Flash evaporation requires the use of an open boat, so that in view of the special requirements mentioned above the rate is limited. The main advantage of the evaporation method is the continuous supply of material. Yet sticking of CdS from the vapor to components of the evaporation device limits the time of evaporation drastically. Coevaporation of Cd and S (three temp. method) leads to highly stoichiometric films with low conductivity wich are not suitable for solar cell applications /8/.

2.2. SPUTTERING

Thin film deposition of CdS by sputtering allows an optimum control of film properties by the discharge parameters and the composition of the gas. Experiments have been reported on cathodic sputtering from CdS cathodes /9/ - which might be doped - in an argon atmosphere. More recent work has been carried out using r.f.-sputtering in reactive atmosphere, containing H_2S or H_2 /10/. These experiments confirm the importance of the parameters substrate temperature ($\sim 200^\circ C$) and the partial pressure of reactive gases. CdS films of 3 to 20 μm thickness have been prepared. The deposition rate was about 0.1 μm/min at a power density of 1.6 W/cm^2.

The electrical conductivity of sputtered CdS films varied according preparation conditions between 10^{-6} to $10^{-2}\Omega^{-1}cm^{-1}$. The size of crystallites was between 0.1 to 0.3 μm and demands further recrystallization processes for successful application in solar cell fabrication.

2.3. SPRAY DEPOSITION

The process of spray deposition presents a convenient technological method for fabrication of low cost solar cells. The spray solution containing f.i. cadmium chloride and thiourea (or other Cd and S compounds respectively) at low concentration is sprayed at rates of about 15 - 30 cm^3/min on a substrate heated up to temperatures between 300 and 400 $^\circ C$. There exists a large variety of starting materials, which have to satisfy the condition a) that they must be soluble in the liquid which may contain the dopants as well and b) that the remainder must be completely volatilized.

The crystallinity is strongly influenced by the substrate temperature. The size of crystallites can be increased by heat treatment in an inert atmosphere. The following reactions lead to the formation of a crystalline film:

$$CdCl_2 + CS(NH_2)_2 + 2H_2O \rightarrow CdS + 2NH_4Cl + CO_2$$

$$Zn(NO_3)_2 + CS(NH_2)_2 + 2H_2O \rightarrow ZnS + 2NH_4NO_3 + CO_2$$

The ammonium salts decompose at elevated substrate temperatures. By appropriate use and composition of the starting materials for the base materials for thin film solar cells, namely CdS and Cd$_x$Zn$_{1-x}$S, have been prepared by different authors /11,12,13/.

In some cases AlCl$_3$ is added to the spray solution to prevent Cu diffusion along the grain bounderies during Cu$_x$S formation process /14/.

2.4. SINTERING

CdS films and thin disks as a base material for the CdS-Cu$_x$S heterojunctions have also been prepared by sintering techniques. The preparation method for sintered films consisted in silk-screen printing on a ceramic substrate followed by a drying and sintering process at temperatures of about 600 °C /15,16/. A different approach is based on compressing CdS powder in self supporting thin disks which are sintered at temperatures between 700 and 850 °C. Grain sizes have been reported in this case of 5 to 10 μm /16/.

Table I summarizes the different methods of preparing thin CdS films. It demonstrates that methods a) to c)

	Depos. rate μm/min	Substr. temp. °C	Grain size μm
a) Evaporation	1	200	5
b) Sputtering	0.1	200	0.1
c) Spray deposition	0.01	300–400	0.5
d) Sintering			(5-10) sint.temp. 600-850 °C

Table I. Deposition methods for CdS films

require substrate temperatures below 400 °C which allows the use of inexpensive substrate materials such as glass. The small grain size obtained by sputtering deposition imposes limitations in photovoltaic operation or requires recrystallization processes.

3. Cu$_x$S THIN FILM TECHNOLOGY

The Cu$_x$S film plays an important role in photovoltaic operation. It acts as an absorbing layer where the minority carriers
are created by interaction with the incident radiation and where
the transport to the junction takes place. Hence the Cu$_x$S film
requires maximum control and reproducibility of optical and transport properties which in turn strongly depend on chemical composition, film thickness, grain size and interface states between
Cu$_x$S and CdS. Hence special interest has to be directed towards
technology and diagnostics of Cu$_x$S films.

3.1. DIPPING PROCESS (WET PROCESS)

The dipping process has been successfully experienced and
yields so far the best results in photovoltaic operation. By
dipping the etched CdS layer in an aqueous acidified CuCl solution a topotaxial reaction takes place, according to

$$CdS + 2CuCl \rightarrow Cu_2S + CdCl_2 .$$

Typical parameters of this reaction are: CuCl concentration
10 g/l, pH-value 3-4, temperature of the solution 90 $^\circ$C, dipping
time 5 - 10 s. Sometimes hydrazine or hydroxylamine is added to
the solution to avoid oxydation of Cu$^+$-ions. In this ion exchange
reaction one Cd-ion is replaced by two single valid copper ions.
Since the S-lattice is not affected by this process, the structure of the Cu$_x$S layer is mainly determined by the structure of
the etched CdS film. The ion exchange reaction occurs preferentially along the grain boundaries which leads to a CdS-Cu$_x$S junction as indicated in figure 5.

The successful photovoltaic operation of topotaxial layers
is based mainly on the fact that the junction is shifted into the
interior of the original CdS crystallites and is not formed on
top of a surface which may exhibit detrimental effects.

Stoichiometry and thickness of the Cu$_x$S layer can be controlled by electrochemical methods during dipping /16,17/ which
provides improvements in homogeneity, reproducibility, and photovoltaic operation.

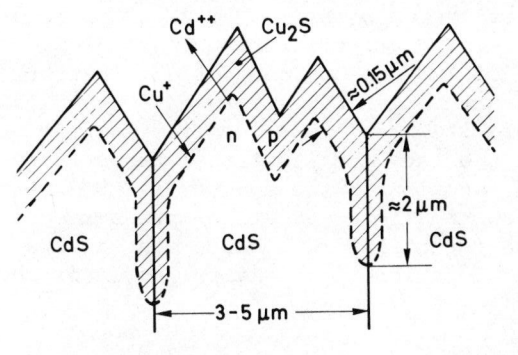

Fig. 5.   Structure of the heterojunction in topotaxially produced
          CdS-Cu$_x$S thin film solar cells.

3.2. EVAPORATION OF CuCl

Evaporation of thin film of CuCl on a CdS substrate followed
by thermal annealing leads to the ion exchange reaction as discussed in section 3.1. This "dry process" results in a topotaxial
growth of the Cu$_x$S film which avoids the deeply penetrating reaction along grain boundaries as observed in the dipping method.
The reaction product CdCl$_2$ has to be washed off by an appropriate
solvent, i.e. water or alcohol. Due to the large number of parameters involved such as purity and stoichiometry of CuCl, partial
pressure of the residual gases, evaporation rate, substrate temperature, temperature gradients during heat treatment, and the
dissolving power of CdCl$_2$, no substantial advantages in photovoltaic operation as compared to the wet process have been reported in literature /18,19/.

3.3. EVAPORATION OF Cu$_x$S

Some authors have used direct evaporation of stoichiometric
Cu$_2$S powder at substrate temperatures of about 250 $^{\circ}$C. The evaporation of Cu$_2$S provides excellent control of thickness and
composition of the resulting Cu$_x$S film. The deposition rates
reported are in the order of 0.5 μm/min /20,21/, which seems to
be sufficiently high in view of the low thickness required in
photovoltaic generators. Whereas vacuum deposition of Cu$_2$S yields
good results in scientific investigations so far the method has

not been applied successfully in fabrication of technical devices.

3.4. SPUTTERING OF Cu_xS

Reactive sputtering in an argon-H_2S-mixture leads to a defined deposition of the Cu_xS layer. The reaction occurs apparently at the substrate. By variation of the partial pressure of H_2S
the stoichiometry of Cu_xS can be controlled. Sputtering rates of
6 μm/h   at   substrate temperatures between 100 and 150 °C have
been indicated /22,23/. Problems arise from the fact that different phases of Cu_xS exist. The electrical resistivity and the
Cu : S ratio as a function of the H_2S partial pressure during
sputtering are shown in figure 6. At low pressure a precipitation
of pure Cu and the formation of Cu cones was observed. In this
Cu rich phase the electrical resistivity increases obviously to
very high values.

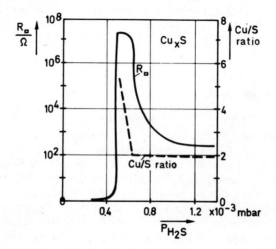

Fig. 6.   Electrical resistivity and Cu : S ratio as a function
          of H_2S partial pressure during reactive sputtering.

Partial pressures above the limits in figure 6 lead to sulfur
rich phases of Cu_xS. The Cu_xS films sputtered on CdS films exhibited epitaxial growth as demonstrated by Armantrout et al. /22/
by electron diffraction and hence showed no higher recombination
velocity at the interface than in topotacially grown junction
layers.

From all methods in Cu_xS film preparation the dipping process
is the one which is most frequently applied. Other methods which
are based on Cu deposition, by evaporation or by electrochemical

methods, and spray processes are considered presently of minor
importance though their technological potential and their compatibility with the corresponding CdS film deposition processes
might lead in the future to a different evaluation and to recon-
sideration of these methods.

4. PROPERTIES OF THE CdS LAYER

The CdS layer acts as a collector for the electrons from
Cu$_x$S during solar cell operation. Furthermore it serves generally
as the basis for the topotaxial formation of the Cu$_x$S layer. Hence
the crystalline structure and the electrical properties of the
film are of major interest. The physical properties of the CdS
films reported in the following sections are characteristic to
films with optimum photovoltaic efficiency.

4.1. CRYSTALLOGRAPHY AND GRAIN SIZE OF CdS FILMS

Crystallites of evaporated CdS films in general exhibit a
hexagonal structure where the c-axis is oriented perpendicular
to the substrate. In some cases also the cubic phase has been observed in the film, depending on the deposition conditions. Stack-
ing faults in the hexagonal lattice are often cubic modifications
of one a,b plane due to very low faulting energy /24/.

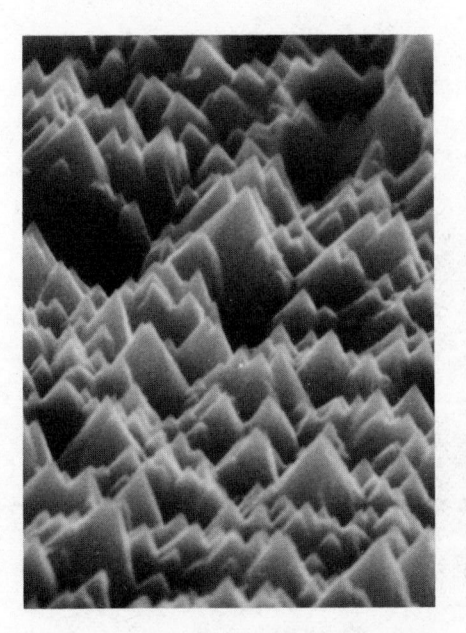

Fig. 7.   SEM picture of a CdS film (left)
and a Cd$_{0.7}$Zn$_{0.3}$S film (right).

The occurance of defects and the orientation of the crystallites both depend on the morphology and the substrate material.
Highly oriented films with few growth anomalies can be achieved
on smooth substrates with vacuum deposited metal layer, e.g. a
silver layer /4/.

Mixed sulfide films $Cd_{1-x}Zn_xS$ exhibit the same crystalline
structure as CdS films. Scanning electron micrographs (SEM) of
both surfaces, etched with HCl, showing hexagonal pyramids do not
indicate any marked difference (see fig. 7).

Under appropriate evaporation conditions a grain size of
2 - 5 μm in films exceeding a thickness of 20 μm on suitable substrates can be achieved. Typical parameters for CdS deposition
are: Substrate temperature 180 ^{O}C, source temperature 1000 ^{O}C,
deposition rate 1.5 μm/min. Larger grains can be obtained by recrystallization of the films. With the method of Gilles and
Cakenberghe /25/ grains of a diameter of one cm can be obtained.
However the orientation of these crystallites is random and precipitation of the activator material at the grain boundaries has
been observed. Furthermore the shrinking process of the film
during recrystallization causes a gap between the individual
grains. A scanning electron micrograph (figure 8) of three adjacent grains shows the different orientations. The 30 μm CdS
film on Ag substrate has been recrystallized by heating in vacuum
for 2 hours at 500 ^{O}C:

Fig. 8.   SEM picture of a recrystallized CdS film.

Due to the different orientations recrystallized films are
not very suitable for solar cell applications. Recrystallization
methods which require even higher temperatures can not be applied
if glass substrates are used /26/. Furthermore the films are incorporating doping materials which affect the performance of the
cells.

4.2. OPTICAL PROPERTIES OF THE CdS FILMS

CdS has a direct band gap at an energy of 2.42 eV at 300 K.
The solar spectrum only contains 20 % of radiation power above
this photon energy. The CdS film therefore contributes only little
to the photocurrent of the cell. Films with good stoichiometry
show a high transmittance at photon energy below 2.4 eV. The
optical gap of Cd$_{1-x}$Zn$_x$S films is shifted to higher energies with
increasing x /27/.

The dependence of the optical band gap on the x-value is
plotted in figure 9.

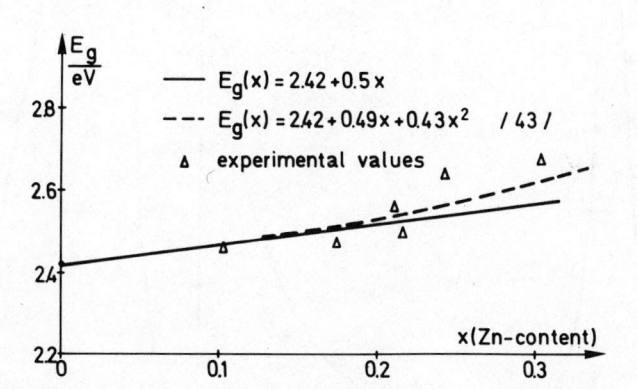

Fig. 9.   Dependence of the band gap on the Zn content of
          Cd$_{1-x}$Zn$_x$S derived from optical transmission
          and luminescence.

4.3. LUMINESCENCE

   Luminescence measurements present a useful tool for the
characterization of the CdS and $Cd_{1-x}Zn_xS$ films. Cathodo- and
photoluminescence have been used in the investigations. Pure films
show a strong band-edge luminescence at about 2.4 eV corresponding to 520 nm. The zinc content of the mixed compound films can
easily be detected by measuring the shift of the edge luminescence. Inhomogeneities of the film compositions become evident
in a broadening of the luminescence peak (fig. 10b). The occurrence of two different peaks corresponds with two different mixtures.
Impurities, namely Ag and Cu, form deep centres in the forbidden
band. Intense luminescence in the red or near infrared is caused
by these impurities (fig. 10a). Other impurities such as In, Fe,
Zn exhibit quenching effects on the luminescence peak in mixed
sulfide films where the Zn : Cd ratio is varying locally.

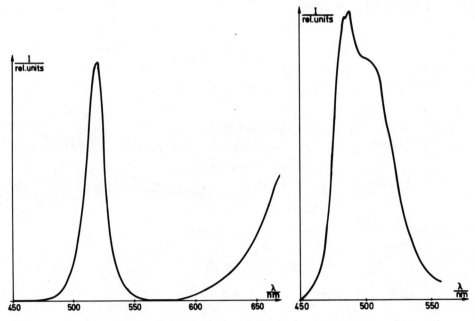

Fig. 10a.   Luminescence of a CdS   Fig. 10b.   Band edge lumines-
            film indicating deep   cence of a $Cd_{1-x}Zn_xS$
            level luminescence   film.
            at λ > 600 nm.

4.4. ELECTRICAL PROPERTIES OF CdS FILMS

CdS films for solar cell applications should have low resistivity in order to reduce the series resistance of the cell. Films which have been evaporated at high rates and at low substrate temperatures are usually n-type and exhibit a resistivity of 1 - 100 Ωcm /28/. Typical values are 10 Ωcm. The residual gases and the purity of the starting material also show an important effect on the resistivity. Low temperature resistivity measurements indicated an activation energy of 25 meV for shallow donor leveld. The carrier concentration is 10^{17} - 10^{18}/cm^3, the mobility is about 10 cm^2/Vs.

It has to be mentioned that these measurements only can be carried out on nonconductive substrates and that the conduction process leads through grain boundaries which strongly influence the electrical transport characteristics. Photovoltaic cells on the other hand are based on conducting substrates, the conversion process occurs mainly in single crystalline regions. The conducting substrates exert a strong influence on the crystallinity of the films. Hence the data quoted above are not necessarily the same as with those films which exhibit optimum photovoltaic performance.

Films with low resistivity can be obtained by In doping /29/. Highly photoconductive films with high dark resistance can be made by Cu or Ag doping. These elements act as deep acceptors at an energy level between 0.8 and 1.2 eV above the valence band and hence cause a fractional compensation of the n-type conductivity /30/.

5. PROPERTIES OF Cu$_x$S FILMS

5.1. STOICHIOMETRY

A phase diagram of the Cu-S system is illustrated in figure 11 and consists of 5 different phases which exist in different crystallographic structures. For the photovoltaic operation the copper rich phases with x $\approx$ 2 are of primary importance. Chalcocite with orthorhombic crystal structure has mainly to be considered. Phase changes with temperatures impose limitations in the operation of solar cells.

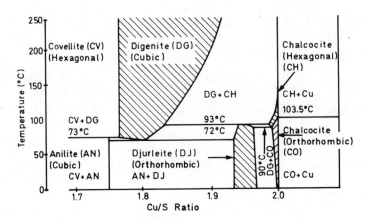

Fig. 11. Phase diagram of the Cu-S-system /31/.

5.2. COULOMETRIC TITRATION

The chemical composition and thickness of the surface layers can be controlled by electrochemical investigation methods. Measurements of the electrochemical potential as a function of time at constant current are recorded in figure 12. At the $CdS-Cu_xS$ film cathodic reactions transform the composition of the film (electrochemical reduction) which is monitored by the electrochemical potential. The reaction rate is controlled by the current /17/.

Thickness and stoichiometry of the layers as measured by this method are indicated in figure 12. Considering the geometrical structure of etched films these values lead to thickness data which agree fairly well with data derived by different diagnostic methods.

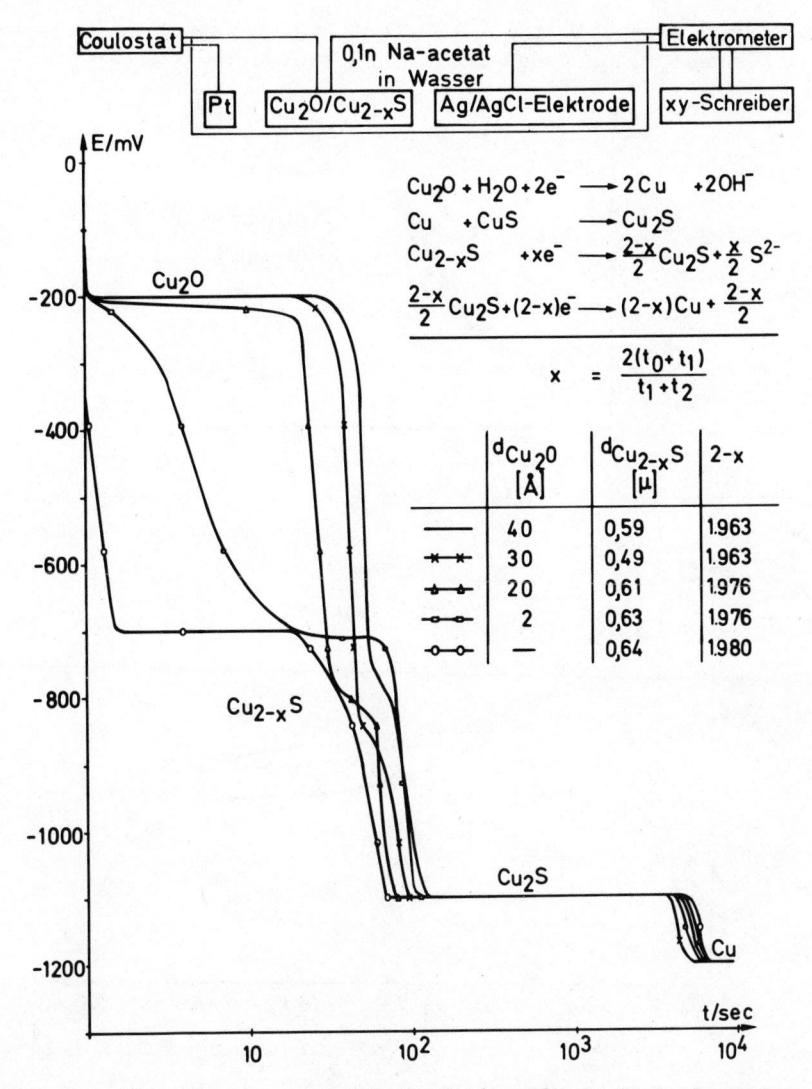

Fig. 12.   Coulometric titration of the CdS-Cu$_x$S-Cu$_2$O system.

5.3 OPTICAL PROPERTIES

The different modifications of Cu$_x$S crystals have been investigated in respect of their optical behaviour. Basing on the known optical constants which vary considerably in the different phases, theoretical values of optical transmission and reflection have been derived /32/ for a large wavelength interval. These values are plotted in figure 13 which clearly indicates that the spectral transmission is sensitively affected by the chemical composition of the Cu$_x$S film, which was assumed to be 0.15 µm thick. This characteristic behaviour presents a sensitive tool to investigate stoichiometry and film thickness. From figure 13 the absorption coefficients of Cu$_x$S can be derived, which are

sketched in figure 14. It is seen that high α-values exist which
indicate direct transitions and a band gap of 1.2 eV.

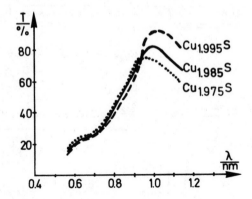

Fig. 13. Optical transmission of different Cu$_x$S films /32/.

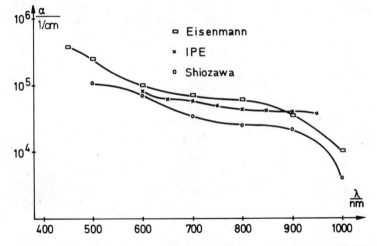

Fig. 14. Spectral absorption coefficient α of Cu$_x$S films.

5.4. ELECTRICAL PROPERTIES

The electrical resistivity of p-type Cu$_x$S films which show optimum photovoltaic performance lies in the range between about 10^{-2} to 10^{-1} Ωcm. This value corresponds to a carrier density of about 10^{20}/cm^3. Mobility measurements indicate values between 1 to 10 cm^2/Vs /33/.

6. PROPERTIES OF THE HETEROJUNCTION

6.1. STRUCTURE OF THE HETEROJUNCTION

In the following pictures obtained by scanning electron microscopy the shape of the CdS-Cu$_x$S heterojunction formed by topotaxial techniques will be illustrated. The structure of the heterojunction and the surface are of major influence on the optical and electrical data of the cell.

Etching in the aqueous HCl solution forms hexagonal pyramids on well oriented CdS films. This textured surface has two favourable properties a) light is increasingly trapped and therefore reflectance is low, b) crystallographic planes with good matching to the Cu$_2$S form the surface of such a pyramid. The surface of an etched CdS film can be seen in the SEM picture of figure 7.

The real shape of the p-n-junction can be revealed by angle lapping and etching techniques /34/. These investigations are essential for the control of the grain size of the CdS films and the thickness and grain boundary penetration of the Cu$_x$S.

A scheme of the cross-section of a sample lapped at a low angle in combination with different etching techniques illustrates the procedure for the analysis of the structure (see fig. 15).

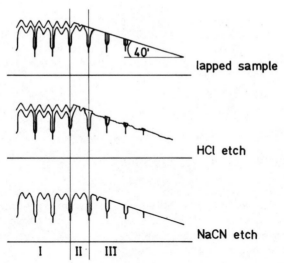

Fig. 15.   Sample preparation techniques.

Mainly three regions of a sample which has been prepared in this procedure can be distinguished:

region	HCl etching	NaCN etching
I surface	–	shape of the junction
II pyramidal region	Thickness of Cu_xS	–
III grain boundary	Cu_xS in grain bound.	grain boundaries

Table II.   Detectable properties of the heterojunction $CdS-Cu_xS$.

In table II the application of the etchants and the possible observation methods are listed.

The following SEM picture show results from such investigations. Fig. 16a is taken from region II after HCl etching. The tips of the pyramids are removed and cross-sections of the Cu_xS walls are revealed. A thickness of 0.15 μm can be derived from this picture. Cells with high photovoltaic response have a Cu_xS layer thickness of about 0.1 μm. The Cu_xS which has penetrated along the grain boundaries is visualized in figure 16b.

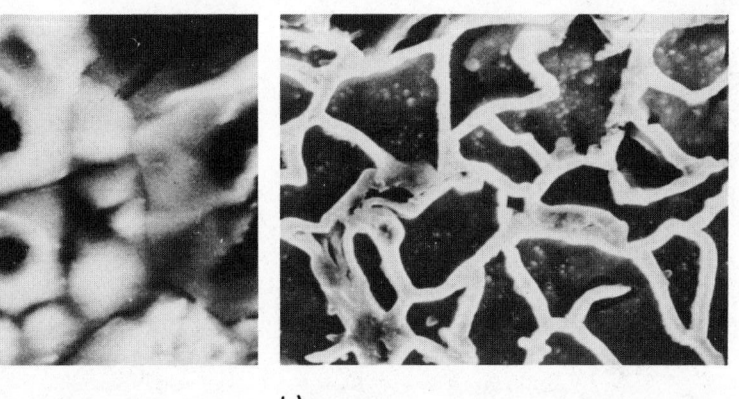

a)

b)

1 μm
d = 0,15 μm   .

1μm
d = 2 x 0,15 μm

Fig. 16.   Scanning electron micrograph of region II a)
            and of region III b)

A dark line indicating the separation of the Cu$_x$S layer of
each grain can be seen in this picture. The dipped polycrystalline
CdS-Cu$_x$S cell can be considered as a number of separate single-
crystalline microcells in parallel, since the thickness of the
Cu$_x$S layer is small compared to the size of the CdS grains (a
factor of 20 - 50).

From such an analysis two pronounced differences between the
wet dip and the dry process can be derived. The wet dip process
leads to a junction in the grains parallel to the surface topology. This formation process is accompanied by reactions along
the grain boundaries. With the dry process these reactions along
the grain boundaries do not occur however, an irregular micro-
cracking at the surface is observed. Furthermore, the junction
profile differs strongly from the surface topology, hence the
thickness of the Cu$_x$S absorber is varying strongly. This behaviour is demonstrated in the scanning electron micrographs of
figures 17a and 17b.

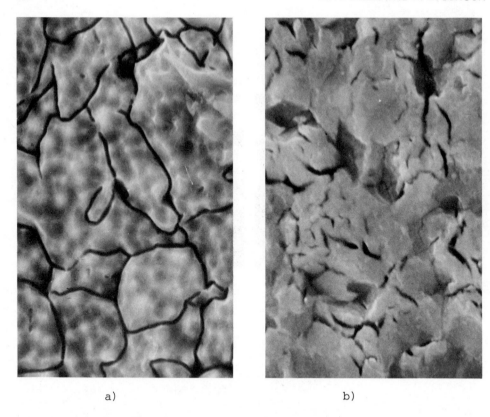

a)   b)

Fig. 17.   SEM pictures of region III
            comparing a) wet and b) dry process.
            The wet process reveals grain boundaries,
            the dry process microcracking.

The irregular structure of the junction formed by the dry
process affects the output current of the cell. In fact, photo-
electric current densities of cells produced by the dry process
are usually lower compared with dipped cells.

Another useful tool for the investigation of the structure
of the heterojunction is the micrograph in the electron beam
induced current (EBIC) mode. Electron-hole pairs generated by
the electron beam generate a current flow across the junction
which modulates locally the current and hereby yields a scanning
image. Figure 18 shows a cross section through the heterojunction
which reveals the crystallite structure of the CdS and the active
collection area in the vicinity of the junction. The image has
been obtained by a double exposure of the secondary emission and
the EBIC image. It is seen that vertical junctions exist along
grain boundaries.

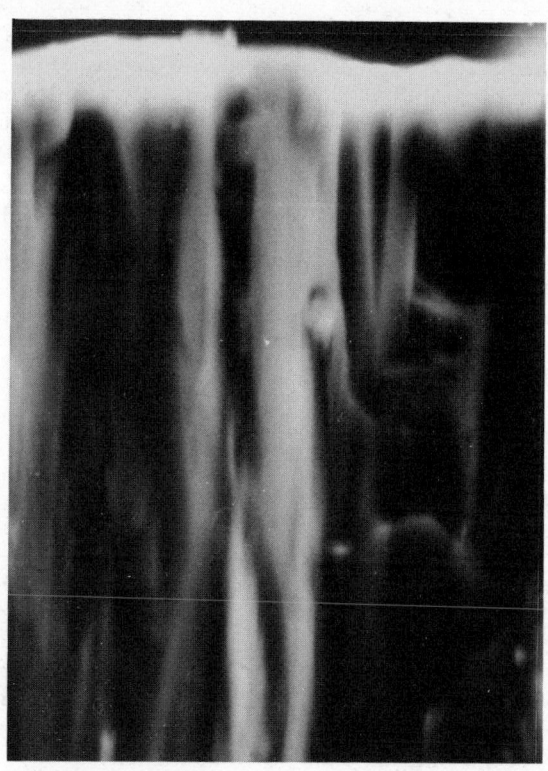

├────────────┤
        10 μm

Fig. 18.   SE and EBIC image of a CdS-Cu$_x$S heterojunction.
           The bright area represents the carrier collection field
           extending from the heterojunction in CdS and Cu$_x$S.

These vertical junctions represent unilluminated diodes
parallel to the active diode areas. An increased reverse current
due to the shunting diodes results in a reduced open circuit
voltage. A detailed study of this phenomena was given by Hewig /35/.
These vertical junctions in some cases extend through the CdS
film hereby causing local short cuts.

6.2. SURFACE EFFECTS OF THE Cu$_x$S FILM

The carrier transport in the very thin Cu$_x$S layer is sensitively affected by surface effects of this layer. The dipping
process does not readily lead to optimum photovoltaic operation
of the Cu$_x$S film. Appropriate treatments have to be applied in
order to improve the properties of the Cu$_x$S and to reduce surface recombination.

A window of a thin Cu_2O layer on the surface reduces surface recombination drastically. This layer however can only be formed on highly stoichiometric Cu_2S or on a metallic copper layer, which can be provided either by the deposition of the additional copper or by the extraction of S from the Cu_xS layer in a H_2 glow-discharge and a subsequent heat-treatment in air /37/. This heat-treatment causes on the one hand copper diffusion into the Cu_xS and thereby an improvement of the stoichiometry, on the other hand the formation of Cu_2O on the surface.

Excessive heating in air results in the formation of CuO which presents no suitable window layer on the Cu_xS surface, hence the surface recombination increases as described in chapter 6.4. The kinetics of the post-dip treatment has been investigated by Pfisterer et al. /37/.

6.3. CAPACITANCE MEASUREMENTS

Much insight in the photoelectrical behaviour of the hetero-junction and in mechanisms of degradation has been derived from capacitance measurements.

By this method detailed information on the doping profile in the CdS film in the vicinity of the heterojunction, the built-in voltage, diffusion of Cu from Cu_xS into CdS and the energy levels of Cu atoms in the CdS band gap has been obtained.

As indicated in section 5.3. the p-type Cu_xS film is highly doped as compared with CdS. Therefore the space charge region extends into the CdS side of the heterojunction where the total diffusion potential is built up. Due to the lattice mismatch at the heterojunction an interface layer originates with a high density of states thus acting like a highly doped interlayer.

During heat treatment of the photovoltaic cell Cu atoms diffuse from the Cu_xS film to the CdS film and hereby compensates in part the n-type conduction mechanism of CdS due to Cu atoms acting as acceptors. The space charge density distribution in the vicinity of the heterojunction is schematically sketched in figure 19a. Figure 19b shows the principal behaviour of $1/c^2$ vs. diode voltage (V). Neglecting the action of the interface layer (N_i), the partly compensated Cu-CdS layer ($N_D^+ - N_A$) extending to x = d and the space charge in the uncompensated CdS layer (N_{D+}) result in the characteristics in figure 19b. The geometrical structure of the junction however and deviations from rectangular space charge profiles lead to modifications of the ideal charac-teristic in figure 19b.

Typical experimental data of capacitance measurements carried out at 100 kHz with solar cells in darkness are shown in figure 19c. The parameter indicated is the duration of heat treatment at 180 $^{\circ}$C which controls diffusion of Cu atoms in the CdS layer. The shift of the $1/c^2$ lines is attributed to an increasing compensation layer both in compensation degree and in extension of the i-CdS layer. This characteristic behaviour is observed only when the heterojunction is "flat" i.e. by avoiding the etching process with a plane geometrical structure.

Figure 19d demonstrates the diffusion process with etched surfaces and a heterojunction with a pronounced geometrical structure which leads to a completely different $1/c^2$ dependence. The different slopes allow the deduction of the geometrical structure which is defined as the ratio between real junction area and the projection of the real junction area on the substrate. This area factor (4 - 8) in structured junctions decreases with duration of heat treatment and with increasing reverse potential due to a smoothing effect of the space charge region (figures 19e and 19f).

Basing on these measurements the band diagram (figure 23) has been derived. The partly compensated i-CdS layer is characterized by depths between 0.5 and 1.0 µm depending on the duration of the heat treatment. The concentration of Cu acceptors derived from these measurements is about $1.2 \times 10^{17}/cm^3$. The Cu solubility and the diffusion coefficient of Cu have also been estimated on the basis of these measurements. The data are summarized in table III /38/.

The data allow an estimation of the Cu diffusion process and its effect on the long time stability. From these evaluations it can be concluded that the diffusion process of Cu into CdS crystallites even at an operating temperature of 80 $^{\circ}$C is not a limiting factor in long time stability and should not cause any substantial degradation within periods of some decades.

Photocapacitance measurements yield information on the deep acceptor level in CdS as indicated in the band diagram figure 23.

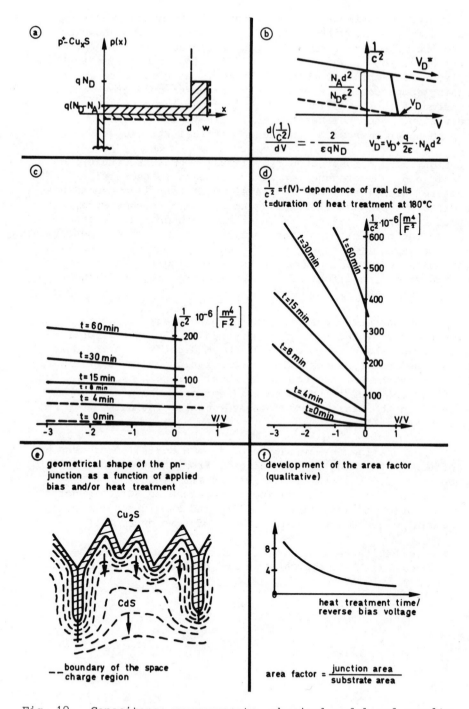

Fig. 19.   Capacitance measurements, physical model and results.

durat. of heat treatm.	N_D (cm^{-3})	d_{min} (μm) d_{max} (μm)	N_{Amax} (cm^{-3}) N_{Amin} (cm^{-3})	N_o (T_1=180°C) max min (cm^{-3})	D_{min} (cm^2/s) D_{max} (cm^2/s)
t=0 min	1.18 x 10^{17}				
t=3 min		0.42 0.44	1.22 x 10^{17} 1.14 x 10^{17}	1.6 x 10^{18} 6 x 10^{17}	1.5 x 10^{-12} 3.2 x 10^{-12}
t=4 min		0.48 0.495	1.21 x 10^{17} 1.15 x 10^{17}	1.6 x 10^{18} 6 x 10^{17}	1.5 x 10^{-12} 3.1 x 10^{-12}

Table III. Values of doping concentration, width of the compensated layer, and parameters of Cu-diffusion from Cu$_2$S into CdS.

6.4. DIFFUSION LENGTH IN Cu$_x$S AND CdS

A very important parameter for photovoltaic operation of CdS-Cu$_x$S cells is the diffusion length of minority carriers especially of electrons in the Cu$_x$S layer. Due to the small thickness of the Cu$_x$S layer and its complex structure measurements of the diffusion length are very difficult. One approach is to prepare special samples with thick Cu$_x$S layers as described by Partain et al. /39/. The diffusion length in polycrystalline CdS sample prepared by lapping (figure 15 region III) has been measured by scanning the electron beam of a SEM across the vertical CdS-Cu$_x$S junction in the grain boundaries /38/. An EBIC image together with a line scan across the junction is shown in figure 20. The shape of the current response I_p according to the formula

$$I_p = I_{p_o} \cdot \exp\left(-\frac{x}{L_p}\right)$$

with x = distance from the junction, indicates a diffusion length in CdS of L_p = 0.2 μm. As the Cu$_x$S layer is too thin, no value for the diffusion length L_n can be extracted from this measurement.

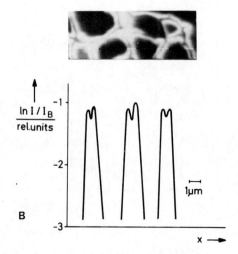

Fig. 20.   Diffusion length measurement in polycrystalline CdS
            a) EBIC image and line scan
            b) normalized current as a function of
               spacing from heterojunction.

     A different method which can also be applied for very thin
assorted layers and which needs no special sample preparation has
been used for the measurement of surface properties and the
diffusion length in the Cu_xS /38/.

     The response of the cell to an electron beam with variable
beam energy has been calculated using the depth dose function of
electron energy in the sample and diffusion equations for the
carrier transport. By matching the parameters of theoretical curves
to measured values the surface recombination velocity, the diffusion length and even the interface recombination velocity can
be derived. Geometrical variations and backscattering losses of
the electron beam affect the accuracy of this method. However,
changes of the properties as a function of post fabrication
treatments can be clearly detected as it is illustrated in figure 21.

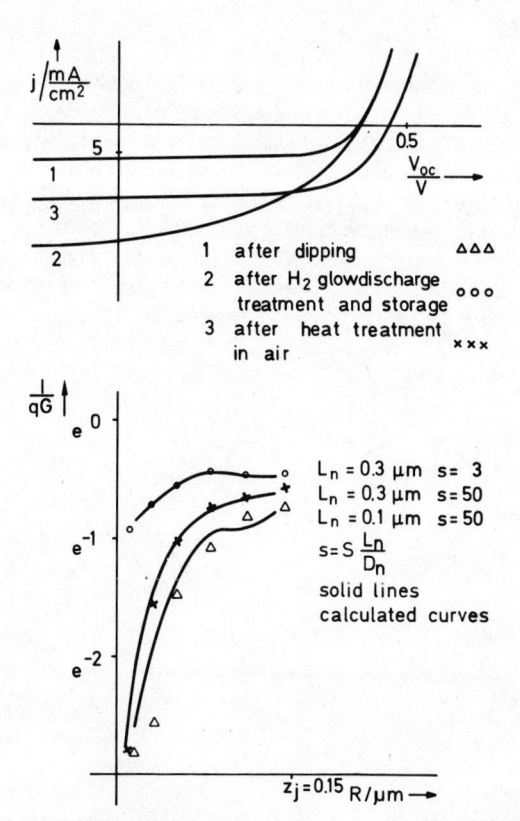

Fig. 21. Variations of transport properties in Cu$_x$S films
          in different stages of processing.

     Immediately after dipping the density of defects in the bulk
Cu$_x$S and at the interface as well as the density of surface recombination centres are very high, resulting in a low current
and voltage output of the cell. After formation of a suitable
surface layer and a copper-rich Cu$_x$S phase by reduction in a
H$_2$ glow discharge, the current increases drastically. In this
stage surface recombination velocity is minimized and the diffusion length increases.

     Annealing in air oxidizes the surface. The open circuit
voltage increases due to the reduction of surface recombination
via tunneling at the junction. The values of diffusion length
remain constant. This has also been observed by Partain /39/.

6.5. SPECTRAL RESPONSE

The CdS-Cu$_x$S junction shows a very complex behaviour when illuminated with light at different wavelengths. These investigations require a special optical equipment which allows illumination with two different light sources: A constant bias light and a chopped primary light causing a chopped current output of the cell which can be detected by lock-in techniques /40/. The quantum efficiency of two cells is plotted in figure 22a and 22b both measured at two different wavelengths of the bias light, causing either quenching or enhancement effects.

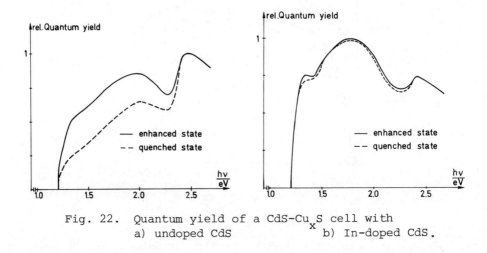

Fig. 22.   Quantum yield of a CdS-Cu$_x$S cell with
            a) undoped CdS   b) In-doped CdS.

It can be seen that the cell on undoped CdS is much more sensitive to the light bias effect which is due to the change of occupation of deep acceptor levels near the junction. These levels modify the space charge and hence the field at the interface layer. Interface recombination is dependent on the electrical field which results from space charge distribution. The effect of interface recombination strongly determines the output characteristics of the cell /44/. The acceptor can be ionized by light in the energy range 1.45 > hν > 0.7 eV and hence causes quenching and reduction of the photocurrent. Light of higher photon energies enhances the photocurrent due to the neutralization of the deep levels. Additional donor doping e.g. by In reduces light bias effect, as demonstrated in figure 22b.

6.6. BAND DIAGRAM

Figure 23 presents an energy band diagram which is based on data derived from the variety of physical investigations. It indicates the structure of the energy bands of the single layers: the Cu$_2$O surface window layer, the Cu$_x$S absorber layer, and the CdS collector layer with deep levels at the interface due to copper diffusion from the Cu$_x$S layer.

This model fits well to all data which have been measured and reported here. Special optical effects which have been observed need a more detailed description of interface recombination phenomena and of light interaction with deep levels.

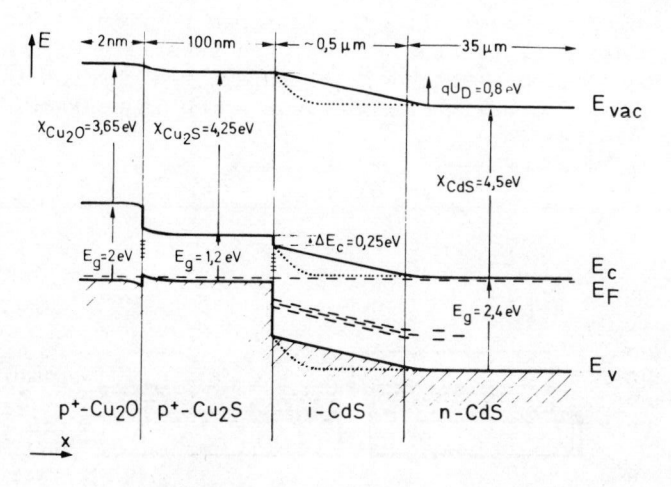

Fig. 23.   Band diagram of a CdS-Cu$_x$S photovoltaic cell.

7. TECHNOLOGY OF CdS-Cu$_x$S PHOTOVOLTAIC GENERATORS

So far the physical realization of the heterojunction and its optimization in photovoltaic efficiency have been discussed. The production process of a photovoltaic generator and its economical use have to satisfy a number of additional criteria which involve a) technological procedures for economic cell and module fabrication allowing large scale production and b) material evaluation for substrates, contacts and encapsulation. These criteria result in a variety of fabrication methods, structures of cells and modules, and materials.

The most promising technologies for CdS film deposition are spraying and vacuum evaporation. In all practical procedures the dipping process which results in a topotaxial growth of the Cu_xS layer has been applied. The cell structures considered are of the frontwall and backwall type. Finally the materials for providing contacts are selected according to their physical and chemical compatibility with the semiconducting films whereas the selection of substrate and encapsulation is based on economy and long time stability and leads mainly to glass.

7.1. CELL STRUCTURES

Cells of the frontwall type exhibit the structure which is shown in figure 24. The incident radiation is transmitted through the front grid and is absorbed in the Cu_xS film. The substrate consists of glass with a conducting film as the counterelectrode. The structure of the backwall cell is shown in figure 25. In this case the radiation penetrates through the substrate with a transparent and contacting electrode, is transmitted through the CdS film and finally reaches the Cu_xS layer which is contacted by non transparent electrodes.

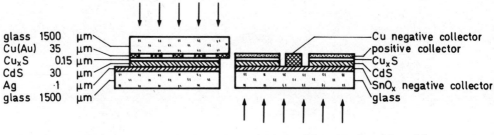

glass	1500	μm
Cu(Au)	35	μm
Cu_xS	0.15	μm
CdS	30	μm
Ag	1	μm
glass	1500	μm

Cu negative collector
positive collector
Cu_xS
CdS
SnO_x negative collector
glass

Fig. 24.   Frontwall cell.   Fig. 25.   Backwall cell.

Both structures provide the potential for large area cell and module production. Frontwall cells have been produced in pilot lines by evaporation, backwall cells by spray technology.

As an example of a pilot line production frontwall cells made by evaporation will be described. This should not over-emphasize the structures and methods used but is attributed to the availability of ample data and experience.

7.2. FABRICATION PROCESS OF CdS-Cu$_x$S CELLS

The fabrication process described here has been realized in a pilot line for cell or module production. The structure of the frontwall glass encapsulated cells has been demonstrated in figure 2. The size of the 1.5 mm thick glass substrate is 7 x 7 cm^2. The integrated generator (module) consists of 8 cells with a total size of 14.5 x 29 cm^2. A flow chart of the processes which have to be applied to the glass substrate and glass plate is sketched in figure 26.

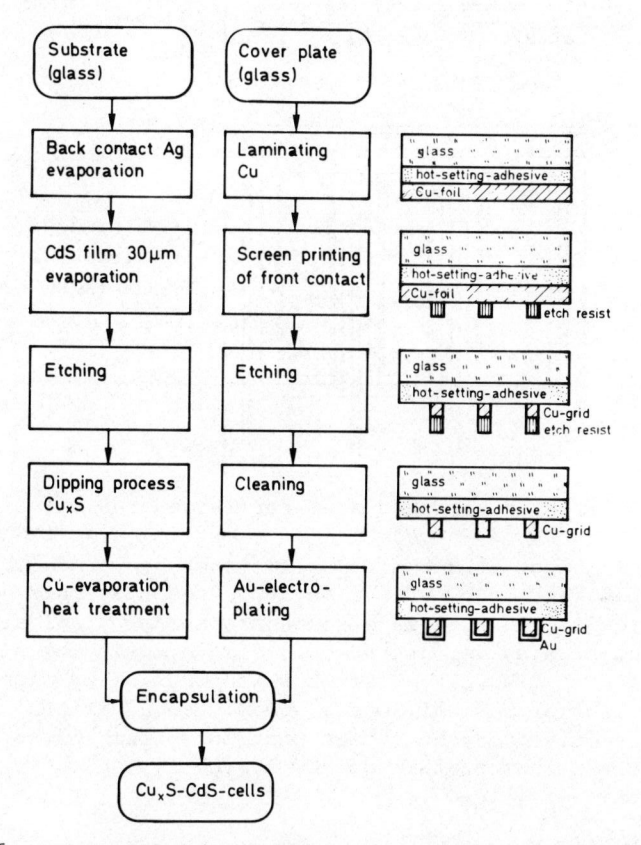

Fig. 26. Steps of fabrication processes of frontwall
          CdS-Cu$_x$S photovoltaic generators.

The back contact on the substrate is produced by evaporating a thin Cr film which acts as an adhesive interlayer and a 1 µm thick Ag film. The geometrical arrangements of the back contact areas have been designed according to the interconnections between the cells in the integrated panel and to the technical requirements (output voltage, output current).

The evaporation system for deposition of a 30 μm CdS film is shown in figure 27. The size of the crucible is 15 x 30 cm^2. The temperature of the crucible during deposition is 1000 ^{o}C, the substrate temperature is 200 ^{o}C. The evaporation system can be used both for cell and module production.

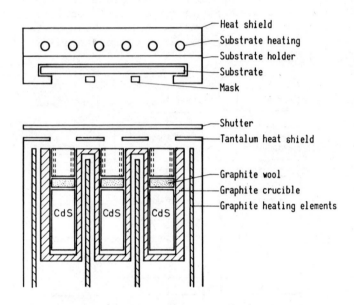

Fig. 27.   CdS evaporation system for large area cells or modules.

After evaporation the CdS film is etched in HCl:H$_2$O solution at a temperature of 60 ^{o}C for about 10 seconds. This etching process results in a surface roughness and a pyramidal structure of 1 - 3 μm size as demonstrated in figure 7. In the next step of fabrication the Cu$_x$S layer is grown topotaxial by dipping in an acid solution of CuCl at 90 ^{o}C. Finally an additional 50 Å Cu film is deposited on the heterojunction system. After heat treatment in air at temperature of 200 ^{o}C the system shows optimum photovoltaic efficiency.

In parallel processing the front contact is produced. The steps, which are shown in figure 26, include laminating of a Cu foil by a hot setting adhesive, screen printing of the front contact pattern, spray etching. The etch resist is removed and the Cu-grid finally is electroplated with gold to provide a compatible ohmic contact with the Cu$_x$S film.

Contacting and encapsulating take place in a press. The principle of operation, schematically sketched in figure 28, includes evacuation, pressing and heating the hot setting adhesive which results in good optical and electrical contact. By this procedure efficient cells and modules have been generated.

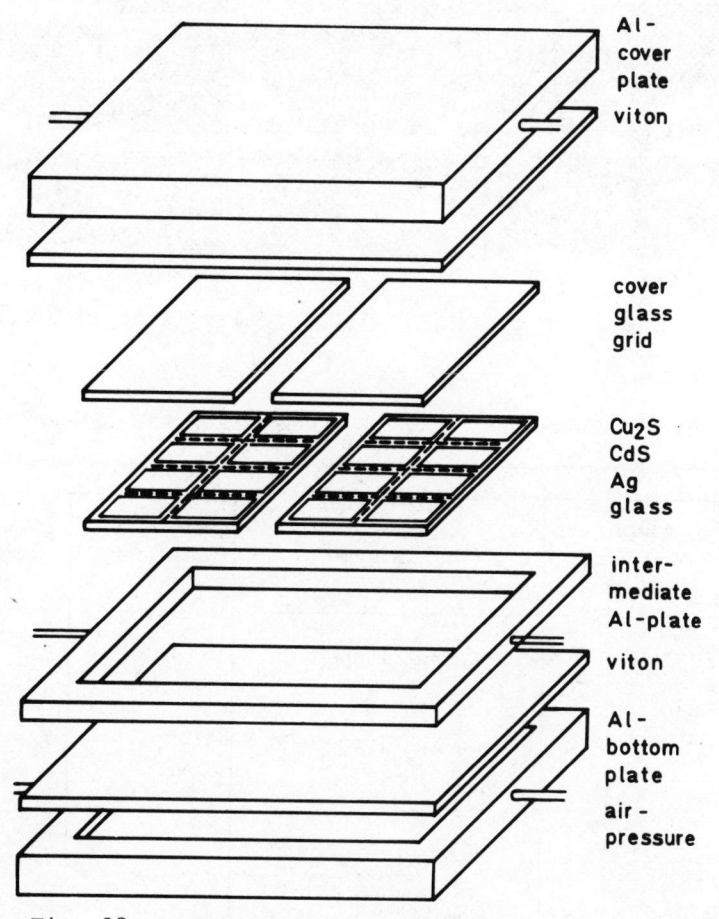

Fig. 28.   Arrangement for module encapsulation.

Figure 29 shows an integrated generator with 18 cells in series connection which is used as a power supply for operation at 6 V. The total material cost according to the processing described amount approximately 20 $/m^2.

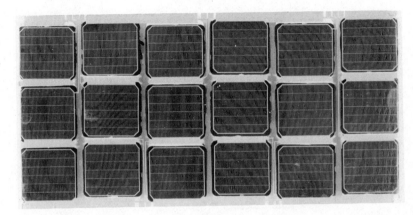

Fig. 29.   Integrated generator with 18 cells (size $14,5 \times 29 \ cm^2$).

8. PERFORMANCE CHARACTERISTICS OF SOLAR CELLS AND GENERATORS

Research group/ institution	ref.	cell type	efficiency %	active area cm^2
Univ. of Delaware	/3/	evap. CdS, dipped Cu_xS	9.15	0.8
Univ. of Stuttgart (IPE)	/4/	evap. CdS, dipped Cu_xS	7.3 4.8	42 8 cells integrated in series (336)
Photon Power, Univ. of Montpellier	/11/	sprayed CdS, dipped Cu_xS	5-6	1
Matsushita	/16/	sintered CdS, dipped Cu_xS	8	4

Table IV.   Comparison of efficiency from different $CdS-Cu_xS$ cells or generators.

In table IV the efficiencies of cells as reported by different groups are listed and the sizes of the active cell area are indicated. The highest efficiency experimentally achieved is 9.15 % with a 0.8 cm^2 cell, which has been obtained by the research group at the University of Delaware. The efficiencies of sprayed cells are somewhat lower (5 - 6 %). Best results on large area devices are 7.3 % for encapsulated 42 cm^2 cell. The IV-characteristic of such a cell is shown in figure 30.

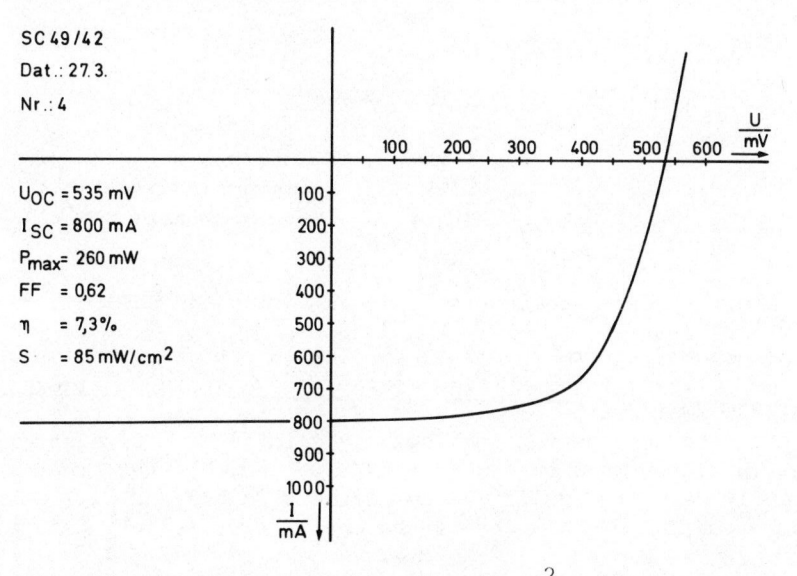

Fig. 30.   IV-characteristic of a 42 cm^2 cell.

An integrated generator with an efficiency of 4.8 % consists of 8 cells (42 cm^2 active area each) in series connection. The IV-characteristic of this integrated generator is shown in figure 31.

These values of efficiencies have to be seen in relation with cells and modules made from single or polycrystalline Si. So far the efficiency is lower by 20 - 30 %. Future efforts and research activities on CdS-Cu$_x$S photovoltaic cells should lead to higher experimental values of the efficiency which is not limited by basic physical effects to efficiencies below 10 %.

A fundamental problem exists, however, in the lack of data and experience of long time stability. Recent developments indicate that degradation processes which have been observed in early investigations are not inherent to the system but can be eliminated by appropriate technological procedures. Hermetically sealed encapsulation and special treatments during fabrication have led to CdS-Cu$_x$S cells and generators which have been in

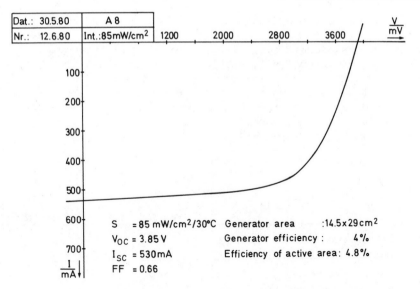

Fig. 31. IV-characteristic of an integrated CdS-Cu$_x$S generator.

operation for more than 3 years. The output characteristics of
some cells during long time operation are shown in figure 32.
A slow degradation of these cells, which according to more recent results have not been optimized in respect with stability,
is indicated in figure 32. It is expected, however, that recently developed cells are by far more stable and will exhibit
no essential degradation of the characteristic output data.

Minor differences in the efficiency between Si cells and
CdS-Cu$_x$S cells however are by far compensated by the advantages
and potentials of thin film solar cells as far as material consumption, energy input for fabrication and production technology
are concerned. The advantages promise that the price goal for
an economic use and large scale production will be achieved in
the near future.

REFERENCES

1. A. G. Stanley, in: Applied Solid State Science, Vol. 5,
    Academic Press, New York, 1975.
2. M. Savelli and J. Bougnot, in: B. O. Seraphin (ed.):
    "Topics in Applied Physics, Vol. 31", Springer, Berlin
    (1979), Chapt. 6.
3. A. M. Barnett, J. A. Bragagnolo, R. B. Hall, J. E. Phillips,
    and J. D. Meakin, Conf. Rec. 13th IEEE Photov. Spec. Conf.,
    Washington, D.C., 1978 (IEEE, New York, 1978), p. 419.
4. W. Arndt, G. Bilger, G. H. Hewig, F. Pfisterer, H. W. Schock,
    J. Woerner, and W. H. Bloss, Proc. 2nd E.C. Photov.
    Solar Energy Conf., Berlin, 1979, (D. Reidel Publ. Comp.,
    Dordrecht, 1979), p. 826.

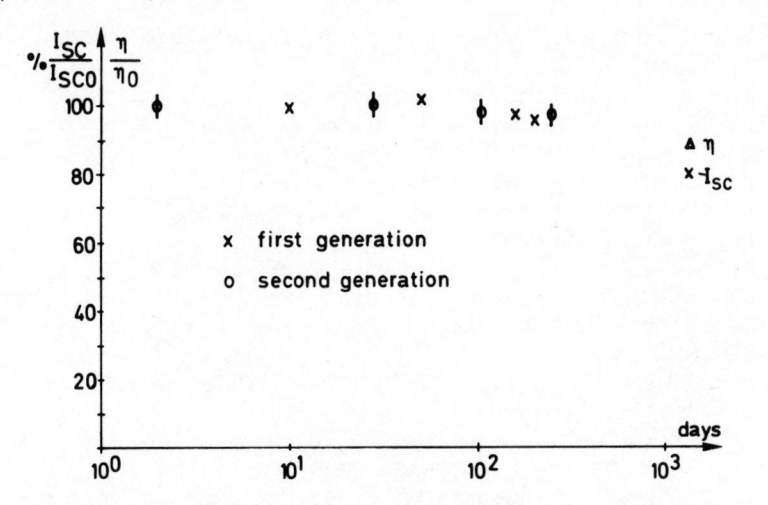

Fig. 32. Long time operation of CdS-Cu$_x$S solar cells.

5. L. Gmelin, Handbuch der anorganischen Chemie, Verlag Chemie,
    Berlin, 1926 ff.
6. L. C. Burton, Solar Cells, 1, 159 (1980).
7. W. Arndt, G. Bilger, W. H. Bloss, G. H. Hewig, F. Pfisterer,
    and H. W. Schock, Proc. 1st E.C. Photov. Solar Energy
    Conf., Luxemburg, 1977, (D. Reidel Publ. Comp., Dordrecht,
    1977), p. 547.
8. J. de Klerk and E. F. Kelly, Rev. Sci. Instrum. 36, 506 (1965).
9. G. Brincourt and S. Martinuzzi, Vide 22, 252 (1967).
10. W. Mueller, H. Frey, K. Radler, and K. H. Schuller, Thin
    Solid Films 59, 327 (1979).

11. M. Perotin, J. Bougnot, J. Marucchi, O. Maris, R. Daures,
    C. Grill, and M. Savelli, Conf. Rec. 14th IEEE Photov.
    Spec. Conf., San Diego, 1980; to be published.
12. A. Banerjee, P. Nath, V. D. Vanakar, and K. L. Chopra, Phys.
    Stat. Sol.(a) 46, 723 (1978).
13. R. R. Chamberlin and J. S. Skarman, Solid State Electronics,
    9, 819 (1966).
14. R. S. Berg and R. D. Nasby, Sadian Labs., Res. Rept.,
    Contract No. E (29-2)-3579, May 1977.
15. S. Vojdani, A. Sharifnia, M. Doroudian, Electron. Lett., 9,
    128 (1973).
16. H. Matsumoto, N. Nakayama, and S. Ikegami, Jpn. J. Appl.
    Phys. 19, 129 (1980).
17. E. Castel and J. Vedel, Analusis 3, 487 (1975).
18. A. N. Casperd, R. Hill, ref. 7, p. 1131.
19. T. S. Te Velde, Energy Conversion 14, 111 (1974).
20. B. Rezig, S. Duchemin, and F. Guastavino, Solar Energy Mater.
    2, 53 (1979).
21. J. J. Loferski, J. Shewchun, E. A. de Meo, R. Arnott,
    E. Crisman, H. L. Hwang, R Beaulieu, and C. C. Wu, Proc.

Int. Conf. Solar Electricity, Toulouse, 1976, (CNES,
     Toulouse, 1976), p. 317.

22. J. Yee, G. Armantrout, E. Fischer-Colbrie, D. Miller,
     E. Hsieh, J. Leong, K. E. Vindelov, and T. G. Brown,
     ref. 3, p. 393.

23. A. D. Jonath and W. W. Anderson, ref. 3, p. 417.

24. L. L. Kazmerski, W. B. Berry, and C. W. Allen, J. Appl.
     Phys. 43, 3515 (1972).

25. J. Gilles and J. van Cakenberghe, Nature 182, 862 (1958).

26. W. Kahle and A. Berger, Phys. Stat. Sol. 14, K 201 (1966).

27. L. C. Burton and T. L. Hench, Appl. Phys. Lett. 29, 612
     (1976).

28. J. L. B. Wilson and J. Woods, J. Phys. Chem. Solids 34,
     171 (1973).

29. F. A. Shirland, F. Augustine, and W. Bower, 2nd Quart. Rept.,
     Contract NAS 3 - 6461, NASA-CR-54302, Clevite Corp., 1965.

30. L. R. Shiozawa, F. Augustine, G. A. Sullivan, J. M. Smith,
     and W. R. Cook Jr., Clevite Corp. Final Rept., Contr.
     AF 33(615) - 5224, 1969.

31. R. W. Potter II, Econ. Geol. 72, 1524 (1977).

32. J. Shewchun, T. Vanderwel, J. P. Marton, A. Kazandjian,
     J. J. Loferski, D. Burk, J. Wu, and R. A. Clarke,
     ref. 3, p. 378.

33. F. Guastavino, H. Luguet et J. Bougnot, Proc. Int. Conf.
     "The Sun in the Service of Mankind", Paris, 1973
     (CNES, Bretigny Sur Orge, 1973).

34. M. K. Mukherjee, F. Pfisterer, G. H. Hewig, H. W. Schock,
     and W. H. Bloss, J. Appl. Physics 48, 1538 (1977).

35. G. H. Hewig and W. H. Bloss, Conf. Rec. XIIth IEEE Photov.
     Spec. Conf. Baton Rouge, 1976 (IEEE, New York, 1976), p. 483.

36. K. Bogus and S. Mattes, Conf. Rec. IXth IEEE Photov. Spec.
     Conf., Silver Spring, 1972 (IEEE, New York, 1972), p. 106.

37. F. Pfisterer, H. W. Schock, and W. H. Bloss, ref. 35,
     p. 502.

38. F. Pfisterer, H. W. Schock, and G. H. Hewig, ref. 4, p. 352.

39. J. J. Oakes, I. G. Greenfield, and L. D. Partain, J. Appl.
     Phys. 48, 2548 (1977).

40. G. H. Hewig, F. Pfisterer, and W. H. Bloss, Proc. Int. Conf.
     Photov. Power Generation, Hamburg (1974), (DGLR, Köln,
     1974), p. 255.

41. L. Eisenmann, Ann. Phys. (6) 10, 129 (1952).

42. L. R. Shiozawa, G. A. Sullivan, F. Augustine, and J. M. Jost,
     Interim Techn. Rept., Contract AF 33 (615) - 5224,
     Clevite Corp., 1967.

43. L. C. Burton, B. Baron, T. L. Hench, and J. D. Meakin, Inst.
     Energy Conversion, Semi-Ann. Rept., July 1978, Contr.
     No. EG-77-C-03-1576.

44. A. Rothwarf, N.C. Wyeth, and J. Phillips, ref. 3, p. 399.

CONVERSION OF SOLAR ENERGY USING TANDEM PHOTOVOLTAIC

CELLS MADE FROM MULTI-ELEMENT SEMICONDUCTORS

Joseph J. Loferski

Division of Engineering
Brown University
Providence, RI   02912

I.   INTRODUCTION

In preceding chapters, the discussion has focused on solar cells in which the photovoltaically active semiconductor is the element silicon or a binary semiconductor like GaAs, Cu_2S, InP, etc. This chapter is concerned with solar cells in which the photovoltaically active semiconductor is based on three or more elements. Specifically, three groups of such semiconductors will be considered, namely, (1) semiconductors which are alloys of binary semiconductors (e.g. $Al_xGa_{(1-x)}As$ or $Ga_xIn_{(1-x)}P_yAs_{(1-y)}$); (2) "true" ternary semiconductors (e.g. $CuInS_2$, $CuInSe_2$) and (3) alloys of these true ternaries (e.g. $CuInS_2Se_{2(1-z)}$, $z < 1.0$).

Why is there interest in solar cells made from these materials which, superficially at least, are more complex than silicon or the binary semiconductors? The basic reason for this interest is that none of the currently well-developed solar cells (Si, GaAs, Cu_xS/CdS) clearly satisfies the requirements for large scale solar energy development, namely low-cost solar cells having acceptable solar energy conversion efficiency ($\geq 10\%$). Furthermore, it is recognized that even if one of these solar cells eventually satisfy the cost and efficiency goals, future generation solar cells will be called upon to operate at even higher efficiencies and even lower costs. The semiconductors discussed in this chapter may play a central role in the attainment of these goals.

The drive toward high efficiency requires the use of the tandem cell (or cascade cell) concept in which solar cells based on photovoltaically active semiconductors having different energy gaps are

arranged in such a way that each of them uses only that portion of
the photons in the solar spectrum which most closely matches its
band gap. As we shall see shortly, tamdem cell systems require
high efficiency cells based on semiconductors whose energy gaps are
distributed over the range from about 1.0 to 2.0eV. If enough
different energy gap semiconductors are available, the theoretical
efficiency of a tandem cell stack with solar energy concentration
approaches 55%. The alloys of binary and ternary semiconductors
provide the only way to obtain materials having essentially continuously varying band gaps and, therefore, these materials make even
the most complex tandem cell systems possible.

The drive toward low-cost cells may also focus on multi-element
semiconductors because certain of these semiconductors have already
produced some of the highest efficiency thin films solar cells ever
made. Specifically, thin film $CuInSe_2$ cells having areas of several
square centimeters and efficiencies in excess of 6% have been reported.[2,3] Thin films of $CuInSe_2$ and other ternary semiconductors
can be deposited by various economically promising methods like
rf-sputtering,[4] vacuum evaporation[2] and chemical spray pyrolysis.[5,6]

The drives toward high efficiency and low cost may intersect
by the fabrication of thin film tandem cell stacks in which the
thin films consist of four-and five-element alloys of the ternary
semiconductors. It has already been demonstrated that thin films
of such alloys can be deposited by rf-sputtering[7] and chemical spray
pyrolysis.[8] It has also been shown that large grained specimens
of these four- and five-element alloys can be used to fabricate
solar cells having efficiencies in excess of 10%[9], i.e. these semiconductor alloys are promising photovoltaic materials.

It is these subjects which we shall discuss in this chapter.
In Sections II through V, we describe some recent developments in
the theory of high efficiency solar cells and some ideas for achieving optimum performance from solar cells based on a single photovoltaically active semiconductor as well as from solar cells based
on a multiplicity of such cells. Section VI summarizes some recent
experimental work on tandem cells using alloys of binary III-V semiconductors. Sections VII and VIII present some recent results on
solar cell applications of ternary semiconductors of the I-III-VI$_2$
type and of their alloys.

II.   INCREASING EFFICIENCY BY RECOURSE TO TANDEM PV CELL SYSTEMS

The maximum AMO or AM1 solar energy conversion efficiency for
a solar cell based on a single photovoltaically acitve semiconductor
is around 25%.[10,11,12] Figure 1 is a bar chart first published by
Wolf[13] which shows the losses occurring in silicon solar cells exposed to AMO sunlight. Two large losses occur because of poor utilization of solar photons. Twenty-four percent of the energy in the

incident beam is lost because the photon energy is too low to produce ionization in silicon, e.e. the photon energy $h\nu < E_G$.

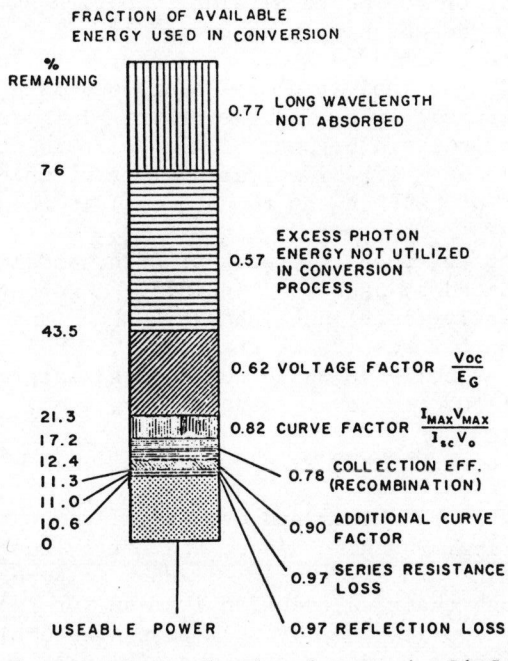

Fig. 1. Bar chart showing losses in Si Solar cells.
From Ref. [13].

Thirty-two and one-half percent of the incident energy is converted into heat inside the silicon because each photon having energy in excess of the band gap can generate only one hole-electron pair; the excess energy it imparts to the ionized carriers ($h\nu - E_G$) is degraded into thermal energy. These are basic physical processes, and the losses associated with them cannot be reduced once the semiconductor has been selected, since it is the energy gap of the semiconductor which determines the magnitudes of these two losses. For the energy gap of Si (1.10eV), the fraction of incident energy remaining after these two losses have been taken into account is 0.43; for the energy gap of GaAs (1.35eV), this remainder is 0.40. The maximum value of this fraction is around 0.44; this means that at least 56% of the solar energy incident on a single solar cell does not get utilized effectively. The magnitude of this loss can be decreased, and the overall sunlight to electricity conversion efficiency can be increased by recourse to tandem cell systems. A tandem solar cell system illustrated schematically in Fig. 2 consists of a group of solar cells made from a number of photovoltaically active semiconductors having different energy gaps so arranged that full sunlight is incident on the cell based on the largest

energy gap material, say E_{G1}. The solar photons having energy
$h\nu < E_{G1}$ are directed to the second solar cell in the chain which
absorbs those photons whose energies lie between E_{G1} and E_{G2} where
E_{G2} is the energy gap of the photovoltaically active semiconductor
in the second cell. Photons having an energy $h\nu < E_{G2}$ are directed
to the third solar cell based on a semiconductor with energy gap
E_{G3}, etc. In Fig. 2, the individual cells are separate p/n junctions, and they are placed one behind the other. This is for illustration only; the cells could be part of a monolithic structure,
or the spectrum could be split by selective filters which reflect
the appropriate band of wavelengths to each cell in the chain.

Some calculation of the efficiency of such tandem cell systems was first conducted by Jackson[14] in 1955. A more complete
calculation was recently (1979) published by Gokcen and Loferski[1],
and we shall focus on the results of that paper. Their calculation
is applicable to solar cells in which the reverse saturation current
I_o is an exponential function of the energy gap, i.e.

$$I_o \sim \exp(-E_G/BkT) \text{ with } 2 > B > 1 \tag{1}$$

This means that I_o must be determined by thermal generation processes, as in the classical Shockley diode, and not by tunneling or
interface recombination processes which do not, in general, depend
on E_G. Thus, although their calculation is made for p/n homojunction solar cells, it applies equally well to p/n heterojunctions,
MIS or SIS cells in which I_o depends exponentially on the energy
gap of the photovoltaically active semiconductor.

The calculation begins with a tabulation of the solar spectrum
like that for AMO shown in Table 1 taken from Reference 15. This

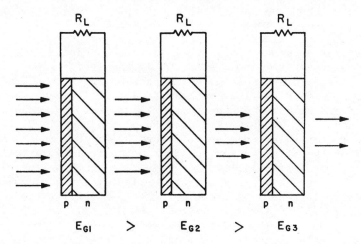

Fig. 2.   Schematic representation of a tandem cell system.

table shows the integrated AMO photon flux $N_{ph}(E)$ for energies between $h\nu = \infty$ and $h\nu = E$

$$N_{ph}(E) = \int_{h\nu = E}^{\infty} n_{ph}(h\nu)d(h\nu) \qquad (2)$$

where $N_{ph}(h\nu)$ is the photon flux at $h\nu$. The table also includes the short circuit current which would flow in a cell having an absorption cutoff at $h\nu = E$ if the collection efficiency were unity, i.e. if every photon having an energy $h\nu > E$ contributed one minority carrier to the short circuit current. Thus

$$I_{sc}(E) = q\, N_{ph}(E) \qquad (3)$$

where q is the charge on the electron. For example, Table 1 shows that for a cell in which the AMO sunlight is incident on a photovoltaically active semiconductor with an energy gap $E_G = 3.00eV$, the limiting value of I_{sc} is 3.993 mA/cm^2.

If a cell based on a photovoltaically active semiconductor having an energy gap E_G has interposed between itself and the solar

Table 1

$E = h\nu$	λ	$10^{-14}N_{ph}(E)$	$I_{sc}(E)$	Solar Energy	$E = h\nu$	λ	$10^{-14}N_{ph}(E)$	$I_{sc}(E)$	Solar Energy
7.000	0.1772	0	0	0	1.900	0.6526	1382.7	22.152	56.61
5.000	0.2480	2.9	0.046	0.25	1.800	0.6888	1568.6	25.134	61.87
4.000	0.3100	31.8	0.510	2.24	1.700	0.7293	1757.3	28.154	67.32
3.900	0.3179	41.1	0.569	2.84	1.600	0.7757	1966.9	31.512	72.94
3.800	0.3263	53.4	0.856	3.59	1.500	0.8273	2198.8	35.227	78.65
3.700	0.3351	68.9	1.104	4.51	1.400	0.8862	2445.0	39.171	84.43
3.600	0.3444	85.9	1.377	5.51	1.300	0.9539	2715.9	43.512	90.34
3.500	0.3542	104.7	1.678	6.57	1.200	1.0353	3020.5	48.391	96.35
3.400	0.3647	125.2	2.006	7.71	1.100	1.1289	3335.6	53.439	102.23
3.300	0.3757	149.3	2.392	8.99	1.000	1.2418	3674.9	58.875	107.89
3.200	0.3875	174.8	2.800	10.30	0.900	1.3789	4011.4	64.367	113.34
3.100	0.3999	203.9	3.267	11.79	0.800	1.5514	4413.5	70.709	118.60
3.000	0.4133	249.2	3.993	14.00	0.700	1.7724	4817.7	77.184	123.42
2.900	0.4275	301.7	4.833	16.46	0.600	200675	5.75.4	82.915	127.14
2.800	0.4428	358.5	5.743	19.09	0.500	2.4803	5513.2	88.327	130.17
2.700	0.4592	433.5	6.945	21.37	0.400	3.0996	5854.2	93.790	132.64
2.600	0.4769	518.6	8.309	25.99	0.300	4.1333	6122.0	98.080	134.14
2 500	0.4959	611.7	9.800	29.78	0.200	6.2249	6325.6	101.343	134.95
2.400	0.5166	711.8	11.404	33.71	0.100	12.4177	6426.9	102.966	135.25
2.300	0.5391	818.9	13.120	37.84	0.000	0.0000	6476.6	103.762	135.30
2.200	0.5636	935.5	14.987	42.08					
2.100	0.5904	1068.3	17.116	46.66					
2.000	0.6199	1223.8	19.606	51.53					

source, a filter which cuts off all photons with energy greater than E, where $E > E_G$, then the limiting short circuit current which this cell can produce is obtained from the relation

$$I_{sc}(E, E_G) = I_{sc}(E_G) - I_{sc}(E) \tag{4}$$

For example, again referring to Table 1, if $E = 3.40eV$ and $E_G = 3.00eV$, then $I_{sc}(3.40eV, 3.00eV) = (3.993 - 2.006) mA/cm^2 = 1.987\ mA/cm^2$.

The following assumptions were made about the junction current-voltage characteristic:

i) It is assumed that the I-V characteristics of the junction can be fitted by a pair of exponential terms of the form

$$I(V) = I_{01}(e^{\frac{qV}{A_1 kT}} - 1) + I_{02}(e^{\frac{qV}{A_2 kT}} - 1) \tag{5}$$

with $A_1 = 1$ and $2 > A_2 > 1$. Further,

$$I_{01} = K_1\ e^{\frac{-E_G}{B_1 kT}} \tag{6}$$

and

$$I_{02} = K_2\ e^{\frac{-E_G}{B_2 kT}} \tag{7}$$

where $2 > B > 1$ and $B \approx A$. The first term in Eq. (5) is the classical Shockley diode term, and the second is usually attributed to recombination and generation in the space charge region. The parameter K_1 depends on parameters of the material like carrier mobilities, lifetimes, etc., and its value can be calculated from the values of these parameters. In their calculation, Gokcen and Loferski assigned a value to K_1 that provides a good fit to the I-V characteristic of an "average" 12% efficient silicon cell, for which at 300°K, $I_{01} = 2.0 \times 10^{-9}\ mA/cm^2$. The value of the parameter K_2 is sensitive to assumptions about the nature and concentration of recombination and generation centers in the space charge region, and calculations of its value are more speculative than are those of K_1. Its value for a particular semiconductor can be obtained by fitting to the observed characteristic. In the case of silicon, at 300°K, a value of $I_{02} \approx 10^{-3}\ mA/cm^2$ provides a reasonable fit. Since the operating point for a silicon cell illuminated by an AM0 spectrum having an intensity equivalent to one or more suns lies in the portion of the I-V characteristic where the first

term of Eq. (5) dominates, it was assumed that this was also true
for cells made from other semiconductors under consideration, i.e.

$$I = I_{01}(e^{\frac{qV}{kT}} - 1) \tag{8}$$

and

$$I_{01} = 6.03 \times 10^9 \, e^{-E_G/kT} \text{ mA/cm}^2 \tag{9}$$

The dependences given by Eqs. (8) and (9) are referred to as the
$A = B = 1$ case.

ii). The voltage at maximum power V_{mp} is obtained from the relation

$$e^{\frac{qV_{mp}}{AkT}} (1 + \frac{qV_{mp}}{AkT}) = \frac{I_{sc}(E,E_G)}{I_0} + 1 \tag{10}$$

The current at maximum power I_{mp} is calculated from the relation

$$I_{mp} = \frac{(qV_{mp}/AkT)}{1 + (qV_{mp}/AkT)} [I_{sc}(E,E_G) + I_0] \tag{11}$$

These expressions for V_{mp} and I_{mp} are based on an equivalent circuit for the illuminated photovoltaic cell in which the internal
shunt resistance R_{sh} is infinite and the internal series resistance
R_s is zero. In "good" solar cells, the losses associated with R_s
can be held to several percent; shunt resistance losses can usually
be neglected.

iii) The AMO solar energy conversion efficiency of a single
cell is given by

$$\eta = \frac{V_{mp} I_{mp}}{135.3C} \times 100\% \tag{12}$$

where C is the concentration ratio, and the AMO solar spectrum intensity is 135.3 mW/cm^2.

Tandem cell systems are of particular interest for concentrator
systems which require high efficiencies if they are ever to become
economically competitive and, therefore, Gokcen and Loferski calculated the efficiency of tandem cell systems as a function of

concentration ratio up to a concentration ratio of 1000. Since the temperature of the cells is likely to increase when they are exposed to concentrated sunlight, they also explored efficiency of tandem cell stacks as a function of temperature. The third parameter of interest in such a study is the number of solar cells based on semiconductors having different values of E_G in the tandem cell system, and this parameter was also factored into the calculation. In their calculation, Gokcen and Loferski set the additional condition that the same number of solar photons should be absorbed in each cell per unit area per unit time, i.e. that the short circuit currents should be equal. This condition must be observed in a monolithic tandem cell structure, but it is not so important if the cells are connected to separate electrical loads as in Fig. 2.

The results of these calculations were published in the form of numberous tables and figures, of which a few particularly important ones are reproduced and discussed here.

First, consider the case of flat plate (concentration ratio unity) tandem cell systems. Table 2 shows the optimum efficiency as a function of number of cells in the stack for "One Sun" and with all cells at 300K. For numbers of cells in excess of four, the optimum efficiency can be approximately reached by more than one combination of energy gap values. For numbers in excess of twelve, the values of E_G can be uniformly distributed over the

Table 2

No. of Cells:	1	2	3	3	4	5
E_G values (eV):	1.5	2.0 & 1.2	2.3, 1.6 & 1	2.4, 1.6 & 1.0	2.5, 1.8 1.3 & 0.9	2.6, 2.0, 1.6 1.2 & 0.8
Efficiency:	23.22	32.48	37.42	37.39	40.45	42.45

No. of Cells:	6	6	6	7
E_G values (eV):	2.7, 2.3, 1.9 1.5, 1.1 & 0.8	2.8, 2.4, 2.0 1.6, 1.2 & 0.8	2.6, 2.2, 1.8 1.4, 1.0 & 0.8	2.9, 2.5, 2.1, 1.8 1.5, 1.1 & 0.8
Efficiency:	43.82	43.67	43.64	44.86

No. of Cells:	7	7	13	26
E_G values (eV):	3.0, 2.6, 2.2 1.8, 1.4, 1.0 & 0.8	2.8, 2.4, 2.1, 1.5, 1.2 & 0.8	0.8 to 3.0 in intervals of 0.2, & 3.4	0.7 to 3.0 in intervals of 0.1 & 3.2 & 3.4
Efficiency:	44.70	44.67	48.13	50.09

range of band gaps involved in the stack. Figure 3 shows the efficiency vs. number of cells having different E_G values for the "One Sun" case with temperature as the running parameter. From this table and figure, we note that the efficiency approaches 50% when the number of cells is 24.

Table 3 shows similar results for concentration ratios of 100, 500 and 1000 suns. Figure 4 shows the efficiency vs. number of cells having different E_G values for a concentration ratio of 1000 with temperature as the running parameter. From Table 3, we observe again that the selection of E_G values can be made rather arbitrarily when the number of cells in the stack exceeds twelve. Furthermore, comparison of Tables 2 and 3 shows that essentially the same combination of E_G values results in the optimum efficiency for all concentration ratios between 1 and 1000. Gokcen and Loferski also showed that if the cells are operating at an elevated temperature, then the optimum combination of energy gaps differs soemwhat from the values corresponding to 300K for small numbers of cells but remains essentially the same for larger numbers of cells. From Fig. 4, we see that the efficiency approaches 60% for 24 cells, 300K and 1000-fold concentration.

Thus we conclude that if we had available about a dozen photovoltaically active semiconductors with E_G values separated by intervals of about 0.2eV from about 1.0eV to 2.4eV and were able to

Table 3

100 Suns:					
no. of cells	1	2	3	6	24
E_G values(eV)	1.4	1.8, 1.0	2.3, 1.5, 0.9	2.6, 2.1, 1.7, 1.3, 1.0, 0.7	0.7 to 3.0 in intervals of 0.1
Efficiency	26.43	37.05	42.52	40.13	55.96
500 Suns					
No. of cells	1	2	3	6	24
E_G values(eV)	1.3	1.8, 1.0	2.2, 1.4, 0.8	2.6, 2.2, 1.8, 1.4, 1.0, 0.7	0.7 to 3.0 in intervals of 0.2
Efficiency	27.62	38.80	44.46	52.26	58.21
1000 Suns					
No. of cells	1	2	3	6	25
E_G values(eV)	1.3	1.8, 1.0	2.2, 1.4, 0.8	2.5, 2.1, 1.7, 1.3, 1.0, 0.7	0.7 to 3.0 in intervals of 0.2
Efficiency	28.18	39.56	45.37	53.21	59.22

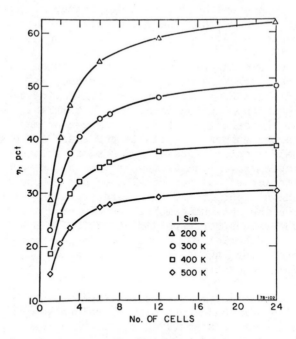

Fig. 3. Variation of efficiency with number of cells
at various temperatures; AMO illumination.

optimize the performance of cells made from each of these semiconductors, then we could combine the cells to produce efficiency optimized tandem cell systems for essentially all concentration
ratios and all temperatures likely to be encountered in practice.
The limiting 300K efficiencies for these stacks would lie between
50% and 60% depending on the concentration ratio used. It is also
interesting to observe that increasing the number of cells beyond
about six does not lead to much further increase in efficiency.

III. DESIGN OF AN OPTIMIZED SOLAR CELL STRUCTURE FOR TANDEM CELL
       SYSTEMS

It is evident from our discussion that realizing the promise
of high efficiency cells requires photovoltaically active semiconductors with band gaps in the range 1.0 to 2.4eV. Figure 5 tabu-
lates the band gaps of some binary and ternary semiconductors which
could serve as the photovoltaically active materials in the cells.
These materials have discrete values of E_G, and while it may be possible to identify a group of them for a specific tandem cell system,

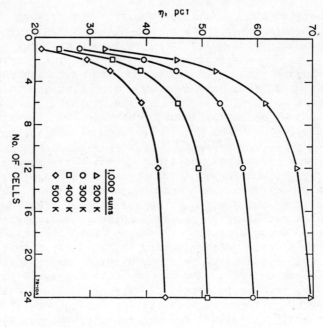

Fig. 4. Variation of efficiency with number
of cells at various temperatures.
C = 1000; AMO spectrum.

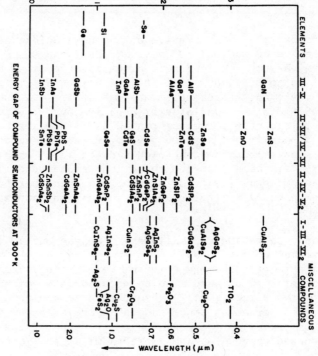

Fig. 5. List of various semiconductors ordered
by their bandgaps, Eg (eV).

this would probably be an impractical way to proceed because the materials are sufficiently different from each other so that a large research and development effort would be needed to lift the efficiency of each one of them close to its limiting value. A better approach is to focus efforts on a particular solar cell structure and materials system in which the required wide range of energy gaps is available. Then the work expended on optimizing a prototype cell of the system is readily transferred to other cells in the system.

First consider the kind of universal cell structure which would satisfy our requirements. To conserve material, the photovoltaically active semiconductors must be direct gap materials. If the solar cells are to be of the p/n homojunction type, excessive surface recombination losses, which are usually encountered at the light receiving surface, must be eliminated if high efficiencies are to be attained. One way to eliminate surface recombination losses is to use the direct gap semiconductor as the light absorbing, photovoltaically active part of a p/n heterojunction in which the other is a semiconductor whose band gap is so large that it cannot absorb any significant fraction of photons from the solar spectrum. A heterojunction device of this type is illustrated in Fig. 6 [15].

There are a number of design factors which must be considered to optimize probability of success. First there is the requirement that the diffusion length of minority carriers in the photovoltaically active material should be as long as possible. Since minority electrons usually have higher mobilities than minority holes, their diffusion lengths are usually larger; this implies that the photovoltaically active semiconductor should be p-type. A second design consideration is related to the effects of lattice mismatch between the heterojunction pair. Lattice mismatch gives rise to dislocations at the interface; dislocations, to dangling bonds; dangling bonds, to interface recombination centers. Arienzo and Loferski[16] showed that these interface states do not affect the short circuit current, but they do strongly affect the current-voltage characteristics, and therefore, the open circuit voltage and fill factor of the resultant photovoltaic cell. Their conclusion was that even in the case of an n-CdS/p-InP cell where the lattice mismatch is only about 0.3%, interface states control the i-V characteristic and, therefore, limit cell performance. Consequently, it is important in selecting pairs of semiconductors for heterojunction cells to reduce lattice mismatch to zero. Another important parameter which affects the band structure and, ultimately, the cell performance is the electron affinity χ of each of the materials. "Spikes" in the conduction band could impede minority electron transport from the small band gap absorber to the n-type wide band gap window. Such a spike can be avoided if the electron affinity of the smaller band

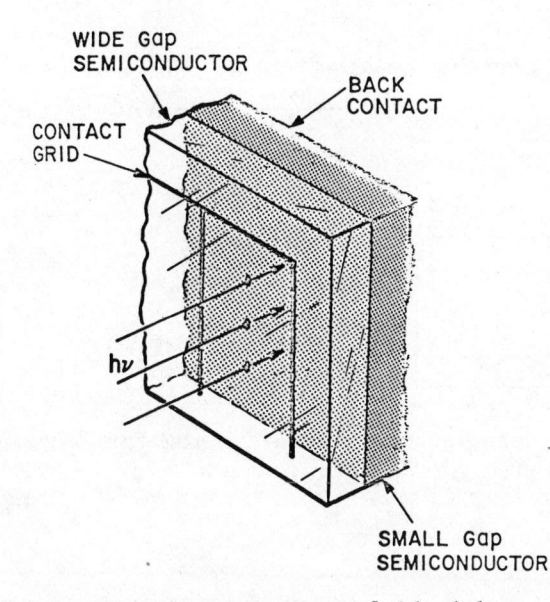

Fig. 6.   Schematic representation of ideal heterojunction
          solar cell.

gap material is less than or at most equal to that of the wide band
gap material. Finally, it is desirable in the cell design if the
space charge region is confined mainly in the photovoltaically active p-type material, since the SCR field will then aid in the collection of minority electrons and relax requirements on the minority
carrier diffusion length in that material. The resulting ideal
heterojunction band gap model incorporating these features is shown
in Fig. 7.

In this structure, the light is incident on the semiconductor
on the left side of the diagram. For this semiconductor pair,
$E_{G1} > E_{G2}$ and $E_{G1} > h\nu_{max}$ (solar) where $h\nu_{max}$ is the highest energy
photon present in the solar illumination. A value of $E_{G1} \gtrsim 2.5eV$
is sufficient to satisfy this requirement. The p-type material in
which light is absorbed is, of course, a direct gap semiconductor.
Secondly, $\chi_1 > \chi_2$ so that there is no spike in the conduction band
to impede the flow of minority electrons from the p-type photovoltaically active semiconductor to the barrier.   In Fig. 7,
$\chi_1 - \chi_2 = \Delta\chi = \Delta E_c$ corresponds to the mismatch of the conduction band

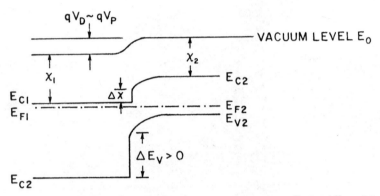

Fig. 7.   Heterojunction band diagram for "ideal" cell.
$\chi_1 > \chi_2$; $E_{G1} > E_{G2}$; $\delta_1 < \delta_2$.

edges at the interface. The mismatch of the valence band edges ΔE_v is equal to $E_{G1} - E_{G2} - \Delta E_c$. Finally, $(E_{C1} - E_F) = \delta_1 < \delta_2 = (E_F - E_{v2})$, i.e. the p-type region is more heavily doped than the n-type region and, therefore, the SCR extends mainly into the p-region.

IV.   SELECTION OF SEMICONDUCTORS FOR TANDEM SOLAR CELL SYSTEMS

As is evident by now, the realization of the full promise of tandem solar cell systems requires that there should be available a selection of solar cells, made from semiconductors having different E_G values, each cell having a solar energy conversion efficiency close to the maximum achievable in its photovoltaically active material. The fabrication of such solar cells requires a number of semiconductors with good photovoltaic properties with band gaps ranging between 1 and 2eV. Let us assume that to simplify fabrication problems the wide band gap semiconductor of the ideal solar cell described above is kept fixed. We have already pointed out that lattice mismatch between semiconductors in the heterojunction must be reduced to zero. This means that the semiconductors needed for the solar cells of the tandem system must all have the same lattice constant as the wide band gap n-type material, even while their band gaps vary over the required 1 to 2eV range. An occasional pair of simple (elemental, binary or ternary) semiconductors, one wide band gap, one having good photovoltaic properties, may exist (for example, CdS and AgInS$_2$ in

Fig. 8), but there does not exist a group of such lattice matched, E_G variable semiconductors. The only way to achieve this requirement is to resort to alloys of more than two semiconductors.

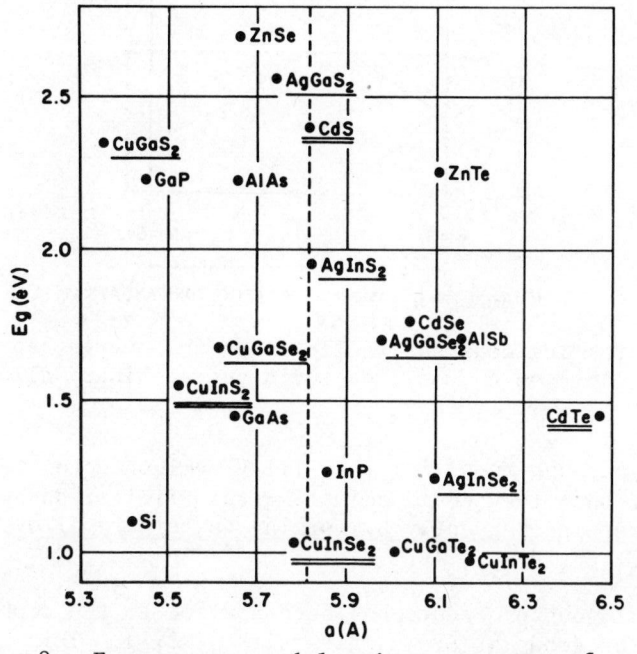

Fig. 8.   Energy gaps and lattice constants of some
          semiconductors of interest for solar cells.

Consider first an alloy formed between two fully miscible semiconductors like $CuInSe_2$ and $CuInS_2$. Figure 9 shows how the lattice constant and energy gap might be expected to behave in such an alloy system. According to Vegard's law, the lattice constant varies essentially linearly with composition. The energy gap varies non-linearly; it is arbitrarily shown to have a variation which is convex downward in the figure. This pseudo-binary alloy will have a lattice constant which matches the lattice constant of a wide band gap material (e.g., CdS) having the lattice constant a_{CdS} for some particular composition. In Fig. 8, this composition would be $MInS_{2(1-x)}Se_{2x}$ to which there corresponds only one value of energy gap. Each such pseudo-binary alloy of pairs of semiconductors provides at most one composition at which lattice matching with another semiconductor is possible and, therefore, only one value of E_G which may or may not be one of the values required for a desired optimum tandem cell system.

The range of bandgaps available can be expanded considerably by adopting a concept introduced by Coleman, Holonyak etal [18]

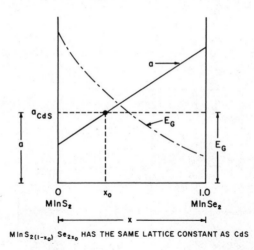

MInS$_{2(1-x_o)}$ Se$_{2x_o}$ HAS THE SAME LATTICE CONSTANT AS CdS

Fig. 9. Lattice constant (solid line) and energy gap (dashed
line) vs composition for a pseudo-binary alloy.

and by Antypas and Moon [19] into III-V semiconductor technology.
They showed that they could adjust, within limits, bandgap and lattice constant independently by forming quaternary alloys of the
type In$_{(1-x)}$ Ga$_x$ P$_{(1-z)}$ As$_z$.

An analogous procedure has been applied to the copper ternary
system by the group at Brown University [7-9]. This results in
five element alloys like Cu$_{(1-x)}$Ag$_x$InSe$_{2y}$S$_{2(1-y)}$. For example,
the alloy with x $\sim$ 0.72 and y $\sim$ 0.70 should have a lattice constant
exactly equal to that of CdS and a band gap around 1.5eV. Figure
9 illustrates how such alloys can in principle be formed among
eight chalcopyrite semiconductors of the type A^I B^{III} C$_2^{VI}$. The
point on the face of the cube having coordinates (x$_o$, y$_o$)corres-
ponds to an alloy having the composition Cu$_{(1-xo)}$ Ag$_{xo}$ InSe$_{2yo}$
S$_{2(1-yo)}$. Points in the interior of the cube correspond to six
element compounds.

Iso-lattice constant, iso-energy gap and iso-electron affinity maps can be constructed for such alloy systems according to
the following procedure developed by Moon etal. Suppose that the
values of lattice constant, energy gap and electron affinity are
known for the particular ternary alloy represented by the point
(x$_o$, y$_o$) on the square shown in Fig. 10. Then the values
of these parameters for alloys represented by a point (x,y) in
the vicinity of (x$_o$, y$_o$)can be represented by a two-dimensional
Taylor series around the point (x$_o$, y$_o$) of the form

$$F(x,y) = \sum_{i=1}^{n} \sum_{j=0}^{i} a_{(i-j)(j)} (x-x_o)^{(1-j)} y^{(j)} \tag{13}$$

If we retain only linear terms, the lattice constant a_o variation assumes the form (Vegard's law)

$$a(x_1, y_1) = \alpha_{00} + \alpha_{10}\, x_1 + \alpha_{01}\, y_1 + \alpha_{11}\, x_1\, y_1 \tag{14}$$

where

$$x_1 = (x - x_o) \text{ and } y_1 = (y - y_o).$$

Quadratic terms need to be included in the expansions for energy gap and electron affinity. The result in the case of the energy gap variation would be an equation of the form:

$$E_G(x_1, y_1) = A_{00} + A_{10}x_1 + A_{01}y_1 + A_{11}x_1y_1 + A_{20}x_1^2 + A_{02}y_1^2 +$$

$$+ A_{21}x_1^2 y_1 + A_{12}x_1 y_1^2 \tag{15}$$

In these equations α_{00} and A_{00} are the lattice constant and the lattice constant and energy gap of the material having a composition $Cu_{(1-x_o)} Ag_{x_o} In Se_{2y_o} S_{2(1-y_o)}$. To illustrate how these would be used to construct the iso-E_G map of the system, let x_o remain constant, i.e., only the ratio of S to Se is varied. Then from Eq. (15), the energy gap of a material corresponding to the coordinates x_o, y would be given by

$$E_G(x_o, y) = E_G(x_o, y_o) + A_{01}\, y_1 + A_{02}\, y_1^2 \tag{16}$$

The parameter A_{02} has been called the bowing parameter by Moon etal [21]. A first approximation to the energy gap topological map of such an alloy system can be constructed by assuming purely linear variations of E_G and a with composition, i.e., by assuming that the bowing parameters A_{20}, A_{02}, A_{21} and A_{12} are all equal to zero. More accurate mapping requires experimental data about the way that parameters of interest vary with composition. To accumulate such data, it is first necessary to synthesize alloys having compositions more or less evenly distributed over the complete range of possible values of x and y. The energy gaps and lattice constants are measured and the results are fitted to Eqs. (14) and (15) to determine the α_{ii} and A_{ii} coefficients. Figure 10 shows such a topological (E_G, a) map for the Cu-Ag-In-S-Se system.

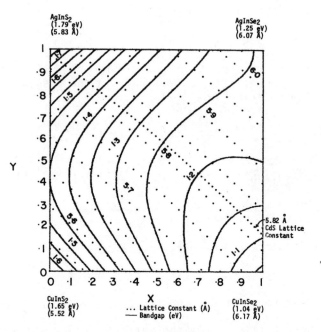

Fig. 10.   Empirical iso-lattice constant and iso-energy gap map
for the $Ag_xCu_{(1-x)}InS_{2y}Se_{2(1-y)}$ system. The isolattice
constant line corresponding to CdS is shown in the
diagram.

When such a map has been constructed, it can be used to select
semiconductors of different bandgap, but the same lattice constant,
so that the lattice matching which is so desirable in a heterojunction can be made between a specified wide bandgap window
semiconductor and a series of photovoltaically active alloy semiconductors. For example, Fig. 10 shows the iso-lattice constant
line with a lattice constant value equal to that of CdS (5.82 eV).
The bandgaps of semiconductors which have this lattice constant
range from about 1.06 eV to 1.80 eV; this range makes the Cu-Ag-
In-S-Se alloy system an attractive possible source of materials to
be used in solar cells of tandem systems. A description of experiments on this and other ternary alloy systems is the subject
of another section of this chapter. However, before we return to
that matter we shall first consider optimum design of individual
cells for incorporation into the system.

V.   OPTIMIZED DESIGN OF DIRECT GAP PHOTOVOLTAIC CELLS

    Once the photovoltaically active semiconductor has been se-
lected for a solar cell, the objective of the solar cell designer
is to extract as much efficiency as possible from a cell made of
that material. Since some of the elements used in the $A^{III} B^V$ and
$A^I B^{III} C_2^{VI}$ semiconductors which are the main subjects of this
chapter are not very abundant, it is also a goal of the solar cell
designer to make cells which are parsimonious of material. Recently, M. Wolf has calculated as a function of cell thickness
and junction position, the theoretical upper limit efficiency of
silicon homojunction solar cells [22] in which homogeneously
doped n- and p-regions are terminated by surfaces having zero surface recombination velocity, i.e., the front and back surfaces of
the cell are covered by minority carrier mirrors (MCM). Wolf's
"ideal design" cell also incorporates an optical mirror on the
surface opposite that on which light is incident. This mirror
doubles the effective optical thickness of the cell, since it permits photons which are not absorbed during their first transit
through the cell to have a second opportunity to be absorbed. For
this cell design, Wolf found that the highest efficiency occurs
for a silicon cell thickness substantially smaller than in cells
which do not employ MCM's and optical mirrors and that the magnitude of this maximum attainable AM1 efficiency is somewhat in
excess of 25%. The thickness of the Si cells in which this efficiency should be achievable is about 75μm

    Spitzer etal [23] have extended these concepts to solar cells
made from the direct gap materials $CuInSe_2$ and GaAs. Their work
focused mainly on $CuInSe_2$ because their model can be applied with
greater confidence to a material having a band gap of 1.0 eV like
$CuInSe_2$ than to a material having a band gap of 1.43 eV like GaAs.
This is because the theory is valid only to solar cells in which
the reverse saturation current I_o results from diffusion of thermally generated minority carriers to the p/n junction. Figure 11
shows the energy band diagram used for their calculations.

    Expressions for the reverse saturation current I_o are cal-
culated from the ambipolar diffusion equation which governs the
transport of the thermally generated minority carriers in the
homogeneous, quasi-neutral regions of the cell. For the one di-
mensional case, this equation has the form

$$D_n \frac{d^2 \Delta n}{d x^2} - \frac{\Delta n}{\tau_n} + G(\lambda) = 0 \qquad (17)$$

where $G(\lambda)$ is the carrier generation function.

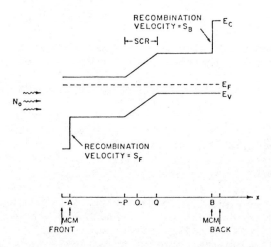

Fig. 11.   The energy band diagram used by Spitzer etal [23] to
calculate the photovoltaic parameters of GaAs and
CuInSe$_2$ homojunctions.   The metallurgical junction
occurs at x = 0.   Q and -P are the coordinates of the
edges of the space charge region.

The surfaces of these homogeneous regions farthest from the junction (at -A and B in Fig. 11.) are commonly characterized by surface recombination velocities S$_A$ for the surface of the cell on
which light is incident and S$_B$ for the back surface of the cell.
This translates into the boundary condition

$$D_n \frac{d\Delta n}{dx}\bigg|_{x = B} = S_B \cdot \Delta n(B) \qquad (18)$$

for the back surface of the cell and an analogous expression for
the front region.

The boundary of the quasi-neutral region at the edge of the
space charge region (SCR) is characterized by the Boltzman boundary condition

$$\Delta n = n_o \left[\exp (qV/kT)-1\right] \qquad (19)$$

where n is the concentration of minority carriers under equilibrium conditions and V is the voltage applied across the junction.

Consider two extreme possibilities, namely S_A and S_B are
both infinite and $S_A = S_B = 0$. In the first case, solution of
Eq.(17) results in an expression for I_o of the form

$$I_o^{(1)} = I_{ono} \coth (w_n/L_h) + I_{opo} \coth (w_p/L_e). \qquad (20)$$

Here w_n and w_p are the widths of the quasi-neutral regions on the
n- and p- sides, respectively; I_{ono} and I_{opo} are the contributions
to the reverse saturation current from the n- and p- regions, respectively, for the ease of thick n-type ($w_n \gg L_h$) and p-type
($w_p \gg L_e$) regions; L_h and L_e are the diffusion lengths of minority holes in the n-region and of minority electrons in the
p-region, respectively. Now coth x $\rightarrow$ 1 as x $\rightarrow \infty$ and coth x $\rightarrow \infty$
as x $\rightarrow$ 0. It is a goal of solar cell design to minimize I_o.
Therefore, if the quasi-neutral regions of the cell are terminated
by surfaces of high surface recombination velocity, I_o (1)is
minimized by recourse to wide ($w \gg L$) quasi-neutral regions. This

minimum value of I_o will be

$$I_o^{(1)} (min) = I_{ono} + I_{opo} \qquad (21)$$

Now consider the case, $S_A = S_B = 0$. Then

$$I_o^{(2)} = I_{ono} \tanh (w_n/L_h) + I_{opo} \tanh (w_p/L_e) \qquad (22)$$

Since tanh x $\rightarrow$ x as x $\rightarrow$ 0, $I_o^{(2)}$ can be much smaller than $I_o^{(1)}$ (min)
This means that the open circuit voltage of a thin ($w_n \ll$
L_h, $w_p \ll L_e$) device should be higher than the open circuit voltage of a thick device since

$$V_{oc} \sim \ln (I_{sc}/I_o) \qquad (23)$$

However, reducing the thickness of the n- and p- regions decreases I_{sc} because thinning down the cell decreases the fraction of solar photons absorbed from the incident sunlight. It is for this reason that an optical mirror is incorporated into the back surface of the cell; it reflects back through the cell those photons which would otherwise have been lost.

Spitzer etal treated the case of p/n homojunctions in CuInSe$_2$ and GaAs. They found that the highest efficiency is to be expected from a device in which the n-type front region is extremely thin, namely about 0.05μm, as shown in Fig. 12. With such a shallow junction, there is virtually no contribution to I_{sc} from the n-type region; most of the sunlight absorption occurs in the p-type base material. As shown in Fig. 13, the addition of a back surface mirror improved the current I_{sc} (and efficiency) by a small amount, only if the total cell thickness was less than 1μm, which contrasts with Wolf's results for the indirect gap semiconductor silicon where the optical mirror plays an important role in maintaining high efficiency. Figure 13 shows that the short circuit current without the optical mirror saturates for cell thicknesses equal to or greater than about 2.5μm. Figure 12 shows that the efficiency saturates at a value of about 26% for a cell thickness of about 1.5μm.

Spitzer et al found that the limiting efficiency was very sensitive to the magnitude of S_B. As shown in Fig. 14, the maximum efficiency achievable in a device with S_B infinite would be about 20% for a device having a thickness of about 7μm, and about 17% for a device having a thickness of 1.5μm. Both short circuit current and open circuit voltage decrease with thickness if S_B is high.

They also considered the effect of varying the acceptor concentration in the p-region on cell performance. This parameter affects several aspects of device performance. In addition to controlling the built-in voltage (V_{bi}), the doping concentrations control the SCR width. Since a significant fraction of the photons are absorbed in the SCR, the SCR width influences J_{sc}. In addition if the SCR width is large compared to the width of the quasi-neutral region, w_n and w_p are reduced assuming that the total thickness of the device is held fixed. This results in a reduction of $I_o(2)$ calculated from Eq. (22). Using reasonable empirical dependences of carrier mobility and lifetime on acceptor concentration, they calculated short circuit current, open circuit voltage and efficiency as a function of carrier concentration for the case $S_B = 0$. The results are shown in Fig. 15. The highest efficiency (26%) occurs for a base acceptor concentration N_A of about 10^{15}/cc. The open circuit voltage passes through a minimum for $N_A = 10^{16}$/cc and increases as N_A decreases. This increase

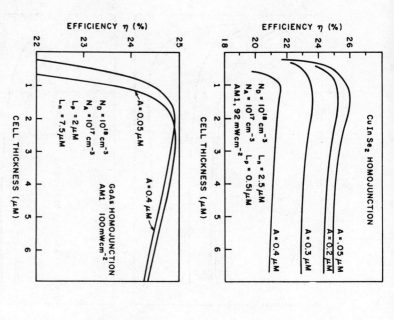

Fig. 12. Efficiency as a function of cell thickness and junction depth A. The upper figure is for CuInSe2, and the lower for GaAs. No optical mirrors are incorporated in these cells but they do have minority carrier mirrors on both external surfaces.

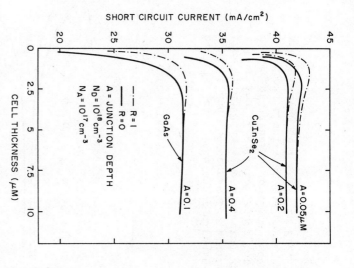

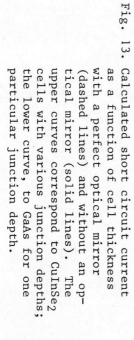

Fig. 13. Calculated short circuit current as a function of cell thickness with a perfect optical mirror (dashed lines) and without an optical mirror (solid lines). The upper curves correspond to CuInSe2 cells with various junction depths; the lower curve, to GaAs for one particular junction depth.

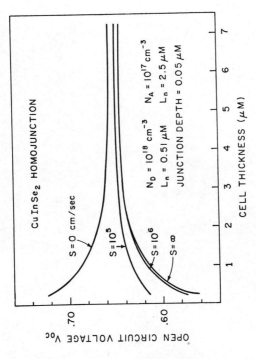

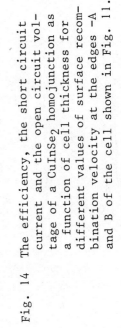

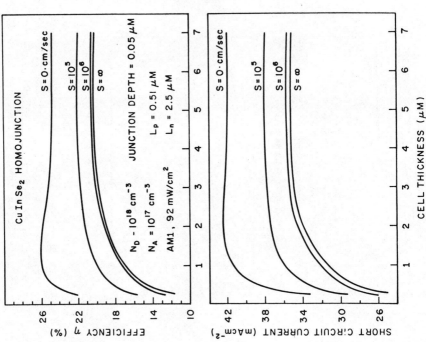

Fig. 14 The efficiency, the short circuit current and the open circuit voltage of a CuInSe₂ homojunction as a function of cell thickness for different values of surface recombination velocity at the edges –A and B of the cell shown in Fig. 11.

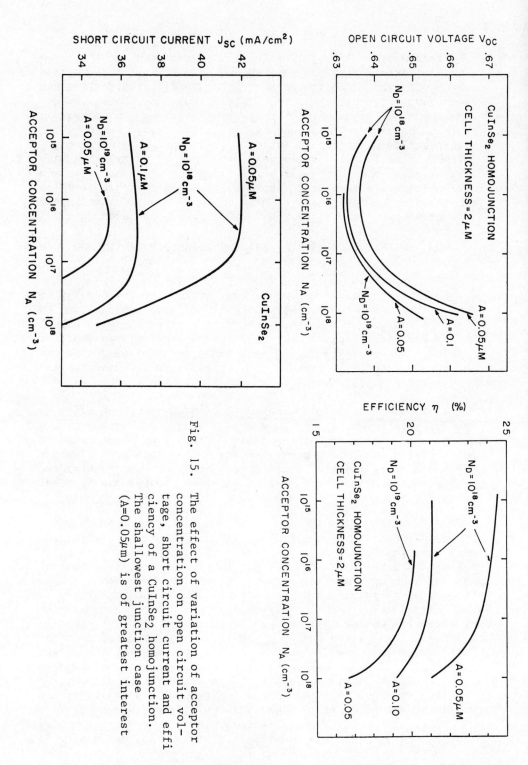

Fig. 15. The effect of variation of acceptor concentration on open circuit voltage, short circuit current and efficiency of a CuInSe2 homojunction. The shallowest junction case (A=0.05μm) is of greatest interest

is related to the drop in $I^{(2)}$ which occurs because w_p decreases as the width of the SCR increases.

It is obvious that minority carrier mirrors have an extremely beneficial effect on device performance. How can such mirrors be realized in actual devices? One way to construct a minority carrier reflecting barrier is to introduce an abrupt change in the doping level at the ends of the quasi-neutral region. In other words, one would construct a device having the configuration $n^+ - n/p$-p^+. As we shall see shortly, such a structure is exactly what is required in monolithic tandem solar cells. Therefore, the improvements in device design described in this section are extremely relevant to tandem cell systems. Another way to realize a surface of zero or low surface recombination velocity is to form a junction with a semiconductor having the same conductivity type but a larger band gap. This is commonly done in p-GaAs/n-GaAs solar cells.

The calculation described here was performed for a p/n homojunction structure. However, it can be readily adapted to a heterojunction structure like the ideal cell of Figs. 6 and 7, since in such a structure the wide band gap window material does not contribute to I_o and, therefore, in Eq. (22), $I_{opo} = 0$. This should result in a small improvement in efficiency with respect to a p/n homojunction having the same p-type base structure.

In conclusion, this theoretical investigation points the way toward optimized performance from a minimum thickness of direct gap semiconductor. It shows that higher efficiency can be expected from slimmed-down solar cells, provided that they incorporate minority carrier mirrors. If these mirrors are of the abrupt junction type (e.g., a p-p^+ junction), they make it easy to incorporate such cells into tandem solar cell systems.

VI   MONOLITHIC AND SPLIT SPECTRUM TANDEM CELL SYSTEMS

The super-efficient, super thin cells described in the preceding section can of course stand alone as the building blocks of solar arrays. However, they can be used as elements in tandem cell systems. There are three principle ways in which they can be assembled to form such systems. In the first method, illustrated in Fig. 2, the cells are placed one over the other but they are electrically isolated so that each bandgap cell is connected to a separate load. Optical losses at the interfaces can be reduced by using a split spectrum system like that shown schematically in Fig. 16. [23] Here mirror M_1 reflects that portion of sunlight which can be used by cell C_1, the largest E_G cell in this three cell tandem system. The dichroic mirror M_2

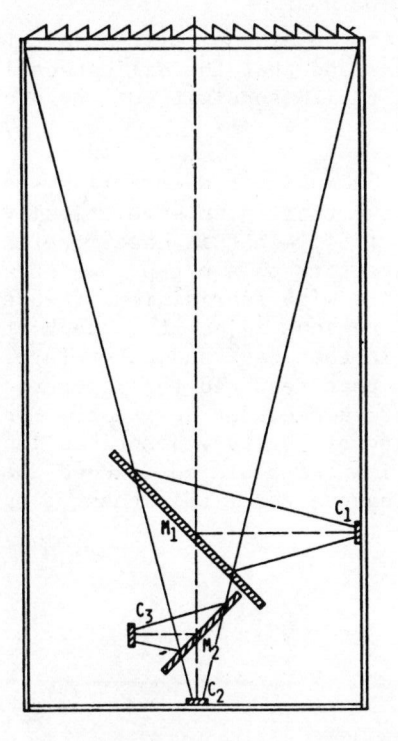

Fig. 16.   Schematic diagram of a three cell split spectrum
            concentrator system using two dichroic mirrors
            M_1 and M_2 and a Fresnel lens for concentrating
            sunlight. From Ref. [24].

performs the same function for cell C_2. The solar photons with
energies less than the energy gap of   cell C_2 are transmitted
to cell C_3. Moon etal [24] investigated such a two cell split
spectrum system in which one cell was a silicon homojunction and
the other a homojunction cell made from a ternary Al–Ga–As alloy
having a band gap of 1.65eV. They fabricated a special dichroic
mirror consisting of 17 layers (a quarter–wave stack with eighth–
wave ends) of alternating ZnS and $Na_3 AlF_6$. This filter reflected
about 99% of the photons having an energy less than 1.65eV and
about 5% of the higher energy photons. When this combination
of cells and dichroic mirror was exposed to concentrated sunlight
having an effective intensity of 165 suns and an AM 1.23 spectrum,
the sum of the power outputs from the Si and the AlGaAs cells
corresponded to a solar energy conversion efficiency of 28.5%.
The authors state that if the dichroic mirror had been a perfect
filter, the efficiency would have been about 31%. Separate

measurements showed that the AℓGaAs cell had an efficiency of 20% when exposed to 400 suns and that the silicon cell had an efficiency of 15% when exposed to 178 suns without the dichroic mirror in place.

The third method for making tandem cell systems is to use a monolithic structure like that shown schematically for and m-cell system in Fig. 17 [25]. In this case, the individual cells have the electronic structure $n^+ - n/p-p^+$, which as we have already noted is compatible with fabrication of super-efficient, thin cells. The n^+/p^+ regions adjacent to each other provide ohmic contact between cells in the stack and, therefore, the doping level must be close to that required for degeneracy. The thickness of a monolithic stack need not be great. For example, if the individual cells are of the type described in the preceding section their total thickness would not exceed about 5µm so that a 12 cell stack would have a total thickness of only 60µm.

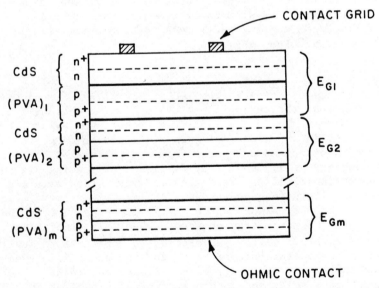

Fig. 17 Schematic representation of an m-junction tandem cell. The left side captions refer to cells like the unit cell of Fig. 6. The right side captions refer to homojunction cells like those discussed in Ref. [25].

At this point, however, we turn to discussions of the current
state of the art of preparation of semiconductors for the ideal
cell of Fig. 6 which can be viewed as the building blocks of
future tandem solar cell systems.

VII SYNTHESIS AND PROPERTIES OF TERNARY ALLOY CHALCOPYRITE
     SEMICONDUCTORS

We have presented what we believe to be convincing arguments
to support the view that realization of tandem cell systems requires a family of semiconductors having the same lattice constant but varying energy gaps. There are a number of such fam-
ilies which can be constructed using copper ternary semiconductors
like $CuInSe_2$, $CuGaTe_2$, $AgInSe_2$ etc. In this section we shall describe the properties of these materials; the methods used to synthesize them; the characterization of the materials; the preparation of large grained materials and thin films from them, and the
fabrication of solar cells from large grained samples of these
alloys.

The ternary semiconductors of the type $CuInSe_2$, $AgInSe_2$, etc.
crystallize in the chalcopyrite structure whose unit cell is compared to that of the zincblend structure in Fig. 18.

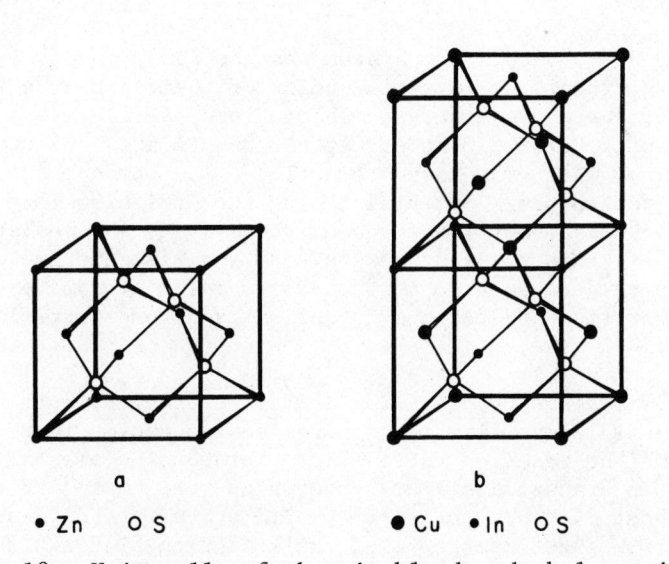

a   b

• Zn   O S   ● Cu ●In   O S

Fig. 18. Unit cells of the zincblend and chalcopyrite
          structures.

The chalcopyrite tetragon consists essentially of two zincblend
cubes; it contains two metal sublattices, one for column I elements (Cu,Ag), the other for column III elements (Aℓ,Ga,In) instead of the single metal sublattice of the zincblend structure.
The x-ray diffraction pattern of the chalcopyrite structure contains the same major lines as that of the zinc blend structure
but it contains additional distinctive "superlattice" lines.

    Table IV lists the energy gaps and lattice constants,
measured at 300K, of ternary chalcopyrite semiconductors which
could be used to prepare ternary alloys [27] . Suppose, for example, that we set as a goal the fabrication of ideal solar cells
like those shown in Fig. 6 with CdS as the wide bandgap n-type
window and four and five-element ternary alloys having the same
lattice constant as CdS as the photovoltaically active semiconductors. Suppose further that, since we expect that in a lattice
matched heterojunction I_o will be determined by thermal generation
across the forbidden energy gap, we wish to fabricate cells whose
photovoltaically active semiconductors have a bandgap near the
maximum of the efficiency versus bandgap curve [10] i.e., $E_G =$
1.5eV. If we were to construct first order iso - E_G, iso-
lattice constant maps based on the values shown in Table IV, we
would find that the group of alloys in Table V would satisfy these
criteria. This was first pointed out by the Brown University
group [19,20] whose research we shall describe in greater detail
in what follows.

    The Cu-Ag-In-S-Se alloy system was the first one explored
by the Brown group. The starting point of their research was the
synthesis of the four ternary semiconductors of this system of
Fig. 10, namely $CuInS_2$, $CuInSe_2$, $AgInS_2$ and $AgInSe_2$ and of five
element alloys made from these ternaries. The goal of the synthesis was to produce a material having the stoichiometric composition and the chalcopyrite structure from which low resistivity
p-type pellets (hopefully large grained) and films could be produced. The powder was used as the source material for resistance
heated evaporation and flash evaporation, and for sputtering targets.

    Synthesis began by preparing a mixture of ultra pure powders
of the elements comprising the desired semiconductor in the appropriate weight fraction ratio. This mixture of elemental powders
was sealed in evacuated silica tubes which were heated at about
350° for about 24 hours to promote solid state interdiffusion of
the elements. The temperature of the resulting sintered mass
was then increased at a rate of about 25° per hour until it became
molten; usually this required a temperature between 700 and 1000°C.
The material was left at this temperature for about 24 hours and
then cooled to room temperature. The reacted material was then

Table IV.   Energy Gaps and Lattice Constants (Measured at 300K)
of Ternary I–III–VI$_2$ Compounds That Could be
Employed in a Pentenary Alloy System

Compound	Bandgap (eV)	Lattice Constant (A°)
1. $CuAlS_2$	5.32	3.49
2. $CuGaS_2$	2.43	5.35
3. $CuInS_2$	1.53	5.52
4. $CuTlS_2$		5.58
5. $CuAlSe_2$		5.60
6. $CuGaSe_2$	1.68	5.61
7. $CuInSe_2$	1.04	5.78
8. $CuTlSe_2$		5.83
9. $CuAlTe_2$	2.06	5.96
10. $CuGaTe_2$	1.2	6.00
11. $CuInTe_2$	0.96	6.17
12. $CuTlTe_2$		
13. $AgAlS_2$		5.70
14. $AgGaS_2$	2.73	5.75
15. $AgInS_2$	1.87	5.82
16. $AgTlS_2$		
17. $AgAlSe_2$		5.96
18. $AgGaSe_2$	1.83	5.98
19. $AgInSe_2$	1.24	6.09
20. $AgTlSe_2$		
21. $AgAlTe_2$	2.27	6.30
22. $AgGaTe_2$	1.32	6.30
23. $AgInTe_2$	0.97	6.42
24. $AgTlTe_2$		

removed from the silica tube and ground in a powder using a
mortar and pestle. The powder was then sealed in an evacuated
silica tube and annealed at a temperature which was about 100°C
below the melting point of the semiconductor for several days to
promote hormogenization. The sample was then ground into a powder
again and an x-ray powder diffraction pattern was recorded to determine that the alloy is single phase. If not, then the material
was once again reground and refired. This process was repeated
until the desired single phase is obtained.

Although compounds prepared by the method just described
exhibited the desired single phase x-ray diffraction pattern, the

Table V. Several Possible Compositions of I–III–VI$_2$
Pentenary Alloy Systems. The Specific Alloy
Composition is That Which (Using a Linear
Estimate) is Lattice Matched to CdS. The
Energy Gaps of Most Materials is Estimated
to be 1.5eV.

Alloy	x	y
$Cu_{(1-y)}Ag_y\ In\ S_{2(1-x)}Se_{2x}$	0.42	0.62
$Cu\ Ga_{(1-y)}Al_y\ Se_{2(1-x)}Te_{2x}$	0.18	0.56
$Cu\ In_{(1-y)}Al_y\ Se_{2(1-x)}Te_{2x}$	0.33	0.50
$Cu\ In_{(1-y)}Al_y\ S_{2(1-x)}Te_{2x}$	0.52	0.20
$Cu\ Ga_{(1-y)}In_y\ S_{2(1-x)}\ Te_{2z}$	0.66	0.66
$Cu\ Ga_{(1-y)}In_y\ Se_2(1-x)\ Te_{2x}$	0.47	0.00*

purity of the material, by semiconductor standards, was not known
and was limited by the purity of the starting elements. Consequently, the crystal structure and band gap of these materials corresponded to intrinsic values for the particular semicondutor,
but the electrical transport characteristics and minority carrier
lifetimes may not have been representative of values achievable in
materials of that composition in a semiconductor-grade purity state.

    The lattice constants of the chalcopyrite material were de-
termined from the x-ray diffraction patterns. Lattice constants
were calculated to at least three significant figures. They were
used to construct actual iso-lattice constant maps like those shown
in Fig. 10 where it is evident that these lines depart from the
strictly linear behavior predicted by Vegard's law. The experimentally deduced curves are then subjected to a least squares fit
to calculate values for the α coefficients of Eq. (14).

    To determine the band gap of the alloys, electron beam catho-
doluminescence measurements were made on pressed bars of the powder. The band gap value assigned to each semiconductor corresponded
to the energy of the maximum in the cathodoluminescence spectrum.

This was done both for consistency and because separate experiments on the ternary compounds yielded energy gap values in good agreement with literature data for those materials [26]. Measurements were made at both room (30°K) and liquid nitrogen (77°K) temperatures.

These measured values of E_G were then used to calculate the "A" coefficients of Eq. (15). Figure 10 is the resulting contour map for the Cu-Ag-In-S-Se ternary alloy system. From this figure it is possible to select a pentenary alloy of any desired combination of lattice constant and energy gap within the limits set by the values of the parameters in the ternary alloys on which the alloy system is based.

The conductivity type of the ternaries and their alloys is another crucial parameter affecting the usefulness of the semiconductor for solar cell applications. Specifically since it was a goal of this research to produce heterojunction solar cells with CdS as the n-type wide bandgap semiconductor, the ternaries and their alloys intended for these cells must be p-type. As shown by previous workers [26], most of the ternary chalcopyrite semiconductors with bandgaps in the range 1 eV to 2 eV can be made either n-or p-type by controlling the chalcogen content. In the case of $CuInSe_2$, an excess of Se leads to p-type samples; a deficiency of Se leads to n-type samples. There are, however, some of the ternary materials in which p-type conductivity has never been reported. One particular such ternary is $AgInS_2$. It is of interest to establish whether the iso-energy gap, iso-lattice constant contour map of an alloy system in which one of the corner point ternaries like $AgInS_2$ is always n-type is divided into two regions, in one of which p-type conductivity is not possible. This question was explored in the Ag-Cu-In-S-Se system. The conductivity type of pressed bar sintered samples which had been saturated with the column VI constituent by adding an excess of the chalcogen to the ampoule in which sintering was to occur was measured using the hot point probe method. It was found that a significant fraction of the area of the square of Fig. 10 was limited to n-type only conductivity. In fact the largest value of E_G for an alloy having the lattice constant of CdS appears to be around 1.2 eV. Because this value was somewhat low, two other possible pentenary combinations listed in Table V were explored.

The first of these was the Cu-Al-Ga-Se-Te ternary alloy system, whose first order topological map is shown in Fig. 19. It indicates that it is possible to obtain lattice match with CdS in materials having E_G ranging between 1.45 and 1.70. Synthesis of powders of this system has, however, proven to be a problem because Al reacts with the commonly used silica reaction vessel. This problem can be eliminated by recourse to alumina reaction vessels but this leads to experimental complications which are

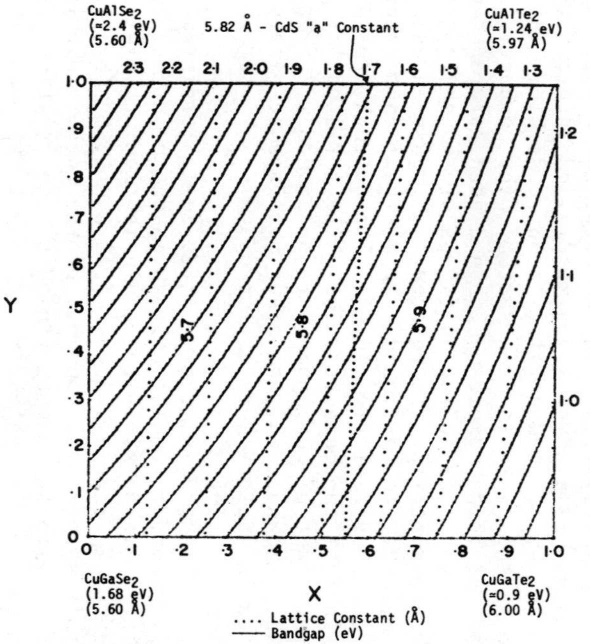

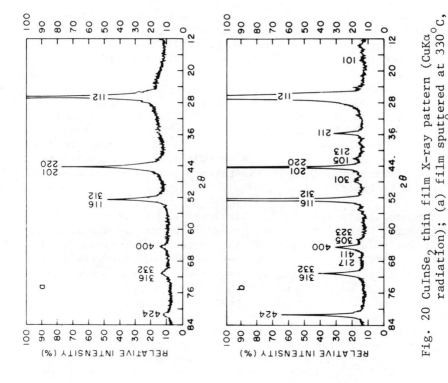

Fig. 20 CuInSe₂ thin film X-ray pattern (CuKα radiation); (a) film sputtered at 330°C, (b) film sputtered at 505°C. From Ref. [4].

Fig. 19 I_so-E_G, iso-lattice constant map of the Cu-Al-Ga-Se-Te alloy system.

best avoided if some other suitable alloy system can be found.

The Cu-In-Ga-Se-Te ternary alloy system turns out to be a good compromise. Based on the first order iso-E_G, iso-lattice constant map of this system, the range of bandgaps of materials which are lattice matched with CdS lies between about 1.02 eV and 1.45 eV. All four ternary semiconductors on which this alloy system is based are known to be available as p-type materials. There are no special problems associated with synthesis of the alloys in this system. This material has been prepared in three compositions which yield energy gaps of 1.25, 1.35 and 1.45 eV.

To test the photovoltaic potential of these alloys, solar cells were fabricated on large grained specimens of the alloy having the composition $CuIn_{0.3}Ga_{0.7}Te_{0.8}Se_{1.2}$. These large grained specimens were prepared from previously synthesized powders either by sintering the powder at a temperature somewhat below the melting point or by slow cooling of molten material.

For sintering, the powder was cold pressed and sintered at 800°C for several tens of hours. The resulting specimen had fused together and appeared to consist of large grains, some of which had dimensions up to several mm^2. Solar cells were prepared on single grains of such specimens by depositing CdS or indium-tin oxide over the pentenary. The CdS was deposited by evaporation in vacuum; the ITO, by argon beam sputtering. The thickness of the CdS was of the order of 5μm; the thickness of the ITO, about 0.3μm. The resistivity of the CdS was about 0.1 ohm cm; the resistivity of the ITO, about 3×10^{-4} ohm cm. Pressure ohmic contacts to the CdS and ITO were made with the help of indium dots.

The resulting cells were illuminated by ELH lamps set to simulate AM1 illumination of 100 mW/cm^2. The efficiency of a 0.2 cm^2 area cell with CdS as the n-type material was 13%; I_{sc} was about 29 mA/cm^2 and V_{oc}, about 0.65V. The efficiency of a 0.75 cm^2 cell with ITO as the n-type material was 13%; I_{sc} was about 33 mA/cm^2; V_{oc}, about 0.55 V and FF, about 0.71.

In another experiment a charge of $CuIn_{0.3}Ga_{0.7}Se_{1.2}Te_{0.8}$ powder was lowered through a Bridgman furnace. Slabs cut from the ingot were large grained (mm^2 in area) but their resistivity as determined by a four-point probe was high. To reduce the resistivity, the platelets cut from the ingot were sealed in evacuated silica tubes containing pentenary powder of the desired composition and annealed at 750°C for periods up to several hundred hours. This treatment reduced the resistivity into the 1 to 10 ohm cm range. A layer of CdS was evaporated over the platelet after it had been mechanically polished and etched. The best resulting device had $V_{oc} \sim 0.43$ V and $I_{sc} \sim 26$ mA/cm^2 but the fill factor was low (FF=0.40) so that the best efficiency was only about 4.5%.

In both cases the measured currents were quite large, close to the theoretical limit for material of this energy gap. This result is similar to that reported by Arienzo and Loferski [16] for heterojunctions involving CdS and various p-type semiconductors. In all cases, I_{sc} was close to the theoretical limit but V_{oc} and FF were the parameter which limited efficiency. The larger V_{oc} observed for cells made on sintered material indicates that there were fewer deleterious interface states in the region between the two materials.

VIII. THIN FILMS OF CuInSe$_2$ AND SOLAR CELLS MADE FROM THEM

There has as yet been no report of thin film photovoltaic cells made from the alloys of ternary semiconductors. However, strikingly successful photovoltaic cells have been made from thin films of CuInSe$_2$. In this section we shall summarize this work.

The first report of significant efficiency in CdS/CuInSe$_2$ thin film cells was published by Kazmerski etal [2]. The films in their devices were produced by vacuum evaporation of previously synthesized CuInSe$_2$ and of commercial CdS from resistance-heated crucibles. In order to ensure that the CuInSe$_2$ would be p-type, Se was co-evaporated with the CuInSe$_2$. Kazmerski etal reported an efficiency of about 6% on cells having an area of about 1 cm^2. These cells were not very reproducible, however, because CuInSe$_2$ decomposes during such vacuum evaporation and, therefore, other ways of preparing thin films have been explored. These include a kind of molecular beam epitaxy by the Mickelson and Chen of Boeing [3], and rf-sputtering and chemical spray pyrolysis by the group at Brown University [7,8].

The Boeing group produces thin films of CuInSe$_2$ by evaporating Cu, In and Se$_2$ from separate resistance heated sources. The evaporation rate of the elements is adjusted so that when vapors of the three elements arrive at the substrate which is heated to a temperature high enough (about 700°C) to promote synthesis, they arrive with the correct ratio of Cu:In:Se to form the compound. The films are deposited on molybdenum coated alimina substrates. The structure is of the p$^+$-p/n-n$^+$ type. The first portion of the CuInSe$_2$ arriving at the substrate contains excess Se because the vapor pressure of the Se source is increased during the first stage of evaporation. After about 1μm of conducting CuInSe$_2$ is deposited, a layer of CuInSe$_2$ having a resistivity in the 1 to 10 ohm cm range is deposited to a thickness of about 2μm. The cell is covered by a layer of CdS followed by a layer of indium doped CdS to form a conducting n$^+$ layer over which the grids are deposited. Devices prepared in this way have been reported to have efficiencies in excess of 9%.

Fig. 21.   SEM micrographs of CuInSe$_2$ films sputtered onto Al$_2$O$_3$
gold-coated substrates at two temperatures: left   ,
330°C, right 505°C. From Ref.[4]

In the Brown University rf-sputtering experiments, a sputtering target is prepared from synthesized CuInSe$_2$ powder by cold
pressing to a pressure of about 7x10^8 Pascals. The targets were
sputtered in argon with a background pressure of 5x10^{-2} Torr. The
total rf-power was 100W; rf-voltage, 800V. The substrates were
maintained at temperatures between 300 and 500°C. Figure 20 shows
the x-ray diffraction patterns of films deposited at roughly these
two substrate temperatures T_s. The film deposited at the lower T_s
exhibits only sphalerite lines in its diffraction spectrum whereas
the film deposited on the higher T_s substrate also shows the chalcopyrite peaks, i.e. the peaks labeled 101, 211, 305 etc.

Figure 21 shows scanning electron micrographs of these two
films. The film deposited on the higher T_s substrate shows well
defined crystallites having dimensions up to about 1μm.

Phtovoltaic cells were fabricated from the higher T_s films
by depositing CdS from a resistance heated evaporator over the
rf-sputtered films. The best a device having an area of 0.2 cm^2

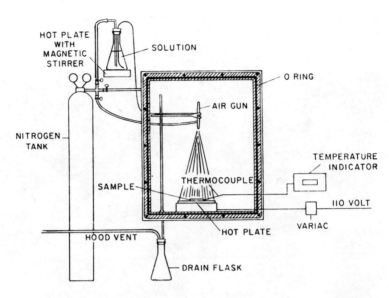

Fig. 22. Schematic representation of experimental setup for depositing copper ternary semiconductors by chemical spray pyrolysis.

had V_{oc} = 0.45V; I_{sc} = 22 mA/cm^2; FF - 51% and efficiency ~ 5.1%.

In the Brown University chemical spray pyrolysis experiments, the CuInSe$_2$ films were deposited using the apparatus shown schematically in Fig. 22 [6]. The spraying solution consisted of Cu$_2$Cl$_2$, InCl$_3$ and N,N-dimethylselenourea with CuInSe$_2$ concentrations in the range from 0.003 to 0.005 M dissolved in dioinized water. A few drops of HCl were added to the solution to increase the solubility of Cu$_2$Cl$_2$. During the spraying process the solution temperature was maintained at about 100°C. The solution was sprayed onto heated glass substrates, using nitrogen as a carrier gas, in a closed system in a continuous flow of nitrogen. Substrate temperatures were varied from 175 to 275 (+10)°C and spraying rates from 7 to 10 ml/min.

The films were homogeneous in appearance and adherent to the substrates. SEM micrographs showed that the as-grown films were polycrystalline, and their grain size was in the range 0.5μm. X-ray diffraction spectra of the films deposited onto substrates in the temperature range 175 - 210°C showed that the films consisted of CuInSe$_2$ of the chalcopyrite structure and of another material, which has been identified as In$_2$O$_3$. As the substrate temperature was increased over the range 220 - 250°C, the intensity of the chalcopyrite peaks decreased and for temperatures in excess of 250°C, only the sphalerite structure was observed. The diffraction pattern corresponding to the In_2O_3 phase also disappeared in films deposited on substrates heated above 250°C.

The resistivity of the films and Hall mobility of the carriers were measured using the four-contact Van der Pauw method. Films deposited in the substrate temperature range from 175 to 250°C were p-type, and their resistivity increased with substrate temperature from 10^{-2} to 3.10^3 ohm-cm. The film deposited at 275°C was n-type with a resistivity about 0.3 ohm-cm, but it was not uniform and EDAX analysis of film composition indicated a strong deficiency of copper.

Transmission and reflection spectra of the films in the visible and near-infrared region were measured at room temperature. In high resistivity samples, the absorption edge was relatively sharp; it corresponded to an energy gap of 1.02 eV which is close to the 1.04 eV energy gap value reported in the literature for $CuInSe_2$ single crystals.

Some samples were subsequently annealed for periods between 0.5 to 1 hour in vacuo or in hydrogen at 400°C. The resistivity of these samples usually increased, and sometimes a change from p- to n-type conductivity was observed. Such short heat-treatments didn't affect either structural or optical properties of the films. Some other samples were "buried" in $CuInSe_2$ crystalline powder, sealed in an evacuated quartz ampoule and annealed for a long period (about 16 hours) at 500°C. Low resistivity samples retained their low resistivity, and the resistivity of high resistivity samples slightly increased, but no conversion to n-type conductivity was observed.

Some n-CdS/p-$CuInSe_2$ heterojunction devices were prepared by evaporating CdS over sprayed $CuInSe_2$ films. Devices made on low-resistivity as-grown films had nearly ohmic I-V characteristics and were not photovoltaic. One device was made by evaporating a comb-shaped layer of CdS over a $CuInSe_2$ layer which had been sprayed on glass and heated for 16 hours in a sealed, evacuated ampoule containing $CuInSe_2$ powder. Its resistivity after heat-treatment was about 100 ohm-cm. To provide ohmic contacts, In and Au were evaporated over the CdS and $CuInSe_2$, respectively. The junction was illuminated by a tungsten lamp having an intensity of about 100 mW/cm^2. The cell exhibited a photovoltaic effect with the following parameters: $V_{oc} \approx 0.3V$, $I_{sc} \approx 1$ mA/cm^2, FF ~ 0.3. The value of I_{sc} was low probably because of the high series resistance attributable to both the high resistivity of $CuInSe_2$ and the lack of backwall contact.

Chemical spray pyrolysis is particularly attractive as a thin film deposition technique because it appears to be an economical way to deposit films. However, the technology is in its infancy and much more experimental work must be done before its potential can be properly assessed.

IX. SUMMARY AND CONCLUSIONS

1. The tandem cell concept provides the framework for achieving the maximum efficiency which is realizable from photovoltaic cells. The concept can be used in both concentrator and no-concentrator systems. It is applicable to both single crystal and thin film cells.

2. Recent developments in solar cell theory provide clear guidelines for extracting the highest efficiency from a given photovoltaically active semiconductor by employing new designs that utilize minority carrier mirrors to increase the open circuit voltage of the cell. Minority carrier mirrors should be equally effective in p/n homojunctions and in p/n heterojunctions.

3. A particularly attractive design for PV cell is a heterojunction consisting of a wide band gap window semiconductor whose energy gap is so large that it absorbs a neglibible amount of sunlight mated with a photovoltaically active, direct gap semiconductor which has incorporated into it a minority carrier mirror, e.g. a back surface field. The two semiconductors comprising such a heterojunction should have matching lattice constants to minimize the concentration of interface states which can control the open circuit voltage if their concentration is high enough. In this ideal-design-cell, the relative doping of the two semiconductors should be so chosen that the space charge region is essentially entirely confined to the photovoltaically active semiconductor. This means that the wide band gap semiconductor should be more heavily doped than the photovoltaically active small band gap material.

4. The photovoltaically active materials to be used in multi-cell (tandem) systems are likely to be quaternary and pentenary alloys of binary and ternary semiconductors since such alloy systems can provide lattice matching with a given wide band gap window material while providing the wide selection of band gaps needed for tandem cell systems.

5. The efficiency of tandem cell systems has already been demonstrated in systems employing alloys of III-V semiconductors. The groundwork has been laid for constructing tandem cell systems from I-III-VI$_2$ materials like CuInSe$_2$ and their alloys. These materials are especially promising for thin film cells of the tandem cell type. Efficiencies up to 10% have already been reported for the CdS/CuInSe$_2$ cell which is a prototype of the "unit cells" required in tandem solar cell systems.

REFERENCES

1. N. A. Gokcen and J. J. Loferski, Solar Energy Materials 1 271
   (1979).
2. L. L. Kazmerski,F. R. White and G. K. Morgan, Appl. Phys. Lett.
   29, 268 (1976).
3. R. A. Mickelson, W. S. Chen, L. F. Buldhaupt and B. Selikson,
   Final Report, Boeing Corporation Contract No. D180-25098-1
   (1978).
4. J. Piekoszewski, J. J. Loferski, R. Beaulieu, J. Beall, B.
   Roessler and J. Shewchun, Solar Energy Materials (1980).
5. M. Gorska, R. Bealieu, J. J. Loferski and B. Roessler, Solar
   Energy Materials 1, 313 (1979).
6. M. Gorska, R. Bealieu, J. J. Loferski, B. Roessler and J.
   Beall, Solar Energy Materials 2 (1980).
7. J. Piekoszewski, J. J. Loferski, R. Beaulieu, B. Roessler and
   J. Shewchun, Final Technical Report, Brown University Contract
   No. SERI-XI-9-8012-1 (1980).
8. M. Gorska, R. Beaulieu, J. J. Loferski and B. Roessler, Private
   communication. Final Technical Report, Brown University,
   Contract No. SERI-XI-9-8012-1 (1980).
9. J. Shewchun, J. J. Loferski, D. Burk, B. K. Garside, R. Beaulieu
   and D. Polk, "Conference Record of the Fourteenth IEEE Photovoltaic Specialists Conference, San Diego--1980" IEEE, New
   York.
10. J. J. Loferski, J. Appl. Phys. 27, 777 (1956).
11. M. Wolf, "Conference Record of the Fourteenth IEEE Photovoltaic
    Specialists Conference, San Diego, 1980" IEEE, New York.
12. M. Spitzer, J. J. Loferski and J. Shewchun, "Conference Record
    of the Fourteenth IEEE Photovoltaic Specialists Conference,
    San Diego, 1980" IEEE, New York.
13. M. Wolf, Proc. IEEE, (1963).
14. E. D. Jackson "Trans. of the Conference on the Use of Solar
    Energy, Tucson, Arizona 1955" p. 22; D. Trivitch and P. A.
    Flinn "Solar Energy Research", F. Daniels and J. A. Duffie,
    Editors, Univ. of Wisconsin Press, Madison, WI, 1955, p. 143.
15. M. Arienzo, Ph.D. Thesis, Brown University, Providence, RI,
    (1978).
16. M. Arienzo and J. J. Loferski, J. Appl. Phys, July 1980.
17. J. J. Coleman, N. Holonyak, Jr., J. J. Ladourse and P. D. Wright.
    J. Appl. Phys. 47 2015 (1976).
18. G. A. Antypas and R. L. Moon, J. Electrochem. Soc. 120 1579
    (1979).
19. J. J. Loferski, J. Shewchun, B. Roessler, R. Beaulieu, J.
    Piekoszewski, M. Gorska and G. H. Chapman, "Conference Record
    of the Thirteenth IEEE Photovoltaic Specialists Conference,
    Washington, DC, 1978", IEEE, New York, p. 190.
20. J. Shewchun, J. J. Loferski, G. H. Chapman, B. K. Garside and
    R. Beaulieu, J. Appl. Phys. 50, No. 11, 6978 (1979).

21. R. L. Moon et al. J. Electron. Mater. $\underline{3}$, 365 (1974).

22. M. Wolf, "Conference Record of the Fourteenth IEEE Photo-
    voltaic Specialists Conference, San Diego, 1980" IEEE, New
    York 563.

23. M. Spitzer, J. J. Loferski and J. Shewchun, "Conference Record
    of the Fourteenth IEEE Photovoltaic Specialists Conference,
    San Diego, 1980", IEEE New York, p. 585.

24. J. A. Cape, J. S. Harris, Jr. and R. Salini, "Conference
    Record of the Thirteenth IEEE Photovoltaic Specialists Con-
    ference, Washington, 1978", IEEE, New York, p. 881.

25. R. L. Moon, L. W. James, H. A. Vander Plas, T. O. Yep, G. A.
    Antaypos and Y. Chai, "Conference Record of the Thirteenth
    IEEE Photovoltaic Specialists Conference, Washington, 1978",
    IEEE, New York p. 859.

26. L. M. Fraas and R. C. Knechtli, "Conference Record of the
    Thirteenth IEEE Photovoltaic Specialists Conference, Washing-
    ton, 1978", IEEE, New York.

27. J. L. Shay and J. H. Wernick, "Ternary Chalcopyrite Semicon-
    ductors", Pergamon Press, New York.

THE PRINCIPLES OF PHOTOELECTROCHEMICAL ENERGY CONVERSION

Heinz Gerischer

Fritz-Haber-Institut der Max-Planck-Gesellschaft
Faradayweg 4-6
D-1000 Berlin 33 West Germany

I.   SUNLIGHT CONVERSION INTO CHEMICAL ENERGY

Visible and near-to-visible infrared light as the sun emits is
a form of energy of very high quality. In thermodynamic language
this means this kind of light is nearly exclusively free energy and
has a very small entropy content. Conversion of light energy into
other forms of useful energy, however, begins in any case with light
absorption and is as a consequence connected with a considerable entropy production. This is a first limitation for the conversion of
light energy into other forms of free energy. Other losses by entropy production follow in consecutive processes as will be pointed
out in the series of lectures given in this Summer School.

The absorption of light quanta converts the light energy into
potential chemical energy in form of excited electronic states.
Since the lifetime of excited electronic states is very short, this
kind of chemical energy can only be preserved if a chemical trans-
formation follows fast enough which leads to chemical situations
with a much longer, in the optimal case with an infinite lifetime.
This consecutive step competes with the quenching processes of the
excited state which would convert all the energy into heat. Such a
fast reaction needs a considerable driving force and in it a considerable part of the stored free energy will be dissipated into
heat. Fig. I.1 shows a principle energy scheme for such an energy
conversion process.

The consecutive reaction which preserves some of the light energy on an intermediate energy level can consist in a change of the
chemical structure of the material. It is very often an ionization
step or a charge transfer reaction. The reasons that the two latter

processes are so common in photoreactions is that electron transfer
is the fastest chemical process, much faster than the rearrangement
of heavy atoms.

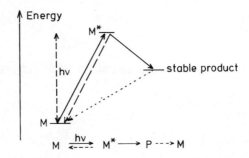

Figure I.1   Scheme of photochemical conversion.

If the main reaction step for energy conservation is an electron
transfer between the excited system and an electron acceptor or electron donor, we have a photoelectrochemical process for energy conversion. Such processes can occur either in homogeneous systems or at
interfaces. A description of the conditions under which such processes occur and how they can be used for solar energy conversion and
storage are the subjects of the series of my lectures at this Summer
School. Fig. I.2 gives a survey of the photochemical processes which
form products of higher energy together with some examples.

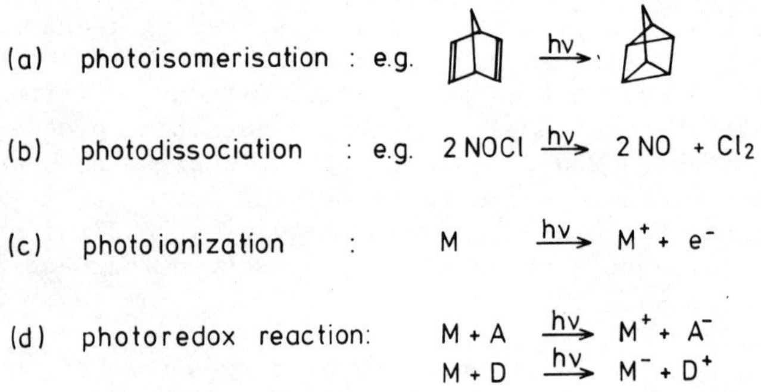

(a)   photoisomerisation : e.g.

(b)   photodissociation   : e.g. $2\,NOCl \xrightarrow{h\nu} 2\,NO + Cl_2$

(c)   photoionization   : $M \xrightarrow{h\nu} M^+ + e^-$

(d)   photoredox reaction: $M + A \xrightarrow{h\nu} M^+ + A^-$
                                          $M + D \xrightarrow{h\nu} M^- + D^+$

Figure I.2   Photochemical processes for energy conversion.

Photoredox Reactions

A redox reaction between an excited state and suitable reagents can be described like a normal redox reaction as an electron transfer between an occupied energy level and a vacant one. The principle of a photoredox reaction is shown in Fig. I.3 for a homogeneous system using an energy level scheme for the excited molecule and the potential reaction partners. In the excited state one electron is transferred to a higher electronic level which was previously vacant

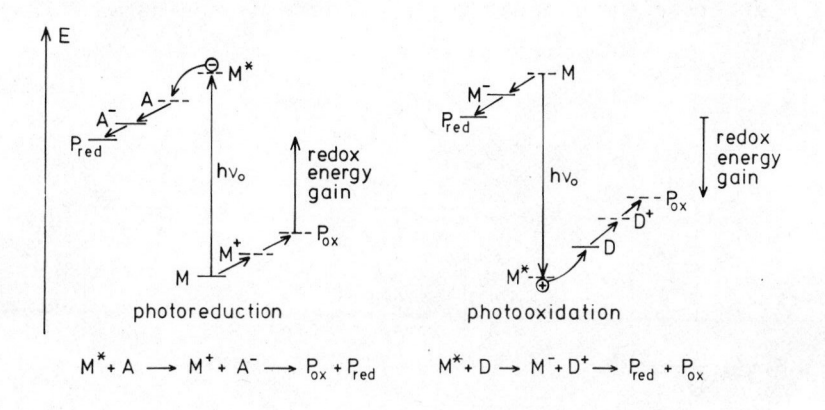

Figure I.3 Homogeneous photoredox reactions for the conversion of light energy with P_{ox} and P_{red} as as stable products.

and the formerly occupied level is now empty. Consequently, an excited molecule can exert as well a reductive step (electron transfer to an acceptor molecule) or an oxidative step (electron transfer from a donor molecule to the excited molecule). This is shown in the energy scheme of Fig. I.3. In order to prevent the reverse electron transfer reactions between the products, the resulting species have to be separated by some means. Homogeneous photoredox reactions will be discussed in another series of lectures (M. Grätzel and J. Albery). I shall concentrate on systems in which the electron transfer reaction being essential for the energy conservation occurs at the interface between a solid and an electrolyte. The excitation can in such cases either happen in the solid or in the molecules adjacent to the solid surface. However, only solids with a particular electronic structure are suitable for such processes. Solids with a metallic character and a continuous absorption spectrum over the whole range of the sun spectrum have a by far too short lifetime of the excited states as to allow a considerable increase of the electronic potential energy by light absorption. They act at the same time as very efficient energy quenchers for excited molecules adjacent to

their surfaces due to energy transfer or compensating electron transfer in both directions. We have to concentrate therefore on semiconducting materials with band gaps above 1 eV.

Fig. I.4 gives a summary of the heterogeneous photoredox reactions at such semiconductor electrodes. If the light is absorbed in
the semiconductor, a photodissociation occurs in most systems generating electron hole pairs. If the electronic states in the energy
bands are highly delocalized, electron hole pairs "forget" their gen-
uine interaction immediately after the act of creation and move
through the solid like independent particles. They disappear after
some time by non-geminate recombination except those carriers passing the interface by a redox reaction. This is indicated in Part a
of Fig. I.4.

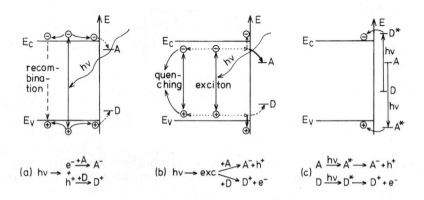

Figure I.4 Photoredox reactions at semiconductor surfaces.

There are solids with a wide band gap, where light absorption
creates so-called excitons which can be considered as strongly interacting electron hole pairs. Such excitons are mobile too and can
react with electron acceptors or electron donors at the interface
between such a solid and an electrolyte leaving a mobile charge
carrier behind in the solid. This is another way of charge separation leading to photocurrents. A description of this process in an
energy scheme is shown in Part b of Fig. I.4.

Finally, the excitation can also occur in adsorbed molecules at
the interface, in which case the solid can act either as the electron
acceptor or the electron donor. This is shown in Part c of Fig. I.4.
All these processes lead to a conservation of part of the light energy in form of redox energy. Redox energy is a form of chemical energy as will be explained in the next section.

Redox Energies and the Scales of Redox Potentials

In order to illustrate in what sense redox energy is equivalent
to chemical energy, we shall recollect a classical example of electrochemistry. Let us consider the reaction between hydrogen and
chlorine forming hydrogen chloride.

(1) $\quad \frac{1}{2} H_2 + \frac{1}{2} Cl_2 + aq \longrightarrow HCl : \quad \Delta G^o_{HCl}$

This chemical reaction can be separated into two electrochemical
steps.

(2) $\quad \frac{1}{2} H_2 + aq \qquad \rightleftharpoons H^+aq + e^-_{U^o_{H^+/H_2}} : \Delta G^o_{H^+} - \mathcal{F} U^o_{H^+/H_2} = 0$

(3) $\quad \frac{1}{2} Cl + aq + e^-_{U^o_{Cl_2/Cl^-}} \qquad \rightleftharpoons Cl^-aq : \quad \Delta G^o_{Cl^-} + \mathcal{F} U^o_{Cl_2/Cl^-} = 0$

(4) $\quad \Delta G^o_{HCl} = \Delta G^o_{H^+} + \Delta G^o_{Cl^-} = -\mathcal{F}(U^o_{Cl_2/Cl^-} - U^o_{H^+/H_2}) = -\mathcal{F} U^o_{NHE}$

In this way, the driving force of the above chemical reaction can be
transformed into an electrical driving force, namely the voltage of
a galvanic cell with a hydrogen and a chlorine electrode inserted
into hydrochloric acid. The voltage difference multiplied with the
Faraday constant $\mathcal{F}$ and the number of electron equivalents in the net
reaction is equivalent to the change in free energy of this chemical
reaction. If one applies the same voltage from outside in opposite
direction to such a cell, no current is flowing. The system is at
equilibrium. We see that the energy difference of the electrons in
the two electrodes of the galvanic cell corresponds to the free energy of the chemical reaction. This is the basis for the scale of redox
potentials in thermodynamic equilibrium which is well known to every
electrochemist. In the conventional scale of redox potentials, the
energy of the electrons in the reversible hydrogen electrode is defined as zero. The free energy of the electrons at equilibrium in any
other redox system is given by the free energy difference of the reaction going on in a complete cell with the standard reference electrode as the counter part. This can be formulated in a general way
in the following reaction equations:

(5) $\quad Ox^+ solv + e^-_{U_{Redox}} \rightleftharpoons Red \cdot solv : \Delta G^o_{Redox} + \mathcal{F} U^o_{Redox} = 0$

(6) $\quad Ox^+_{ref} + e^-_{U_{ref}} \rightleftharpoons Red_{ref} : \Delta G^o_{ref} + \mathcal{F} U^o_{ref} = 0$

(7) $\quad Ox^+ solv + Red_{ref} \rightleftharpoons Red \cdot solv + Ox^+_{ref} : \Delta G^o = \Delta G^o_{e^-}$

$$(8) \quad U^o_{Redox} - U^o_{ref} = -\frac{1}{\mathcal{F}}(\Delta G^o_{Redox} - \Delta G^o_{ref})$$

We have used in Fig. I.1 to I.3 a spectroscopic scheme for the description of the electronic energy levels in order to show how electronic excitation leads by electron transfer to the conversion and storage of light energy. We can also describe redox energies in the same way and define a scale of redox potentials for the electrons in redox systems in which the systems with a higher electron energy act as electron donors while systems with a lower electron energy act as electron acceptors. In order to get the closest connection with spectroscopic usage, the electron in the vacuum is defined as the zero point of this scale. This is also in accordance with the description of electronic energy levels in solids and the choice of the vacuum level as the reference point for their energies. In such a scale, the redox reactions are written as shown in the following equations:

$$(9) \quad Ox^+ \cdot solv + e^-_{vac} \longrightarrow Red \cdot solv \quad : \quad \Delta G^o_{Redox}$$

$$(10) \quad H^+ solv + e^-_{vac} \longrightarrow \tfrac{1}{2} H_2 + solv \quad : \quad \Delta G^o_{H^+/H_2}$$

$$(11) \quad E^o_{Redox} = \frac{1}{N_L} \cdot \Delta G^o_{Redox}$$

$$(12) \quad E^o_{H^+/H_2} = \frac{1}{N_L} \cdot \Delta G^o_{H^+/H_2} \approx -4.6 \pm 0.1 \ [eV]$$

$$(13) \quad E^o_{Redox} = -e U^o_{Redox} + E^o_{H^+/H_2} \ [eV]$$

The connection between this "absolute" scale of redox potentials and the conventional scale is given by a linear shift of the zero points (shifted by the energy level of the electrons in the standard hydrogen electrode versus the vacuum level) and by an inversion of the sign of this scale. For photoelectrochemical reactions it is convenient to use the absolute scale even if one measures the redox potentials as usual in the conventional scale versus an arbitrarily chosen reference electrode.

For the understanding of the redox potential scales it is helpful to consider a thermodynamic cycle which indicates what properties of a system control the position of a redox potential. This is shown in Fig. I.5 in terms of the absolute scale of redox energies for the hydrogen electrode and for a model redox system. The absolute energy level of the electrons in the hydrogen electrode is not very exactly known. However, it must be located according to various derivations between $- 4.5$ and $- 4.7$ eV. In Fig. I.6 the two redox potential scales are compared with each other assuming a value of $- 4.5$ eV for the hydrogen electrode in the absolute scale.

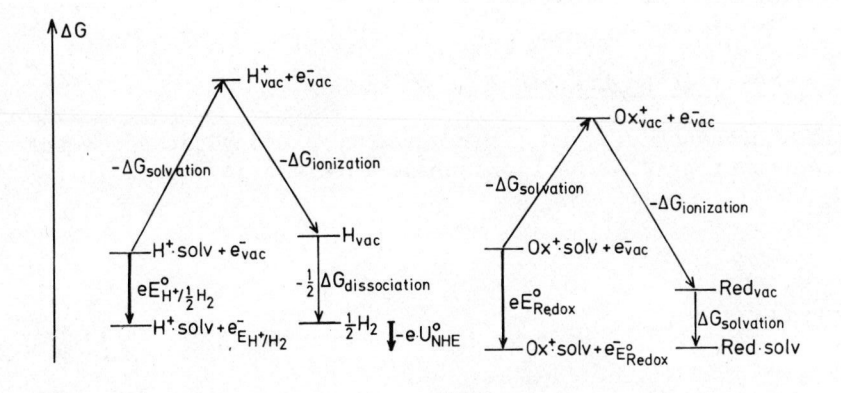

Figure I.5 Redox energy scales derived by thermodynamic cycles.

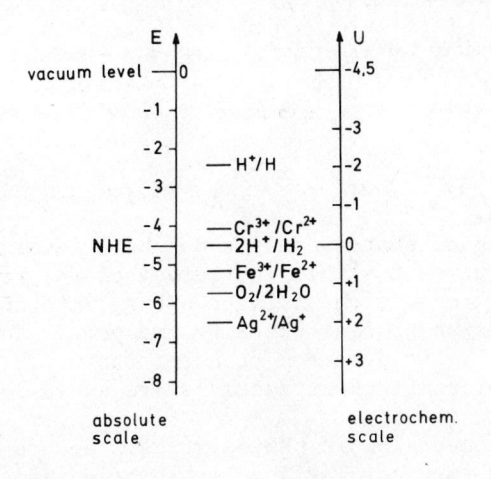

Figure I.6 Correlation between absolute and electrochemical scale of redoxpotentials.

One point should be mentioned before we turn to the discussion of real systems for light energy conversion. In the spectroscopic description of electronic energies the statistical entropy term is usually not considered. The term scheme describes potential energies of electrons in a single molecule. If we have to deal with ensembles of molecules or with solids, a statistical entropy term appears in the free energy. In the redox potential scales this statistical term is accounted for by the concentration dependent term $kT/e \cdot \ln(C_{ox}/C_{red})$ which reduces to zero at equal concentrations of the oxidized and reduced species. If the standard states are differently defined, the standard potential contains a different entropy contribution. It should be kept in mind that such an entropy term has to be added to all descriptions of redox energies, if the system is not kept under standard conditions.

Photosynthesis as an Example

In photosynthesis, CO_2 is reduced to carbohydrates and water is oxidized to oxygen. A reaction scheme is given in Fig. I.7.

Net reaction: $CO_2 + H_2O + light \longrightarrow \frac{1}{n}(CH_2O)_n + O_2$

Photoredox reactions

(1) $Chl_I + hv_1 \rightarrow Chl_I^* \xrightarrow{+A_1} Chl_I^+ + A_1^-$

(2) $Chl_{II} + hv_2 \rightarrow Chl_{II}^* \xrightarrow{+A_2} Chl_{II}^+ + A_2^-$

Consecutive redox reactions

$A_1^- + CO_2 \xrightarrow[+H^+]{catalysts} A_1 + \frac{1}{n}(CH_2O)_n + \frac{1}{2}H_2O$

$Chl_{II}^+ + H_2O \xrightarrow{catalysts} Chl_{II} + \frac{1}{4}O_2 + H^+$

coupling reaction between (1) and (2): $Chl_I^+ + A_2^- \longrightarrow Chl_I + A_2$

Chl_I, Chl_{II} = chlorophyll containing reaction centers; A_1, A_2 = electron acceptors

Figure I.7   Reaction scheme of photosynthesis.

The mechanism of photosynthesis has been investigated into much detail. Fig. I.8 gives a simplified survey of the redox reactions which perform the conversion of light energy into chemical energy. The energy conversion process contains two photoredox reactions coupled in series by an electron transfer chain. The energy correlations in this series of redox reactions are shown in Fig. I.9.

The energy conversion in photosynthesis occurs in localized centers which are very complex and highly organized. These energy conversion centers are imbedded into membranes of bilayer dimensions. The membrane has the function to keep the products separated. Such localized centers however have a very limited light absorption. To improve the light absorption, nature has invented an antenna system of other chlorophyll molecules which are not active as energy conversion centers but absorb sunlight and transport the light energy via energy transfer to the reaction centers.

The transfer of electrons through the thylakoid membrane leads to a charge separation and the resulting electric field would prevent

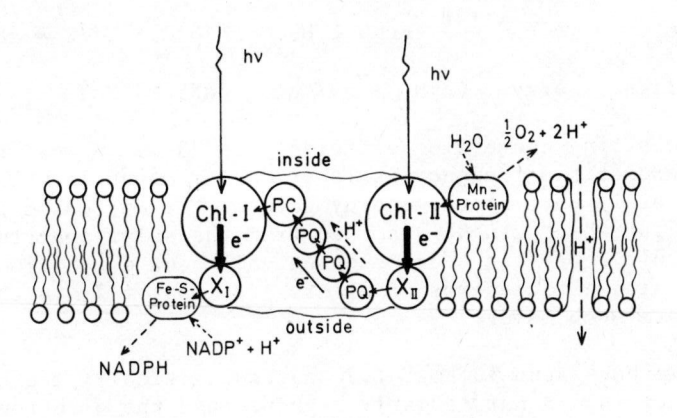

electron tranport   through   Thylakoid   membrane
Chl = Chlorophyll, X = primary   electron acceptor
PQ = Plastoquinone, PC = Plastocyanin
NADP$^+$= Nicotinamide - adenine - dinucleotidephosphate

Fig. I.8

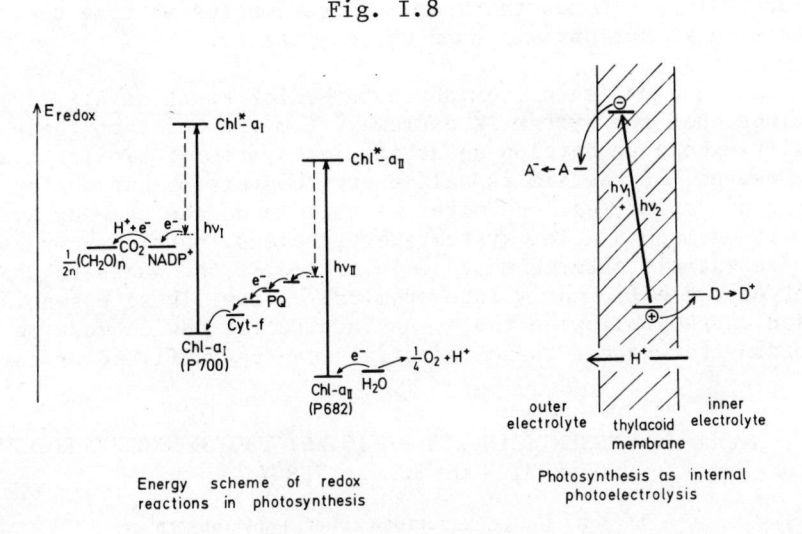

Energy   scheme of redox
reactions in photosynthesis

Photosynthesis as internal
photoelectrolysis

Fig. I.9

the continuation of the electron transfer. This field is used in
photosynthesis for a proton transfer parallel to the electron transfer and in this way the electron current is internally compensated.
Nature uses this proton transfer process also as part of the energy
conversion.

We are interested in the efficiency of the light energy conversion which can be calculated for monochromatic light at the maximal
absorbance of chlorophyll with optimal use of the light quanta from
the ratio between the stored and the absorbed energy:

$$6\ CO_2 + 6\ H_2O \xrightarrow{\ +48\ h\nu\ } C_6H_{12}O_6 + 6\ O_2\ :\ \Delta E = 2800\ kJ/Mol$$

light energy: $48\ N_L \cdot h\nu/Mol = 8435\ kJ/Mol$

The resulting conversion efficiency is 33 %. In reality, even the monochromatic efficiency is lower due to metabolism which consumes immediately part of the energy gained. The optimal efficiency for all light which can be absorbed from the solar spectrum related to the whole solar light can theoretically not exceed 10 %. The best conversion efficiencies of plants are about 4 %; in the average, they are much lower.

We have seen in Fig. I.9 that energetically the conversion efficiency is not particularly high because the electrons lose a great deal of their energy which they have in the excited chlorophyll molecules in the consecutive electron transfer steps. It appears that these energy losses are needed in order to make the electron transfer steps fast enough. Furthermore, two light quanta are needed for the transfer of one electron through the membrane. From a theoretical analysis one comes therefore to the conclusion that the energy conversion is not optimized in photosynthesis.

What can we learn from photosynthesis? First of all, one must conclude that the system is extremely complex and that there will be little hope to develop an artificial system of similar complexity. However, if the aim is only energy conversion but not synthesis of organic molecules, one might be able to do the same or even better with a much simpler system. Nevertheless, nature shows us that some essential features must be fulfilled if one wants to convert light energy efficiently into chemical energy. These essentials are listed in the following table. Any artificial device must be judged according to whether these essentials can be fulfilled or not.

TABLE I - ESSENTIALS FOR EFFICIENT PHOTOELECTROCHEMICAL
              CONVERSION OF SOLAR ENERGY

1) Fast charge separation after light absorption

2) separation of products in order to prevent
   reverse reactions

3) efficient absorption of solar light with
   minimal entropy production

4) adjustment of the redox potentials of the
   excited states to the redox reactions which
   store the energy

5) long term stability (or continuous reproduction)

Artificial Systems for Energy Conversion

Many attempts have been made and are still being investigated to construct systems in which the photoelectrochemical energy conversion can technically be used. The simplest system would be a homogeneous photoredox reaction leading to products which do not react with each other, at least not during the time before they can be separated from each other. The stored energy can be used afterwards either in form of heat by introducing a catalyst for the reverse reaction between the products or by inserting selective electrodes which interact only with one of the products of the redox reactions and convert the stored energy in this way into electrical energy of a galvanic cell. The scheme of suitable photoredox reactions was already given in Fig. I.3.

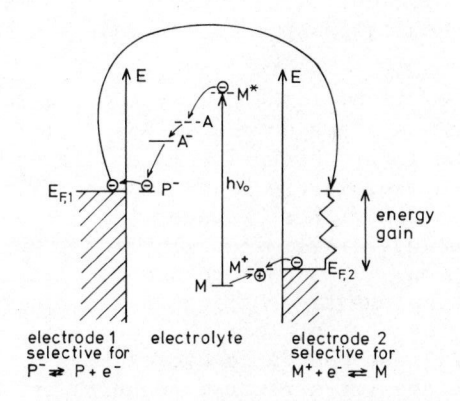

Figure I.10   Energy schemo of a photogalvanic cell
with reaction $M^* + A \rightarrow M^+ + A^-$.

Fig. I.10 shows the principle of a photogalvanic cell with two selective electrodes which interact separately with the products of the photoredox reactions in the homogeneous electrolyte. These homogeneous photoelectrochemical processes will be discussed in much more detail in the series of lectures by M. Grätzel and J. Albery. I shall concentrate on systems where the redox reactions with the excited states occur at interfaces, particularly at the contact between an electrode and an electrolyte.

The system coming closest to photosynthesis is shown in Fig. I.11. It consists of a bilayer membrane in which photoactive molecules are imbedded which undergo photoredox reactions at the contact to the neighboring electrolytes. Such systems have been investigated by Tien and Berns and others. The membrane separates two compartments which contain two different redox couples. The system cannot act as a storage device because charge transfer through the membrane would create electrical fields through the membrane, if the excess charge is not immediately picked up at the two electrodes of this cell.

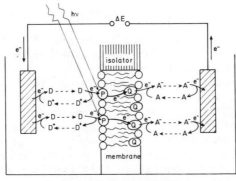

P = pigment   Q = electron acceptor

Figure I.11   Redox cell with asymmetrical membrane for photochemical energy conversion.

The energy relations in such a system are schematically shown
in Fig. I.12. These systems have very low efficiency by two reasons.
One is the limited light absorption in a bilayer which cannot be improved by using thicker layers because such layers would have a too
high internal resistance. Furthermore, the concentration of the sensitizer in such a layer is limited what means a further restriction
for light absorption. The second reason is that it is extremely difficult to make the properties of such a membrane asymmetric enough
in order to move the charge, left on the sensitizer molecule after
electron transfer, fast enough to the other side of the membrane.
Therefore, the, chance for a reverse reaction is very large. Efficiencies reached with such devices have been extremely low (below
0.1 o/oo).

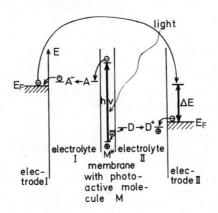

Figure I.12   Well with photoactive membrane.

The system where also monolayers of dyes are used but the asymmetry for electron transfer reactions is very well granted, is shown in Fig. I.13. In these systems, the dye molecules are adsorbed on a semiconductor and can in the excited state inject either electrons (Fig. I.13 a) or holes (Fig. I.13 b) into the semiconductor. These

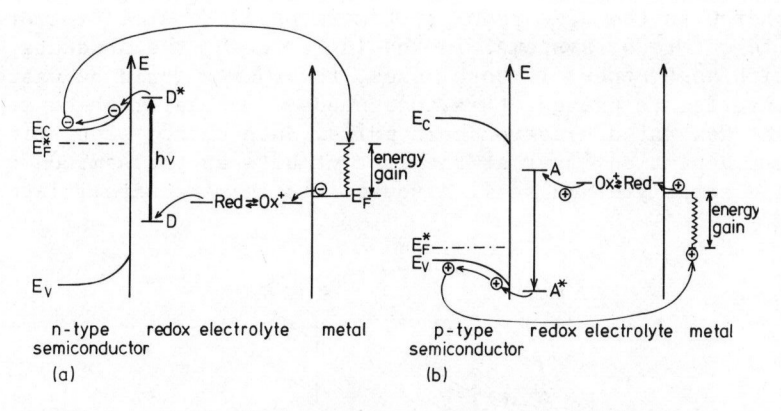

Figure I.13   Regenerative photocells with:
                   (a) sensitized electron injection
                   (b) sensitized hole injection.

charge carriers are removed from the interface very quickly, if the n-type or p-type semiconductors used in such systems form a depletion layer in contact with this electrolyte. Such depletion layers contain an electric field which drives the injected charge carriers to the bulk of the semiconductor. We shall come back to these situations in the following lectures. This is indicated in Fig. I.13 by a downwards or respectively upwards bending of the characteristic energy levels of the semiconductor, the edges of the conduction band and the valence band. Electrons move downwards in this scheme while holes move upwards. In order to regenerate the oxidized or reduced dye molecules which have to absorb the sunlight, the electrolyte must contain a redox couple which can restore the oxidation state of the sensitizer molecules. This is indicated in these two figures. If only this redox couple is present in the solution, the potential of the counter electrode is controlled by its redox potential. The injected charge carriers in the semiconductor however let increase the free energy of electrons or holes in these electrodes providing in this way electrical energy for external consumption. The efficiency of these devices is again restricted by the limited light absorption of monolayers. Otherwise, they could be very useful.

The most efficient systems for photoelectrochemical energy conversion at present are based on light absorption in the electrodes themselves. Semiconductors with a suitable band gap often have a very strong light absorption above this threshold and, at the same

time, the mobility of the generated charge carriers can be very
high. If they reach the contact with an electrolyte, they can undergo redox reactions there as was shown in Fig. I.4 a.

If both carriers would reach the interface with equal probability, they would not only recombine there or in the bulk but also react either with the same redox system in opposite direction or with
two different redox systems. In the latter case, the products would
react with each other. In both cases, the energy would be wasted as
heat. In order to prevent this, one needs a mechanism which separates the generated electron hole pairs. Such charge separation occurs in a depletion layer at the contact between the semiconductor
and the electrolyte. We shall discuss this in more detail later.

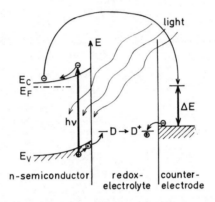

Figure I.14 Cell with semiconductor electrode as voltage generator.

For this introduction, only the principle of energy conversion
by such an electrode is shown in Fig. I.14, where an n-type semiconductor is used for illustration. The redox couple of the electrolyte
reacts under illumination with the holes generated by light absorption. The electrons are left in the semiconductor and their accumulation in the bulk leads to a negative voltage difference between
the semiconductor and the counter electrode. The latter remains more
or less in equilibrium with the redox system in solution. The electric energy of the electrons can be used for external consumption as
in a photovoltaic solid state device.

If the counter electrode is not in contact with the same redox
electrolyte but with a different one which is separated from the
first one by a membrane as shown in Fig. I.15, such systems can be
used as redox batteries in which the light energy can be stored directly. The membrane is necessary in order to prevent the direct reaction between the two redox systems. This membrane therefore must

be impermeable for the components of the redox systems but conductive enough for the electrolytic current.

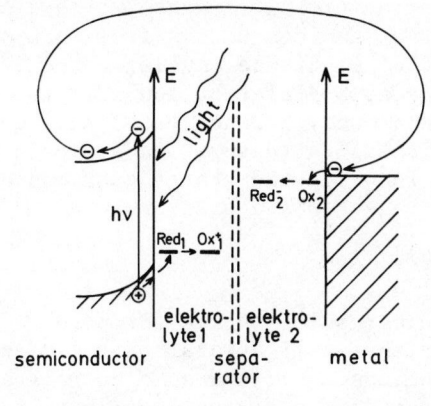

Figure I.15 Photoelectrolytic storage cell.

We shall concentrate in the following on devices in which semiconductor electrodes are used as the primary light energy converters and therefore we shall have to discuss next the general features of semiconductor electrodes in order to understand the principles of charge transfer reactions at such materials. On this basis we can then proceed to the mechanism of solar energy conversion and the problems involved.

References to Lecture for Further Reading:

1.   M. Archer, J. Appl. Electrochem. 5, 17 (1979).

2.   J. M. Bolton (Ed.), "Solar Power and Fuels", Academic Press: New York 1977.

3.   H. Gerischer, Faraday Discussions 58, 219 (1974).

4.   H. Gerischer, J. Electrochem. Soc. 125, 218C (1978).

5.   H. Gerischer, in: "2nd EC Photovoltaic Solar Energy Conference" (Eds. R. van Overstraeten, W. Palz), D. Reidel Publ. Comp.: Dordrecht-Boston-London 1979, p. 115.

6.   Govindjee (Ed.), "Bioenergetics of Photosynthesis", Academic Press: New York 1975.

7.   A. Heller (Ed.), "Semiconductor Liquid-Junction Solar Cells", Proceedings 77-3, The Electrochemical Society: Princeton 1977.

8.   G. Porter and M. Archer, Interdisciplinary Science Rev. 1, 119 (1976).

II. FUNDAMENTALS OF SEMICONDUCTOR ELECTROCHEMISTRY

There are two principle differences which distinguish the electrochemical behavior of semiconductors from that of metal electrodes. First, the by orders of magnitude smaller concentration of mobile charge carriers leads to a different double layer structure and to different kinetic relations. Secondly, the existence of an energy gap in the electronic band structure leads to a distinction between electrode reactions in which the charge carriers of either the conduction band or the valence band are involved.

The Space Charge Layer

The first feature results in the formation of a space charge layer of variable extension if excess charge is accumulated at the contact between a semiconductor and an electrolyte. Fig. II.1 gives a schematic comparison of the charge distribution and the course of potential at the contact between an electrolyte and a metal or a semiconductor.

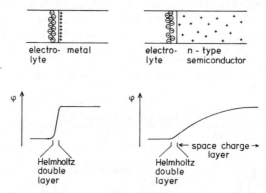

Figure II.1   Charge and Potential distribution at metal/electrolyte and semiconductor/electrolyte contact.

For a description of electrode reactions it is very useful to describe the contact properties by the relation between the electronic energy levels in the electrolyte and in the solid. This is done in Fig. II.2 for a metal electrode, an n-type and a p-type semiconductor electrode for three different situations of the charge distribution at the interface. On the left hand side, we have the zero point of charge (no excess charge on the electrodes). In the middle, we have a negative charge on the electrode and, on the right hand side, we have a positive charge. As characteristic energies for the solids are shown the Fermi energies and the position of the band edges in the semiconductors, all related to a reference energy in the electrolyte. If the semiconductor has a negative excess charge, the bands are bent upwards from the surface to the bulk. If it has a positive excess charge, the energy bands are bent downwards. We see in this figure that the whole electronic band system is shifted upwards and downwards with the excess charge in a metal electrode, while in the case of a semiconductor the energy position at the band edges remains practically constant at the interface if the majority carriers are removed from the surface, but are shifted upwards or downwards if they are accumulated.

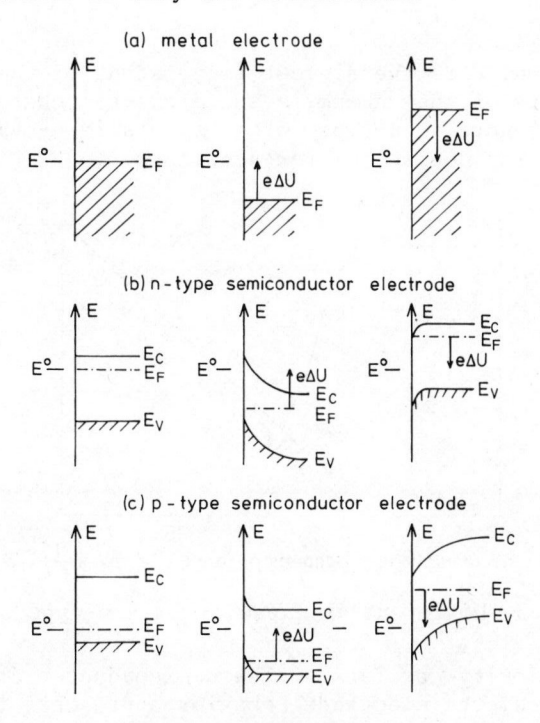

Figure II.2   Electric double layer structure and position of electronic energy levels.

The local concentration of the electronic charge carriers is given in such pictures by the distance between the Fermi level and the band edges:

$$(1) \qquad n(x) = N_C \cdot \exp\left(-\frac{E_C(x) - E_F}{kT}\right)$$

$$(2) \qquad p(x) = N_V \cdot \exp\left(-\frac{E_F - E_V(x)}{kT}\right)$$

These properties can be checked experimentally by capacity measurements. The most important situation for a semiconductor in connection with energy conversion is the depletion layer, i.e. a positive excess charge on an n-type or a negative excess charge on a p-type semiconductor. In this case, the differential capacity, provided that there are no slow relaxation processes inside the semiconductor and the position of the band edges at the surface of the semiconductor does not vary with the applied voltage, can be represented by the so-called Mott–Schottky equation which relates the capacity to the potential difference between the surface and the bulk of a semiconductor, ΔV_{sc}

(3) $\qquad C^{-2} = \dfrac{2}{\varepsilon \varepsilon_o e N_i} \left(z_i \, \Delta V_{sc} - \dfrac{kT}{e} \right)$

In this equation, ε is the dielectric constant of the material, ε_o the permittivity of the vacuum, e the elemental charge, N_i the concentration of donors in n-type, of acceptors in p-type semiconductors, z_i is +1 for donors, −1 for acceptors.

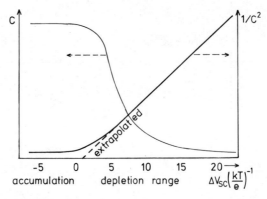

Figure II.3   Differential capacity of a n-type semiconductor.

If the majority carriers of the semiconductor are accumulated at the interface, the electrode behavior approaches that one of a metal because now the excess charge remains very concentrated at the interface. Fig. II.3 gives a picture of the capacity behavior of an n-type semiconductor against the potential drop across the interface in a linear representation and also in a Mott–Schottky plot.

Although the potential difference at the interface cannot be directly measured, the variation of this potential difference goes parallel with the electrode potential measured against a reference electrode. Therefore, one can derive the position of the zero point of charge, which is called the flat band potential, from measuring the space charge capacity in dependence of the electrode potential applied. The flat band potential is identical with the position of the Fermi level in the uncharged semiconductor. Since at the flat band potential the distance between the Fermi level and the band edges is the same on the surface as in the bulk, one obtains in this way also the position of the band edges at the interface. This is shown in the following equations for an n-type and a p-type semiconductor:

(4) $z_i \, \Delta V_{sc} = U - U_{fb}$

(5) $E_C = - eU_{fb} - kT \ln \dfrac{n_o}{N_C} + E^o_{ref}$

(6) $E_V = - eU_{fb} + kT \ln \dfrac{p_o}{N_V} + E^o_{ref}$

E^o_{ref} is the position of the Fermi level in the reference electrode.

The position of the Fermi level in an uncharged semiconductor relative to the electrolyte depends also on surface dipols and dipolar layers at the interface. The latter can be formed by excess charge, directly accumulated at the surface in electronic surface

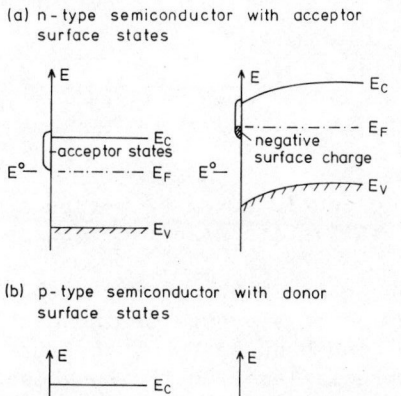

Figure II.4   Influence of surface states on position of electron
energy levels in relation to the electrolyte.

states or in ionic states at the surface being generated by interaction with the electrolyte or by attachment from the electrolyte. Such surface charges create a Helmholtz double layer like on a metal. They influence the differential capacity, if they respond to a variation of the position of the Fermi level in the semiconductor what can be the case for electronic surface states. The ionic charge depends on the chemical composition of the electrolyte, if charged components interact with the surface. There are plenty examples for both cases. A consequence of a variation of the charge in surface states with a variation of the Fermi level in the semiconductor is a shift of the position of the band edges at the semiconductor surface relative to the energy levels in the electrolyte. This is de-

monstrated in Fig. II.4 for two semiconductors, one having surface
states of acceptor character, the other surface states of donor
character.

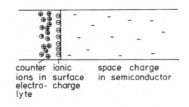

counter ionic   space charge
ions in surface   in semiconductor
electro- charge
lyte

$\varphi \uparrow$ Electric Potential

$U-U_{fb} > 0$

U

$U-U_{fb} < 0$

distance

Figure II.5   Electric double layer at p-type semiconductor electrode
              with negative ionic surface charge.

Attached surface charge interacting with the electrolyte is
indicated in Fig. II.5 for a p-type semiconductor. It is assumed
that the position of the band edges at the interface is practically
not changed, if the Fermi level in the semiconductor is varied by
an applied voltage. The figure shows the course of the electrical
potential in the space charge layer for different voltages applied.
A shift of the band edge positions can only be caused by a variation of the potential drop in the Helmholtz double layer, ΔV_H. This
occurs often when the chemical composition of the electrolyte is
changed. A common example is the electrical double layer between
oxides and aqueous electrolytes, where protons can be attached or
dissociated from the surface particularly after hydrolysis of the
surface layer. A similar situation occurs at sulfides inserted into
an electrolyte containing sulfide ions. Such cases can be treated
as chemical equilibria between the semiconductor surface and the
electrolyte with charge transfer through the Helmholtz double layer.
The following example may illustrate such situations.

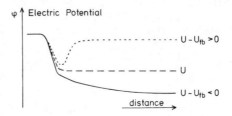

(7) $$\Delta V_H = \frac{kT}{e} \ln \left(\frac{a_{H^+} \, a_{O^-_s}}{a_{OH_s}} \right) + \text{Const.}$$

The consequence is a variation of the potential drop in the Helm-holtz double layer which has the form of the Nernst equation for the pH dependence, if the activity coefficients of the species on the sur-face do not change with their concentration. This assumption is not fulfilled at higher charge densities and leads to deviations from the Nernst equation.

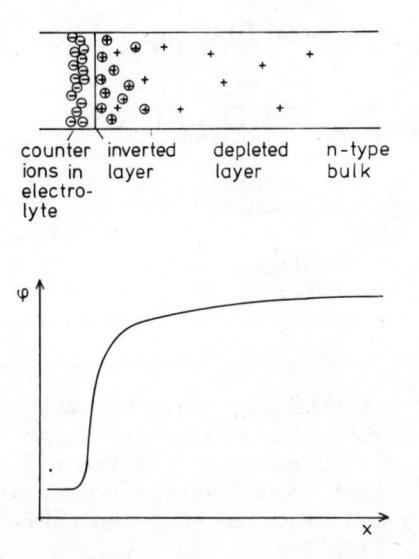

counter ions in electro-lyte   inverted layer   depleted layer   n-type bulk

Figure II.6   Charge and potential distribution at a n-type
             semiconductor/electrolyte contact with inversion
             layer.

Finally, there is another situation which can be important for the discussion of semiconductor electrode behavior. Instead of a depletion layer obtained by removing the majority carriers from the interface as we have assumed in Fig. II.2 and Fig. II.3, an inver-sion layer can be formed. This occurs if in some way minority car-riers are generated in the space charge region, for instance by thermal excitation or by injection from the electrolyte or, as we shall discuss later, by illumination. This situation means that the excess charge on the semiconductor is now not only formed by the immobile donors or acceptors but in a very small region close to the surface also by the minority carriers. Fig. II.6 shows this for an n-type semiconductor at positive bias. This situation leads to a much steeper potential gradient at the interfacial region than for a normal depletion layer and affects more or less the potential drop in the Helmholtz double layer, which means a considerable shift of the band edge positions at the interface.

Kinetics of Electron Transfer Reactions

Electron transfer can generally be described as an exchange of electrons between occupied and vacant energy levels. Since electron transfer is a very fast process, fast compared to the relaxation times of most other energy modes, the Franck-Condon principle has to be applied on a description of this process with respect to the structural conditions and, besides this, the energy has to be conserved. This leads to a description of the rate of electron transfer as shown in the following equations:

$$(8) \quad \psi_i(E_i)_{occ} + \psi_f(E_f)_{vac} \longrightarrow \psi_i(E_i)_{vac} + \psi_f(E_f)_{occ}$$

$$(9) \quad \upsilon = \nu_{ET} \cdot \int \varkappa(E) \, D_{occ}(E) \cdot D_{vac}(E) \, dE \qquad \text{with} \quad E_i = E_f$$

At interfaces, the electron transfer occurs between occupied and vacant states at both sides of the interface. The distribution of electronic states in metals as well as in semiconductors is given by their band structure. If we know the position of the band edges relative to the electrolyte, we know in principle the average state density and energy distribution.

The energy levels in the electrolyte are given by the localized electron acceptors which are the oxidized species of a redox system and the electron donors which are the reduced species. Their density of states is simply given by their concentration. What is their energy position and distribution? The energy of an electron in the occupied state is given by the energy needed to remove it from the reduced species and to bring it to a reference state, for instance the vacuum level. It corresponds to the ionization energy. Likewise, the position of the vacant level is given by the electron affinity of the vacant state, for instance the energy gained in bringing an electron from the vacuum level to this state. The ions in solution interact with their surroundings very strongly and the interaction energy will very much depend on the electric charge of the species. Therefore, we must expect a drastic difference in the position of the energy levels for the occupied and vacant state of the same redox system. If we represent the interaction between the species and their surroundings by a generalized reaction coordinate, we can derive a distribution function for these energy levels. Since the interaction is fluctuating with the thermal movements of the solvent and the interacting species in the electrolyte, such a distribution function represents a statistical average over time. The result is a Gaussian distribution of the energy states which is shown in Fig. II.7 in a one-dimensional representation. It contains as characteristic parameter the so-called reorganization energy E_R and can be described by the following equations:

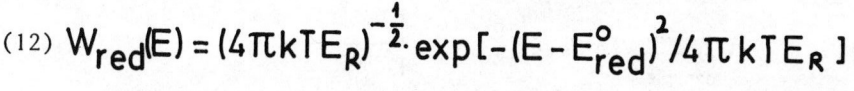

(10) $D_{occ}(E) = c_{red}W_{red}(E); \quad D_{vac}(E) = c_{ox} \cdot W_{ox}(E)$

(11) $W_{ox}(E) = (4\pi kTE_R)^{-\frac{1}{2}} \cdot exp[-(E - E_{ox}^o)^2/4\pi kTE_R]$

(12) $W_{red}(E) = (4\pi kTE_R)^{-\frac{1}{2}} \cdot exp[-(E - E_{red}^o)^2/4\pi kTE_R]$

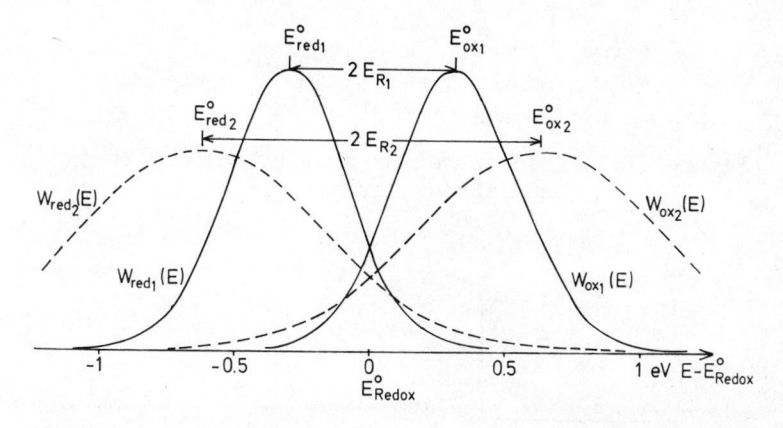

Figure II.7   Distribution functions of redox energy levely for two redox systems with $E_{R_1} = 4\pi kT$ and $E_{R_2} = 8\pi kT$ (T=298°K).

The energy positions of the maxima of these distribution functions are related to the standard redox potential identical with the standard Fermi level of the redox system, E_{Redox}^o, by the following equations.

(13) $E_{red}^o = E_{Redox}^o - E_R$

(14) $E_{ox}^o = E_{Redox}^o + E_R$

The energy level of the standard redox potential can be reached with equal probability by a thermal fluctuation of the occupied or the vacant redox species as equ. (11) and (12) indicate.

Applying the general formula for the electron transfer reactions one can describe the electron transfer at a metal electrode as shown in Fig. II.8 and II.9. At equilibrium, the Fermi levels in the redox electrolyte and in the metal coincide and, as one sees, electron transfer occurs at energy levels in the close neighborhood of the Fermi level. If one applies a voltage, the Fermi level in the electrode is shifted upwards (cathodic) or downwards (anodic) relative

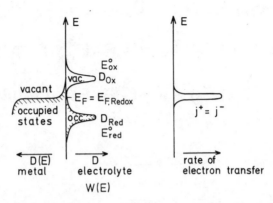

Figure II.8 Redox reaction at metal electrode in
equilibrium situation.

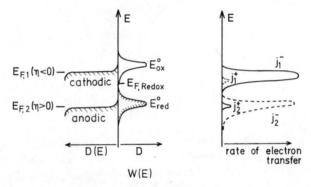

Figure II.9 Redox reaction at metal electrode at cathodic
$(E_{F,1}$ and $j_1)$ and anodic $(E_{F,2}$ and $j_2)$.

to the energy levels in the electrolyte and the consequence is, as
shown in Fig. II.9 that now the electron transfer in cathodic direction or in anodic direction exceeds the reverse transfer rate. The
electron transfer still occurs in the energy range around the Fermi
level of the metal because the change from full occupation to full
vacancy occurs in the metal much more abruptly than in the electron
state distribution of the electrolyte.

The modification of electron tranfer processes by the band gap
of a semiconductor is very obvious. The Fermi level in the absence
of degeneracy is located in the semiconductor within the band gap.
Therefore the energy range in which electron transfer occurs at a
metal electrode is excluded from this process. Fig. II.10 indicates
what one has to expect by comparing the position of the band edges
of a semiconductor with various positions of the electronic energy

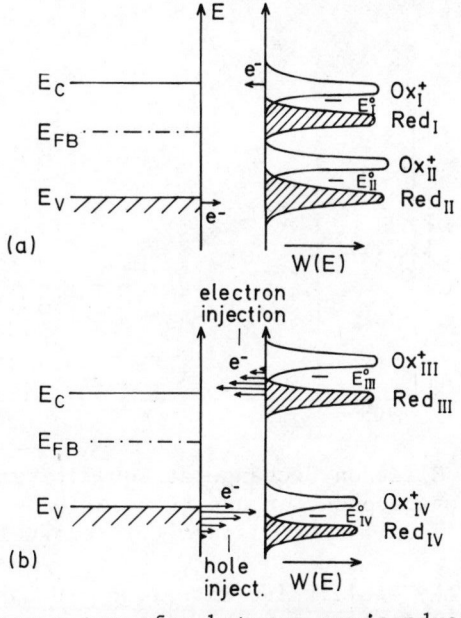

Figure II.10   Electron transfer between semiconductor and different redox couples at flat band situation.

levels in redox electrolytes. The semiconductor in this case is supposed to be at the flat band potential. The figure indicates that electron transfer occurs only on the energy levels close to the band edges. If the energy levels of the redox system overlap energywise more with the conduction band, we have a preferential electron transfer between this band and the redox system, if we have a larger overlapping with the valence band we have an electron transfer in this energy range which can better be described as an exchange of holes.

If we insert a semiconductor into a redox electrolyte and electrons can be exchanged, equilibrium will be achieved by a shift of the Fermi level in the bulk of the semiconductor to the same height as in the redox electrolyte. The consequences are shown in Fig. II.11 for an n-type semiconductor in contact with two different redox systems. The first system (left side) has a redox Fermi level closer to the conduction band edge. In equilibrium, a depletion layer is formed with a moderate band bending. The exchange current density is small and electron exchange occurs only in the energy range of the conduction band as shown in this picture. If the contact is made with a very oxidizing redox system having a redox Fermi level at the valence band edge or close to it, an inversion layer is formed by injection of holes into the surface from the redox system. The ex-

change current depends on the equilibrium concentration of holes at
the surface and can be very large. This is shown on the right hand
side of Fig. II.11.

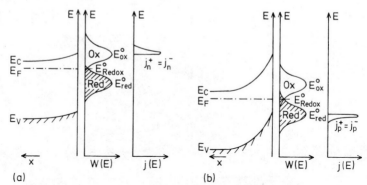

Figure II.11   Electron Exchange at Equilibrium between n-type
semiconductor and reducing:
(a) or oxidising (b) redox electrolyte.

If we leave the equilibrium situation and apply a voltage to
such semiconductor electrodes in contact with redox electrolytes,
the position of the band edges at the interface will normally not
vary to a considerable extent but the space charge layer varies and
with it the surface concentration of electrons or holes. Fig. II.12

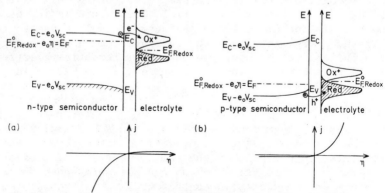

Figure II.12   a) Electron transfer via conduction band at n-type
semiconductor, b) Electron transfer via valence band
at p-type semiconductor, and current characteristics.

shows two situations, left an n-type semiconductor in contact with
a redox system where electron transfer occurs via the conduction
band and right a p-type semiconductor in contact with a redox system where electron transfer occurs via the valence band.

In the simplest cases are the reaction rates in one direction exclusively controlled by the concentration of the electronic charge carriers at the interface while the rate of the reverse reaction is constant. The following equations are then valid:

$$(15) \quad j_c = j_{c,0}\left[\frac{c_{red}}{c_{red,0}} - \frac{c_{ox}}{c_{ox,0}} \cdot \frac{n_s}{n_{s,0}}\right]$$

$$(15a) \quad \frac{n_s}{n_{s,0}} = \exp\left(- \frac{e\eta}{kT}\right)$$

$$(16) \quad j_v = j_{v,0}\left[\frac{c_{red}}{c_{red,0}} \cdot \frac{p_s}{p_{s,0}} - \frac{c_{ox}}{c_{ox,0}}\right]$$

$$(16a) \quad \frac{p_s}{p_{s,0}} = \exp\left(\frac{e\eta}{kT}\right)$$

$$(17) \quad \eta = U - U_0$$

In these equations c_0, n_0, p_0 mean the concentrations at equilibrium, j_0 are the exchange current densities and U_0 is the equilibrium potential. Fig. II.12 shows also the resulting current voltage curves.

One sees that the current voltage curves have rectifying character. The reverse current is extremely small, if the distance between the redox Fermi level in the electrolyte and the respective band edge is large. From an analysis of the current voltage behavior one can usually directly decide which energy band is responsible for the electron transfer reaction.

If the energy position of the band edges varies with the applied voltage, which occurs in the presence of surface states, or if the attached surface charge is modified by the electrode reaction, the simple rate equations given above must be modified and the influence of the variation of the band edge position must be taken into account. This can result in very complex current voltage curves, since the shift of the band edge position will normally have a non-linear correlation with the applied voltage. If we assume a linear correlation between the potential drop in the Helmholtz double layer and the applied voltage, one gets the following modified rate equations neglecting the concentration polarization ($c_{red}/c_{red,0} = c_{ox}/c_{ox,0} = 1$):

$$(18) \quad j_c = j_{c,0}\left[\exp\left(\alpha \frac{e\eta_H}{kT}\right) - \exp\left(-\frac{e(\eta - \alpha\eta_H)}{kT}\right)\right]$$

$$(19) \quad j_v = j_{v,0} \left[\exp\left(\frac{e(\eta - \beta\eta_H)}{kT}\right) - \exp\left(-\beta \frac{e\eta_H}{kT}\right) \right]$$

$$(20) \quad \eta_H = \Delta V_H - \Delta V_{H,0} ; \quad \beta = 1 - \alpha$$

Fig. II.13 gives a summary of the current voltage curves which may be found at semiconductor electrodes with various electron transfer mechanisms. The current voltage curves are plotted in a logarithmic current scale against the overvoltage for the particular reaction.

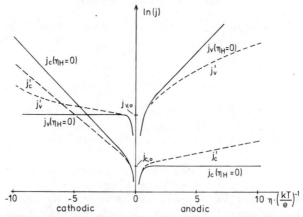

Figure II.13   Current voltage curves for redox reactions at semiconductors via the conduction band (j_c) or valence .e
band (j_v): j with $\eta_H = 0$; j^1 with $\eta_H \neq 0$.

If an inversion layer is formed by hole injection, as assumed in Fig. II.11 on the right hand side, the current voltage curve will mainly be controlled by the electron transfer through the inversion layer and to a lesser degree by the rate of the redox reaction at the interface. In such a case, one has to apply with some modifications the current voltage behavior of p-n junctions.

The picture of electron transfer reactions given here does not adequately describe electron transfer reactions in which a very large amount of energy can be dissipated. For example, if we consider the reduction of a redox system with a very large difference between the position of the redox Fermi level in the electrolyte and the conduction band by electrons from the conduction band, as shown in Fig. II.14, the rate of electron transfer is still high, although the peak position of the vacant energy levels is far below the band edge. In this case, Franck-Condon factors for transitions into vibrationally excited final states have to be taken into account in

the theory which have not been considered in the derivation given
before which was based on a classical approximation of the description of energy levels. Experimental experience with homogeneous
electron transfer reactions has shown that electron transfer remains
fast, even if large energies have to be dissipated in such a step.

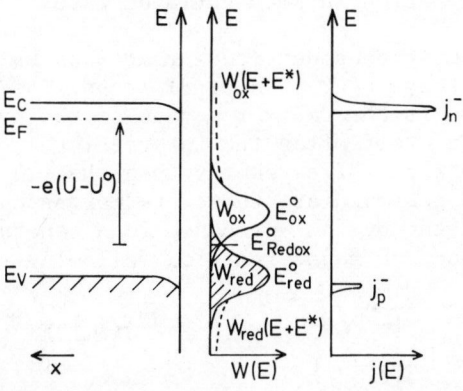

Figure II.14   Electron transfer forn n-type semiconductor electrode
to oxidising redox system at large overvoltage.

This means that in Fig. II.14 after electron transfer the resulting
final states of the occupied species will vibrationally be highly
excited including the modes of the surroundings (polarised solvent
molecules). The excess energy is afterwards dissipated very quickly
as heat. Further, one should not forget that even the classical rate
equation (cf equ. (9,11,12)) for electron transfer contains a very
large frequency factor, ν_{ET}, in the order of 10^{10} to 10^{12} and that
therefore the rate can be high even if the exponential term has a
very small value. The additional probability to obtain a vibrationally excited final state in such an electron transfer step is indicated in Fig. II.14 by an extension of the distribution function.
The same situation is found for a hole transfer from the semiconductor to a redox electrolyte with a very large excess energy of the
hole over the most probable state of the electron donor in the solution.

A further complication of the electron transfer at semiconductors can occur if electronic surface states are present. If these
surface states have energy positions within the band gap they can
take part in the electron transfer reactions. This is indicated in
Fig. II.15 for electronic surface states below the conduction band
edge, and in Fig. II.16 for electronic surface states with donor
character close to the valence band edge. If the surface states are
located too far away from the band edges they can still exchange

electrons with redox systems, however, they will often not be able to exchange electrons with the respective energy bands. This has the consequence that their state of charge and the position of the Fermi level in these surface states will be controlled by the redox electrolyte and will not or very little be affected by the position of the Fermi level in the bulk of the semiconductor. In such cases kinetics are little influenced by the presence of such surface states. The only effect is a constant shift of the band edge positions if the charge concentration in such surface states is considerable.

Finally, we must consider redox reactions which are complicated by the fact that they occur in several steps. The classical example of such a reaction is the redox reaction of the hydrogen ion or the reduction of water. Very often the intermediate formed in such a reaction has a much higher free energy than the final products. The formation of this intermediate therefore corresponds to a high activation energy for the overall process. In a generalized form we shall discuss here a two-step redox reaction following the scheme

(21) $\quad A^+ + 2e^- \rightleftarrows B^- \quad : \quad E^o_{Redox} = E^o_{A^+/B^-}$

(22) $\quad A^+ + e^- \rightleftarrows I \quad : \quad E^o_{A^+/I}$

(23) $\quad I + e^- \rightleftarrows B^- \quad : \quad E^o_{I/B^-}$

The standard redox Fermi energies of the two-one electron transfer step processes are connected with the Fermi level of the overall redox reaction by the relation

(24) $\quad E^o_{A^+/B^-} = \frac{1}{2}(E^o_{A^+/I} + E^o_{I/B^-})$

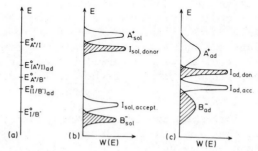

Figure II.17   (a) Standard redox potentials in absolute scale at fermi energies for 2-step redox reaction: $A^+ + 2e-$ $A^+ + 2e- \rightleftarrows B^-$ with intermediate I: in solution and adsorbed state. (b,c) Electron exchange term distribution in solution (b) and in adsorbed state (c).

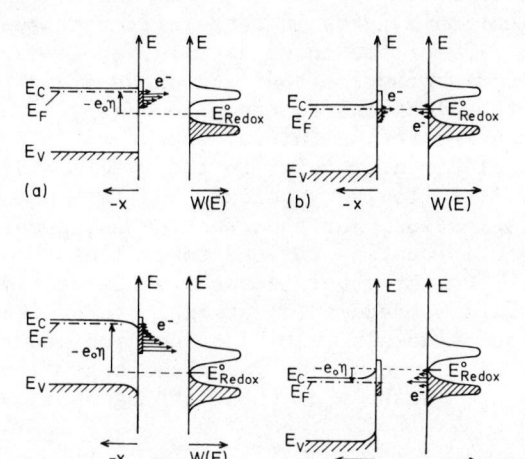

Figure II.15   Electron energy levels of n–type semiconductor with
               surface states (around E_C) in contact with redox
               couple; energy spectrum of electron transfer; (a)
               flat band (cathodic); (b) equilibrium; (c) strongly
               cathodic: (d) anodic.

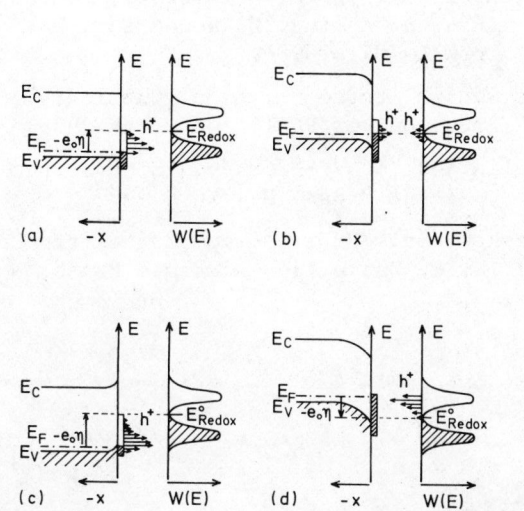

Figure II.16   Electron energy levels of p–type semiconductor with
               surface states (around E_V) in contact with redox
               couple; energy spectrum of hole transfer:
               (a) flat band (anodic); (b) equilibrium; (c) strongly
               anodic; (d) cathodic.

In order to discuss the kinetics one has to consider the distribution of electronic terms in these different species. This is shown in Fig. II.17 for the above example. It is very obvious from this energy term distribution that one needs a high overvoltage in order to bring the electrons to the high energy level needed for the electron transfer to the initial acceptor. On the other hand, the consecutive electron transfer to the intermediate forming the final product will follow very quickly and is connected with a very large energy dissipation. For the reaction in opposite direction we have the analogous situation for the removal of the electron from the product B^-. The energy barriers can be decreased if the energy of the intermediate is reduced by strong interaction with the interface. This is the principle feature of heterogeneous catalysis for such processes. The consequences for the energy term distribution in the intermediate is shown on the right hand side of Fig. II.17.

References

1.   P. J. Boddy, J. Electroanal. Chem $\underline{10}$, 199 (1965).

2.   H. Gerischer, in: "Physical Chemistry", Vol. IX A: "Electrochemistry" (Eds. H. Eyring, D. Henderson, W. Jost), Academic Press: New York 1970, p. 463.

3.   H. Gerischer, in: "Electrocatalysis on Non-Metallic Surfaces" (Ed. A. D. Franklin), NBS Special Publication $\underline{455}$, Washington 1976, p. 1.

4.   V. G. Levich, in: "Physical Chemistry", Vol. IX B: "Electrochemistry" (Eds. H. Eyring, D. Henderson, W. Jost), Academic Press, New York 1970, p. 986.

5.   R. Memming, in: "Electroanalytical Chemistry", Vol. $\underline{11}$ (Ed. A. J. Bard), Marcel Dekker Publ.: New York 1979.

6.   V. A. Myamlin and Yu. V. Pleskov, "Electrochemistry of Semiconductors", Plenum Press: New York 1967.

7.   Yu. V. Pleskov, in: "Progress in Surface and Membrane Science" Vol. $\underline{7}$ (Ed. J. F. Danielli), Academic Press: New York 1973, p. 57.

III. THE SEMICONDUCTOR ELECTROLYTE CONTACT UNDER ILLUMINATION AND PHOTODECOMPOSITION REACTIONS

Distribution of Electrons and Holes under Illumination

The absorption of light quanta with above-band-gap energy generates electron hole pairs in semiconductors. Consequently, the concentration of electrons and holes is increased beyond their equilibrium concentration until the recombination reaction compensates the rate of generation by light absorption. Light quanta with higher energies than the band gap generate electron hole pairs with excess translational energy. However, these excess energies are very quickly dissipated and the excess electronic carriers will reach the thermal distribution of their translational energy in their respective energy bands within a time which is much shorter than their usual lifetime. One can therefore treat the individual electronic charge carriers within their respective energy band as being in thermodynamic equilibrium in spite of the fact that the overall electronic system of the semiconductor is not in equilibrium. The free energy of the individual charge carriers is expressed then as a so-called quasi-Fermi level, E_F. The following equations describe the electronic free energies at equilibrium and in quasi-equilibrium:

(1) $$E_F = E_c + kT \ln \frac{n_0}{N_c} = E_v - kT \ln \frac{p_0}{N_v}$$

(2) $$n_0 \cdot p_0 = N_c \cdot N_v \cdot \exp\left(-\frac{E_{gap}}{kT}\right)$$

(3) $$_nE_F^* = E_c + kT \ln \frac{n^*}{N_c} \quad ; \quad n^* = n_0 + \Delta n^*$$

(4) $$_pE_F^* = E_v - kT \ln \frac{p^*}{N_v} \quad ; \quad p^* = p_0 + \Delta p^*$$

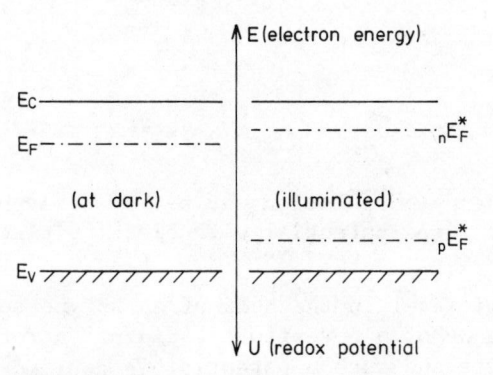

Figure III.1 Fermi and quasi-fermi levels in n-type semiconductor.

While at real thermal equilibrium the Fermi levels of electrons and holes coincide, they differ in the steady state of illumination. This is shown in Fig. III.1 for an n-type semiconductor. One sees that the quasi-Fermi level deviates drastically for the minority carriers, while the effect on the majority carriers is small.

The light absorption is not homogeneous because the light intensity increases with increasing distance from the surface. Consequently, the excess charge distribution is also inhomogeneous in the steady state. If a space charge layer exists underneath the illuminated surface, the charge distribution in the space charge layer will also be affected so more so higher the illumination intensity. The most drastic changes are found in a depletion layer in which minority carriers now accumulate at the surface due to the forces exerted by the electric field in this region. The increase of the concentration of minority carriers at the interface has drastic consequences for all the reactions in which minority carriers participate. The energy gain in photoelectrolysis relies on such reactions. We shall therefore concentrate our discussion on illuminated depletion layers.

Two different situations must be considered. One in which photoelectrolysis is done with an externally fixed potential of the semiconductor electrode. In this way, photocurrent voltage curves are normally measured in galvanic cells with semiconductor electrodes. The other situation corresponds to the photoelectrochemical cells for voltage generation. These contain semiconductor electrodes in contact with a redox system in which a photovoltage is obtained under illumination.

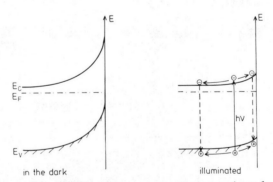

Figure III.2   Energy correlations at n-type semiconductor electrode at fixed potential with blocking electrolyte contact.

If the Fermi level in the bulk of an n-type semiconductor is controlled by means of a potentiostat against a reference electrode in the electrolyte and such a potential is applied that a depletion layer is formed in the dark, the generation of electron hole pairs

and the charge separation in the depletion layer leads to an accumulation of holes at the surface. If the minority carriers cannot leave
the semiconductor because there is no redox system present which can
consume these charge carriers, this charge separation leads to a
counter voltage in the semiconductor which is opposite to the voltage
drop along the space charge layer. It therefore reduces the band
bending. On the other hand, the accumulated charge of the minority
carriers at or very close to the surface attracts a counter charge
from the electrolyte and in this way, the potential drop in the Helmholtz double layer increases. This compensates for the loss of potential drop in the space charge layer. Fig. III.2 shows this situation
for an n-type semiconductor in the dark and under illumination.

If a redox reaction with some components of the electrolyte can
occur, the accumulation of minority carriers and the shift of the
band edges due to their accumulation goes on until the photocurrent
generated by the electron hole pair separation is compensated by the
rate of the surface reaction. Such a situation is shown in Fig. III.3,
where part of the photocurrent is lost by recombination via the anodic
and cathodic process of the redox reaction.

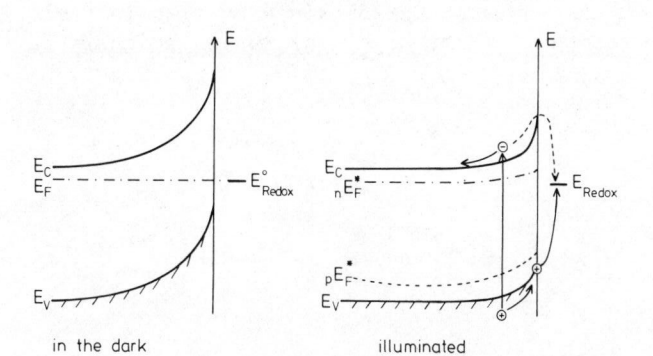

Figure III.3 Energy correlation in n-type semiconductor electrode
              at fixed electrode potential with redox reaction.

The situation is quite different if the electrode potential is
not kept constant and the electrolyte contains a redox system. The
equilibrium in the dark is shown in Fig. III.4 for an n-type semiconductor. The electron hole pair separation in the depletion layer
now cannot lead to such an accumulation of the minority carriers because they can react with the redox system and leave the semiconductor. The electrons remain in the semiconductor as long as it is not
connected with a counter electrode. The remaining counter charge in
the semiconductor also decreases the band bending. However, since
this excess charge accumulates in the bulk and the counter charge is
transferred to the electrolyte, the Helmholtz double layer is affected much less than at fixed potential. Due to the decreased band bending the Fermi energy of the majority carriers in the bulk is increas-

ed as shown in Fig. III.4. This figure also shows the quasi-Fermi

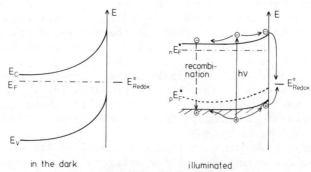

in the dark   illuminated

Figure III.4   Energy correlations in n–type semiconductor electrode
in contact with redox electrolyte at open circut.

level of the minority carriers which deviates largely from the Fermi
level of the electrons in the bulk. The difference between the Fermi
level in the bulk of the semiconductor and the Fermi level of the redox system is the photovoltage obtained by illumination. This is the
free energy gained from the light absorption which is available for
work.

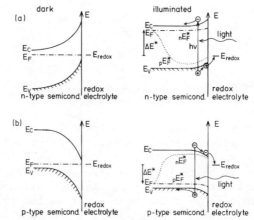

Figure III.5   Generation of a driving force for photoelectrolysis
at n–type (a) and p–type (b) semiconductor electrodes.

If the semiconductor is connected with a counter electrode, a
photocurrent is generated and photoelectrolysis occurs without an
external voltage. This situation is delineated in Fig. III.5 for an
n–type and a p–type semiconductor. An exact calculation of the distribution of the charge carriers at such an illuminated semiconductor electrolyte contact is very difficult. Some approximations can

be obtained on the basis of a simplified model. This goes back to a treatment of Gärtner and has been improved recently by Wilson, Rice and Reichman taking into account surface recombination and/or slow charge transfer at the semiconductor-electrolyte interface. These treatments are based on the model of the boundary layer shown in Fig. III.6.

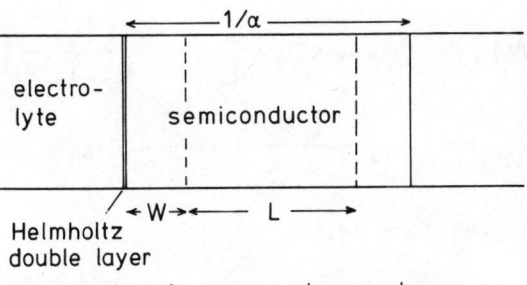

W = width of space charge layer,
L = mean free diffusion length of
minorities, α = absorption coefficient

Fig. III.6

If the mean diffusion length and the penetration depth $1/\alpha$ of the light exceed largely the extension of the space charge layer, the flux of the minority carriers to the surface outside the space charge layer can be calculated from the following differential equation:

$$(5) \quad D\frac{d^2c}{dx^2} - k_r c + I_o \alpha \exp(-\alpha x) = 0$$

where D = diffusion coefficient; c = concentration of the minority carriers; I_o = light intensity at the surface; $k_r = k_r^o \cdot m$ = rate of recombination (with m = concentration of the majority carrier); $k_r = \frac{1}{\tau}$; τ = life time of the minority carriers.

The boundary conditions are:

$$(6a) \quad x = \infty \quad : \quad c = 0$$

$$(6b) \quad x = W \quad : \quad c = c_W$$

c_W is the concentration at x = W and has to be related to the surface concentration by an additional condition.

Gärtner solved this equation with the assumption that the concentration of the minority carriers at the end of the space charge

layer is practically zero assuming that the minority carriers are removed very quickly through the space charge layer and disappear immediately at the surface. Reichman had taken into account the possibility that the minority carriers accumulate at the surface and therefore a finite concentration will be found at the end of the space
charge layer. The latter solution is given in the following equations:

$$
(7) \quad c(x \geqslant W) = c_W \cdot \exp(-\frac{x}{L}) + \frac{\alpha I_W}{k_r - \alpha^2 D}\left(\exp(-\alpha x) - \exp(\frac{x}{L})\right)
$$

$$
(8) \quad D\left(\frac{dc}{dx}\right)_{x=W} = I_W \frac{L}{L + \frac{1}{\alpha}} - \frac{D}{L} c_W
$$

In these expressions is:

$$
(9) \quad I_W = I_0 \exp(-\alpha W)
$$

$$
(10) \quad L = \sqrt{\frac{D}{k_r}} = \sqrt{D\tau}
$$

At large band bending caused by for example an externally applied
voltage, the minority carriers indeed disappear at the space charge
layer very quickly and Gärtner's solution with $c_W = 0$ is valid. This
gives the saturation current. At lower band bending where the saturation current is not yet reached, the theoretical description is much
more difficult. Reichman has assumed that through the space charge
layer the minority carriers remain in thermal equilibrium and therefore the concentration of the minority carriers is controlled by
their concentration at the surface needed to obtain the same rate of
the redox reaction as the flux of minority carriers from the bulk to
the space charge layer together with the photocurrent generated by
light absorption within the space charge layer. For the latter contribution recombination in the space charge layer is neglected in a
first approximation. The continuity equation for the current through
the interface is then given by the following relation:

$$
(11) \quad D\left(\frac{dc}{dx}\right)_{x=W} = \frac{|j_{photo}|}{\mathcal{F}} - I_0 [1 - \exp(-\alpha W)]
$$

Together with equ. (8) one can obtain from equ. (11) the concentration c_W if j_{photo} and the other parameters are known.

The redox reaction at the interface needs a definite concentration of the minority carriers for a definite rate. We have seen be-

fore that the rate of a redox reaction at a semiconductor electrode
is approximately proportional to the concentration of the electrons
or holes. The surface concentration is a function of c_w, the band
bending and the current passing the space charge layer or being generated therein. Without going into further detail one sees that a
redox reaction which needs a high concentration of minority carriers
to proceed in forward direction can cause a considerable increase in
concentration of the minority carriers at the end of the space charge
layer, so more so smaller the band bending is. This effect is under-
estimated in Reichman's treatment who has made the assumption that
the minority carriers remain practically in equilibrium throughout
the space charge layer,

$$(12) \qquad c_S = c_W \cdot \exp\left(- \frac{ze\Delta V_{sc}}{kT}\right)$$

with $z = +1$ for holes and -1 for electrons.

In reality for getting a fast enough flux of the minority carriers through the space charge layer, there must be a decrease in
their free energy which has the consequence that the increase of the
minority carriers at the end of the space charge layer is higher than
assumed in Reichman's calculation.

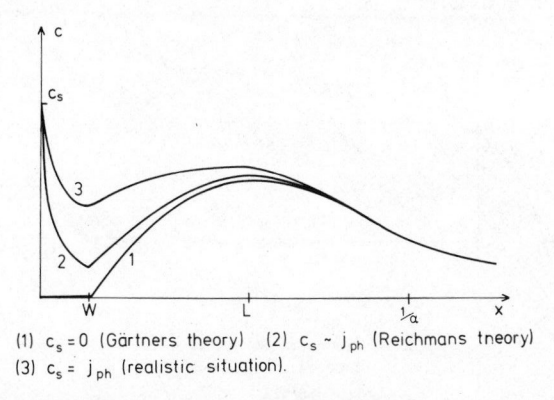

(1) $c_s = 0$ (Gärtners theory)   (2) $c_s \sim j_{ph}$ (Reichmans theory)
(3) $c_s = j_{ph}$ (realistic situation).

Figure III.7   Distribution of minority carrier concentration in
illuminated semiconductor/electrolyte contact for
different surface concentrations.

Fig. III.7 illustrates the situation as assumed in the treatments of
Reichman and Gärtner and also indicates the tendency of the real distribution of the concentration in the space charge region. So higher
the concentration of the minority carriers at the end of the space
charge layer, where the electric field becomes the driving force for
the transport of the minority carriers, so larger will be the loss
of electron hole pairs by recombination in the bulk of the semiconductor, because the driving force for diffusion to the surface is
decreased. This discussion shows that the rate of the redox reaction
can be very important for the photocurrent yield at low band bending.

A theoretically much more difficult situation arises if the
light absorption occurs mainly within the depletion layer. At large
band bending and fast reaction of the minority carriers at the surface, practically all generated electron hole pairs will contribute
to the photocurrent. At smaller band bending recombination and reverse currents decrease the quantum yield and the charge distribution will considerably be changed. No calculations for the correlation between photocurrent and photovoltages are available at present under these conditions.

Further complications in comparing theoretical calculations
with practical results arise from the presence of surface states
which can take part in the redox reactions and affect the charge
distribution if their occupancy is changed under illumination.
They can also contribute to recombination which has been taken
into account in a calculation of Wilson. However, there are many
real systems in which the photovoltage obtained at open circuit is
not controlled by a recombination in the semiconductor or recombination via surface states, but by the reverse redox current with the
majority carriers of the semiconductor. This is a kind of additional
surface recombination via electron transfer to the redox system in
both directions occurring via both energy bands simultaneously. Such
a situation was indicated in Fig. III.4.

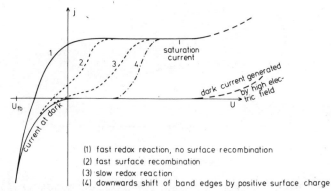

(1) fast redox reaction, no surface recombination
(2) fast surface recombination
(3) slow redox reaction
(4) downwards shift of band edges by positive surface charge

Figure III.8   Current voltage curves for n-type semiconductor electrode
under illumination.

Finally, Fig. III.8 shows a summary of the type of current
voltage curves obtained under illumination with different processes
controlling the decay of photocurrent for an n-type semiconductor.
At high enough voltages a saturation current is reached in all cases.
However, a high field strength in the space charge layer often induces the generation of minority carriers by internal ionization
processes. This happens if the field strength exceeds a critical
value and depends very much on impurities and defects which also let
increase the recombination rate very much.

    In the potential range where the saturation current is observed
one can measure the photocurrent spectra for the particular semiconductors which correspond to the absorption spectrum of the material
if the absorption coefficients are so small that the penetration
depth of the light is larger than the mean diffusion length plus the
extension of the space charge layer. Otherwise, the quantum yield depends little on the absorption coefficient since all light would be
absorbed in a region where electron hole pair separation is very efficient. Fig. III.9 gives some examples for such photocurrent spectra.

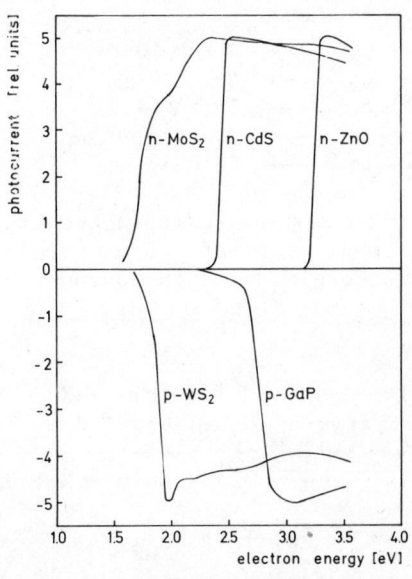

Figure III.9 Photocurrent spectra for electrochemical reactions
              at semiconductors in the saturation range.

Photodecomposition of Semiconductors

    The most serious problem of all photoelectrochemical devices is
the susceptibility of semiconductors in contact with electrolytes to
photodecomposition. Such decomposition reactions are electrochemical
redox processes in which the semiconductor is involved and they can
occur already in the dark. The electronic charge carriers of the semiconductors are direct reaction partners in these processes. The next
table gives a summary over a number of decomposition reactions in
which either holes or electrons participate. Since these reactions
are electrochemical processes their thermodynamics can be described
in terms of redox potentials at which these processes would be in
equilibrium.

TABLE III   ELECTROLYTIC DECOMPOSITION OF SEMICONDUCTORS

Anodic processes (p-type specimen or photodecomposition of n-type materials)

$$Si + 4h^+ + aq \longrightarrow SiO_2 \cdot aq + 4H^+ \cdot aq$$
$$GaAs + 6h^+ + aq \longrightarrow Ga^{3+} \cdot aq + AsO_3^{3-} \cdot aq + 6H^+ \cdot aq$$
$$CdS + 2h^+ + aq \longrightarrow Cd^{2+} \cdot aq + S$$
$$ZnO + 2h^+ + aq \longrightarrow Zn^{2+} \cdot aq + \tfrac{1}{2}O_2$$
$$Cu_2O + 2h^+ + aq \longrightarrow 2CuO + 2H^+ \cdot aq$$
$$MoS_2 + 14h^+ + aq \longrightarrow MoO_3 \cdot aq + 2SO_3^{2-} \cdot aq + 18H^+ \cdot aq$$
$$GaSe + 7h^+ + aq \longrightarrow Ga^{3+} \cdot aq + SeO_3^{2-} \cdot aq + 6H^+ \cdot aq$$

Cathodic processes (n-type specimen or photodecomposition of p-type materials)

$$Si + 4e^- + aq \longrightarrow SiH_4 + 4OH^-$$
$$CdS + 2e^- + aq \longrightarrow Cd + S^{2-} \cdot aq$$
$$ZnO + 2e^- + aq \longrightarrow Zn + 2OH^- \cdot aq$$
$$Cu_2O + 2e^- + aq \longrightarrow 2Cu + 2OH^- \cdot aq$$
III
$$GaSe + 2e^- + aq \longrightarrow Ga + Se^{2-} \cdot aq$$

The two following examples indicate how the redox potential can be derived by a combination of two electrode reactions, one of them being the reference electrode in aqueous solution the standard hydrogen electrode:

Anodic decomposition:

$$MX + 2h^+ + solv \longrightarrow M^{2+} \cdot solv + X$$
$$2H^+ \cdot solv \longrightarrow H_2 + solv + 2h^+$$
$$\overline{MX + 2H^+ \cdot solv \longrightarrow M^{2+} \cdot solv + X + H_2 ;\ \Delta_p G^o_{decomp}}$$

(13)
$$_pE_{decomp} = -\frac{\Delta_p G^o_{decomp}}{RT} + E^o_{NHE},$$

what corresponds to a surface concentration of holes:

$$p_{s,decomp} = N_V \cdot exp\left(-\frac{_pE_{decomp} - E_V}{kT}\right).$$

Cathodic decomposition:

$$MX + 2e^- + solv \longrightarrow M + X^{2-} \cdot solv$$
$$H_2 + solv \longrightarrow 2H^+ \cdot solv + 2e^-$$
$$\overline{MX + H_2 + solv \longrightarrow M + X^{2-} \cdot solv + 2H^+ \cdot solv ;\ \Delta_n G^o_{decomp}}$$

(14)
$$_nE_{decomp} = +\frac{\Delta_n G^o_{decomp}}{RT} + E^o_{NHE},$$

corresponding to a surface concentration of electrons:

$$n_{s,decomp} = N_C \cdot exp\left(-\frac{E_C - _nE_{decomp}}{kT}\right).$$

E^o_{NHE} is the Fermi level of the standard hydrogen electrode in this solvent.

The free energy difference of such a cell reaction gives the redox potential difference versus the reference electrode, i.e. the position of the equilibrium redox potential in the respective redox potential scale. At constant composition of the semiconductor and the electrolyte, the only variable in such a system is the free energy of the electronic charge carriers. We have seen in previous equations that the free energy of the electrons and the holes is controlled by the position of the band edge besides their concentration. If one knows the position of the band edges of an individual semiconductor in contact with an electrolyte and can calculate the position of the equilibrium potential of the decomposition reaction from thermodynamic data, one can predict the concentration of the respective electronic charge carrier at the surface which is needed to reach equilibrium. If these concentrations are exceeded, we have a driving force in the forward direction of the decomposition reactions. We see that decomposition will be favored so more so smaller the concentration of these charge carriers is in the equilibrium situation, i.e. so farther away the position of the decomposition potential is from the position of the band edges of the reacting charge carriers. Since under illumination the quasi-Fermi levels can at most approach the position of the band edges, decomposition will normally not occur if the decomposition potentials are located in an energy range below E_V for the anodic or above E_c for the cathodic process. Assuming that illumination cannot shift the position of the band edges but only varies the position of the quasi-Fermi levels within the band gap, one can from a thermodynamic analysis conclude, whether a semiconductor should be inert against decomposition or not. A scheme of the situations which can be found for a semiconductor in contact with an electrolyte is shown in Fig. III.10.

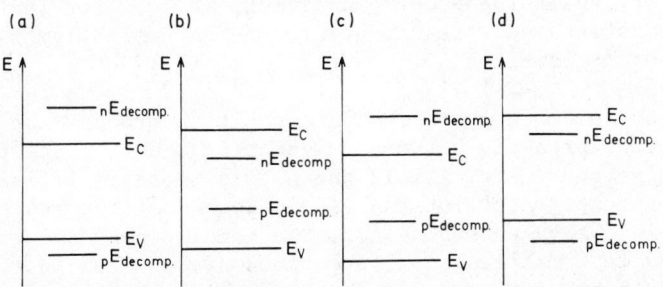

Figure III.10   Stability criteria for electrolytic decomposition of semiconductors: (a) stable, (b) instable, (c) cathodic stability only, (d) anodic stability only.

Fig. III.11 gives a summary for a number of semiconductors for which the position of the band edges and the decomposition potentials are known. One sees that apparently, at least in contact with aqueous solutions, none of these semiconductors is thermodynamically stable against anodic photodecomposition, while some are indeed stable against cathodic decomposition.

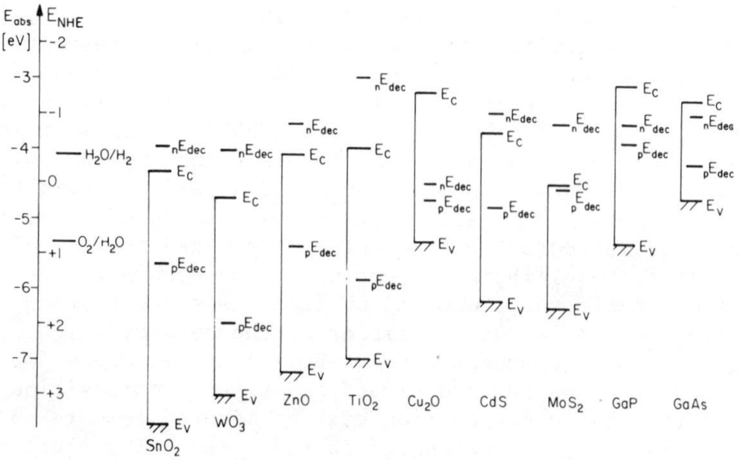

Figure III.11 Positions of band edges and decomposition potentials of various semiconductors in electrolytes of $_p$H 7.

In reality, the situation is fortunately better with respect to photocorrosion than thermodynamics seem to tell us. The reason is that the decomposition reactions are complicated processes involving a series of consecutive reaction steps. These reaction steps lead to intermediates which often have a much higher energy and therefore form a reaction barrier for the process. To overcome this barrier one needs a much higher free energy of the electronic charge carriers than thermodynamically predicted and this can let increase very much the stability against photodecomposition. This is particularly the case if competing redox reactions are present which limit the increase of the minority carriers.

In most semiconductors the decomposition process can be considered as a bond breaking between neighboring atoms by the presence of holes or electrons. Fig. III.12 shows this in a simplified crystal model in its upper part for the anodic decomposition reaction. The electrons in the valence band form the bonding states in these more or less covalent materials. If electrons are missing in these bonding states, the bonds are weakened. Due to the positive charge, such an electronic defect can easily be attacked by a nucleophilic reagent. In this way, a new chemical bond is formed with one of the atoms while the other remains in a radical state. The electron in this radical state will have a higher energy, and therefore this intermediate will either act as an efficient trap for holes or it might interact with another intermediate forming a new chemical bond. Fig. III.12 shows the first case in which now one atom is left with a much weaker bond to the surface which will further be easily attacked and in this way, the atoms are stepwise removed from the surface by an oxidation due to an accumulation of holes.

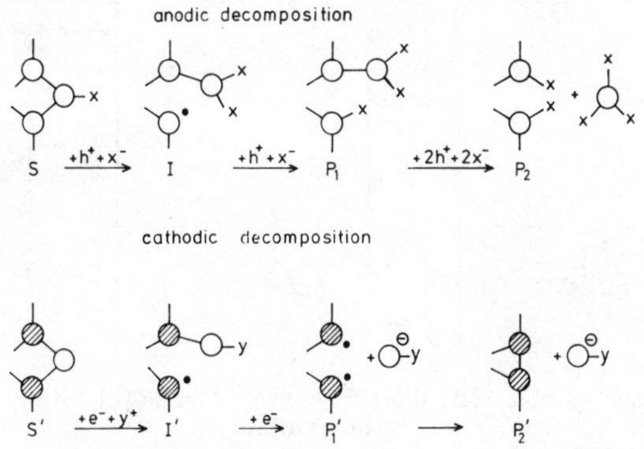

Figure III.12   Models for the attack on semiconductor bonds by accumulation of holes or electrons at the contact with an electrolyte.

The cathodic decomposition process can be discussed in a very similar way. This is shown on the lower part of Fig. III.12. Accumulation of electrons in the conduction band which contains usually non-bonding or anti-bonding orbitals will favor the interaction with electrophilic reagents and in this way cause the interruption of the chemical bonds as shown in this figure. The intermediate states with unpaired electrons generated on the surface by this process are supposed to interact with each other which occurs for example in the case of the reduction of compound semiconductors where a surface layer of the metal is formed.

Including the stepwise decomposition reaction according to this model one can again describe the conditions for stability in a quasi-thermodynamic way. This is done in Fig. III.13, where the equilibrium potentials for the formation of the intermediate and the consecutive products as well as the final products are schematically represented for different cases. The reference state is again the position of the band edges which controls the potential energy of the electronic charge carriers. In this picture it is distinguished between an absolute thermodynamic stability where the free energy of the final product is higher than the initial situation with the electronic charge carriers at the energy of the band edges. The second row shows a characterization of a so-called kinetic stability where the first intermediate has a higher standard free energy and can be formed therefore only very slowly or practically not under normal conditions of illumination without an externally applied voltage. The most unfavorable situation is where all reactions including the intermediates

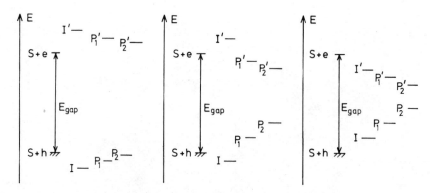

(a) thermodynamic stable   (b) thermodynamic instable(c) totally instable
                                              kinetic stable

Figure III.13   Energy diagram for anodic and cathodic decomposition
                        reactions;

$$S\xrightarrow{h^+}I\rightarrow P_1\rightarrow P_2; \qquad\qquad S\xrightarrow{e^-}I'\rightarrow P_1'\rightarrow P_2'.$$

are going downhill free energywise as shown in the third row of this
picture.

Even in this unfavorable case there is a chance to stabilize the
semiconductor by a competing redox reaction. A description of this
situation in quasi-thermodynamic terms is outlined in Fig. III.14.
The reaction equations given in this figure illustrate the meaning
of these energy levels which characterize again the standard free energies of the different reaction steps. If the redox reaction is fast

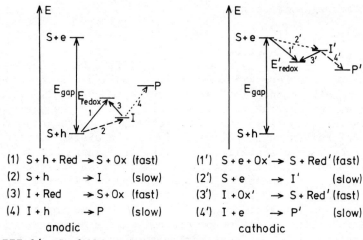

(1) $S+h+Red \rightarrow S+Ox$ (fast)   (1') $S+e+Ox' \rightarrow S+Red'$ (fast)

(2) $S+h \qquad\rightarrow I$ (slow)   (2') $S+e \qquad \rightarrow I'$ (slow)

(3) $I+Red \quad\rightarrow S+Ox$ (fast)   (3') $I+Ox' \quad \rightarrow S+Red'$ (fast)

(4) $I+h \qquad\rightarrow P$ (slow)   (4') $I+e \qquad \rightarrow P'$ (slow)

        anodic   cathodic

Figure III.14   Stabilisation of semiconductors by competing
                        redox reactions.

enough to prevent the second step, namely the forming of a new bond with the radical, but can restitute the previous state of the surface, the semiconductor can even be stabilized after the first step of the decomposition occurred.This is certainly an extremely simplified description of the situation. However, it qualitatively shows the essential features and demonstrates how important the driving forces in the individual reaction steps are for what is going on in reality. All practical systems which are used in photoelectrochemical devices appear to be stable due to the kinetic competition of redox reactions with the decomposition process.

References

1.   A. J. Bard and M. S. Wrighton, J. Electrochem. Soc. 124, 1706 (1977).

2.   A. J. Bard, J. Photochem. 10, 59 (1979).

3.   W. W. Gärtner, Phys. Rev. 116, 84 (1959).

4.   H. Gerischer, J. Electroanal. Chem. 82, 133 (1977).

5.   H. Gerischer, J. Vac. Sci. Technol. 8, 1422 (1978).

6.   H. Gerischer, in: "Topics in Applied Physics", Vol. 31: "Solar Energy Conversion" (Ed. B. O. Seraphin), Springer Verlag: Berlin-Heidelberg 1979, p. 115.

7.   J. Reichman, Appl. Phys. Lett. 36, 574 (1980).

8.   H. Reiss, J. Electrochem. Soc. 125, 937 (1978).

9.   R. H. Wilson, in: "Semiconductor Liquid-Junction Solar Cells" (Ed. A. Heller), Proceedings 77-3, The Electrochemical Society: Princeton 1977, p. 67.

10. and loc. cit. II, 2, 5, 6.

IV.   PHOTOELECTROCHEMICAL CELLS AND THEIR PROBLEMS

Regenerative Cells

The photovoltage generated in a depletion layer of a semiconductor electrode can be used for direct generation of electric power. The principle was already shown in Fig. I.14. The redox reaction at the counter electrode reverses the chemical change at the
semiconductor electrode. Since such cells should operate for a very
long period of time, even a very small side reaction can ruin the
performance after some while. If the parasitic reaction occurs only
within the redox electrolyte, the failure can easily be repaired by
an exchange of the electrolyte. If however it is a corrosion of the
semiconductor, the lifetime of the whole cell is limited. Therefore
the redox reaction at the semiconductor should proceed very fast in
order to prevent corrosion.

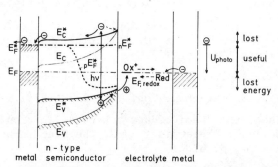

Figure IV.1   Regenerative photoelectrolysis cell with n-type
              semiconductor in the dark and under illumination.

Fig. IV.1 shows schematically the energy relations in such a
cell in the dark and under illumination. It also indicates the losses of free energy for the conversion of light with a quantum energy of the band gap whereby it is assumed that the redox reaction at
the metallic counter electrode causes practically no energy loss.
One sees, however, that the redox reaction at the semiconductor itself contributes very much to the loss in energy, if it has a redox
potential which keeps the surface concentration of holes small
enough in order to prevent corrosion.

Fig. IV.2 shows the analogous cell with a p-type semiconductor.
At equilibrium the surface concentration of holes is minimized. If
the semiconductor is stable against cathodic decomposition one can

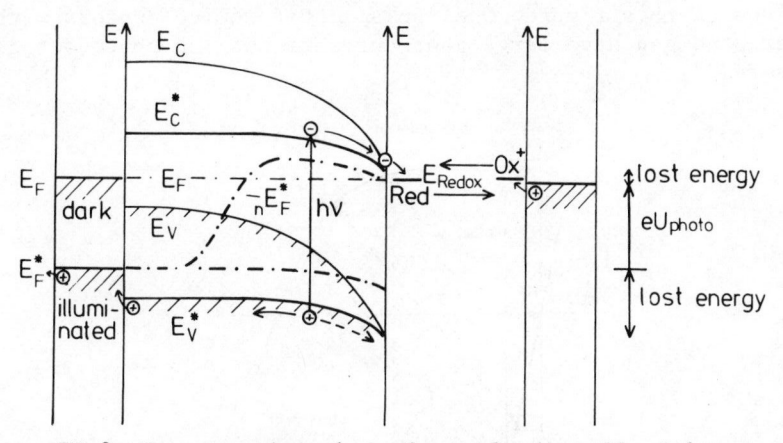

Figure IV.2   Regenerative photoelectrolysis cell with p-type
              semiconductor in the dark and under illumination.

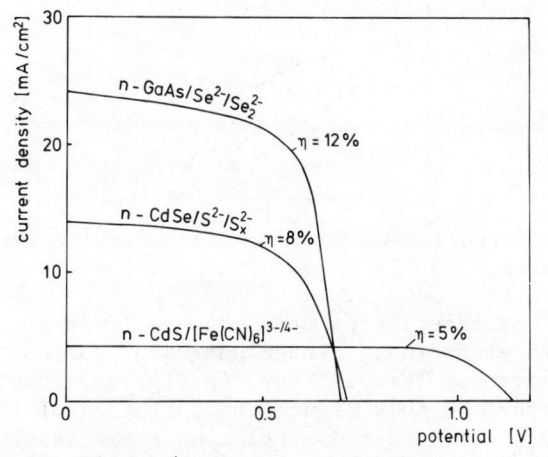

Figure IV.3   Power characteristics of 3 regenerative solar cells.

use reducing redox systems which cause a very large band bending or
even generate an inversion layer. The anodic corrosion problem is
however still critical under illumination, since at reduced band
bending the concentration of holes at the surface will become considerable and may exceed the critical limit for decomposition of
the semiconductor. The figure also indicates what energy is lost
for light quanta with band gap energy.

The efficiency of such cells is characterized by its photocurrent photovoltage correlations which are obtained if the two electrodes are connected via an external resistance. This gives the so-
called power characteristics for which Fig. IV.3 gives three examples. The conversion efficiencies for solar light of these systems

are given in this figure. One can see that semiconductors with a
smaller band gap have large photocurrents but low photovoltages and
vice versa.

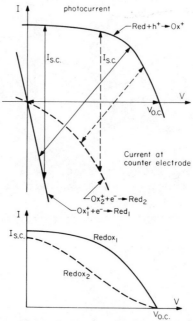

Figure IV.4   Construction of power characteristics for regenerative
              cells.

   The power characteristics of a photocell can be constructed
from the individual current voltage curves of the photoelectrode and
the counter electrode. This is shown in Fig. IV.4, again for an n-
type semiconductor. In this figure the influence of the counter electrode is indicated. If the redox reaction there is slow, i.e. if it
needs large overvoltages, the power characteristics are very much
worsened. A similar effect has the photoelectrode if the photocurrent
does not increase very steeply below the open circuit photovoltage
and does not reach the saturation current very soon.

   The optimal conversion efficiencies can be obtained if the band
bending is largest at equilibrium in the dark and the redox reaction
is very fast. The latter condition is usually fulfilled for one elec-
tron-transfer-step reactions. It is not the case for more complex reactions like the hydrogen evolution or oxygen evolution process. The
maximum band bending is however limited by the band gap. In order to
see what redox systems can be used for particular semiconductors one
has to know the energy position of the band edges in the redox energy
scale. Fig. IV.5 gives a few examples. One sees that some redox systems have position of band edges in a range where redox systems in
aqueous solution are not stable, because they decompose the solvent.

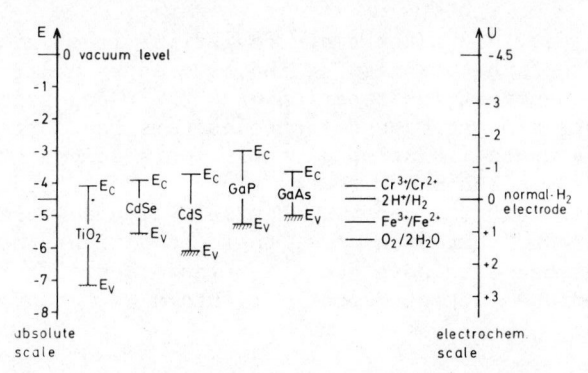

Figure IV.5   Energetic positions of energy bands of semiconductors
             at the contact with the electrolytes in absolute and
             electrochemical scale of redox potentials (some redox
             systems for comparison).

This is a limitation depending on the solvent which can be overcome
to some extent by a variation of the solvent. Different solvents behave also differently with respect to photodecomposition. It can
therefore be attractive to employ redox systems in non-aqueous solvents for such type of solar cells.

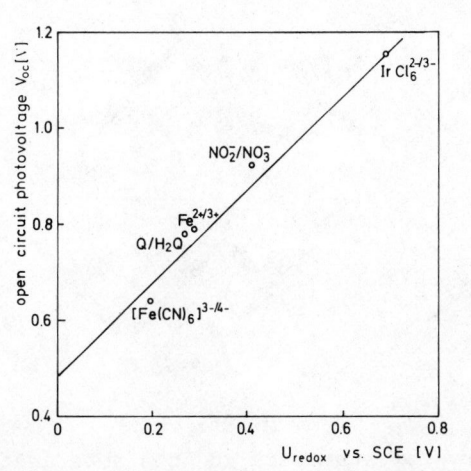

Figure IV.6   Dependence of open-circut-voltage of Cds-electrode in
             contact with redox systems on their redox potential.

The photovoltage obtained from a semiconductor electrode at the
same illumination intensity should approximately be proportional to
the Schottky barrier height in the dark. Fig. IV.6 shows that in the
case of cadmium sulfide this can be fulfilled, although the semiconductor corroded in this example in contact with the most oxidizing
redox systems.

However, very often there are deviations from this linearity
due to a shift of the position of the band edges under illumination
or as a consequence of the reaction with the redox system. In order
to prevent corrosion, compound semiconductors have often been employed in a saturated solution of their ionic components like sulfides in sulfide solutions with the redox system polysulfide/sul-
fide or selenides in the redox system polyselenide/selenide. However, it was found that even under these conditions the crystalline
semiconductors can decompose and are transformed either into a surface with numerous lattice defects or into a different polycrystalline material.

It is obvious that crystal imperfections must have an enormous
influence on the performance of photoelectrochemical cells with semiconductors. The situation is exactly the same as for other solid
state semiconductor devices. A very impressive example is given by
semiconducting materials with layer structure which were supposed
to be particularly stable against photodecomposition. The study of
such materials for photoelectrochemical energy conversion has been
started by Tributsch and a number of interesting investigations
have been performed recently.

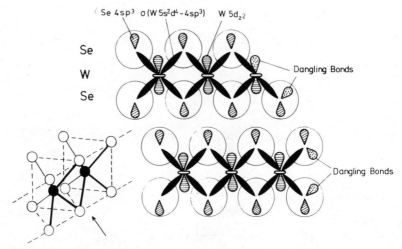

Figure IV.7   Crystal structure and characterisation of valence
band bonds in WSe_2.

Fig. IV.7 shows the crystal structure and the electronic structure of a material like tungsten sulfide or selenide and molybdenum
sulfide or selenide. These materials have band gaps in the order of
1 to 1.2 eV and therefore should have similar properties for solar
energy conversion as silicon. Their stability against photodecomposition is caused by the fact that the highest valence band and the

lowest conduction band is predominantly formed by the d-electrons of
the metal atom which are very well shielded against interaction with
their surroundings by the two layers of the chalcogen atoms. Therefore the corrosion reaction which needs a formation of new bonds
with components of the solvent are prevented at the intact layers
while electron transfer which needs very little electronic interaction proceeds in a normal way. However, at steps and edges of the surface as well at structural defects (dislocations), the electronic
states of the metal atoms are widely exposed to the interaction with
the surroundings and corrosion can proceed rapidly at such sites. This
is clearly shown if one compares the power characteristics of regenerative cells with layered semiconductors of different surface performance. Fig. IV.8 shows power characteristics for molybdenum selenide
and tungsten selenide electrodes with different surface qualities.

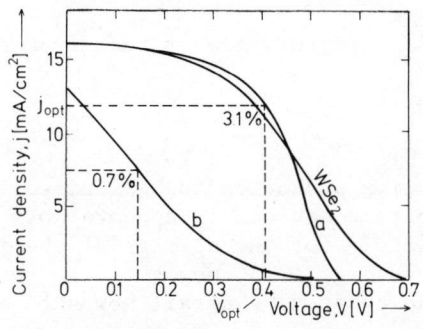

Figure Iv.8 Power characteristics of regenerative solar cells with
             MoS$_2$ and WSe$_2$. Curves (a) and (b) MoS$_2$ - electrodes
             with different surface perfection.

    A real device for practical use must be built very compactly
in order to minimize the energy losses inside the cell. Fig. IV.9
shows a possible construction. In order to get illumination of the
semiconductor from the front one needs a transparent counter electrode. Glass covered with doped and highly conducting transparent
oxide layers of tin or indium have been used for this purpose. It
appears, however, that for high current densities and large areas
the conductivity of these materials is insufficient. Therefore, in
addition to these conductive layers one needs a metallic grid. Very
important is also a rapid compensation of the concentration changes
which occur at both electrodes. If the compensation must be done by
diffusion alone, the electrolyte layer should be kept extremely thin.
It appears however difficult to reach this. Fortunately, in a somewhat thicker electrolyte layer, thermal convection contributes very
much to the equilibration of the electrolyte. Geometric conditions
have to be optimized in order to reach a maximum transport of the
products from one electrode to the other. Very little is known at

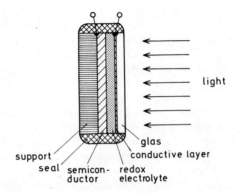

Figure IV.9   Model of a photovoltaic cell with a semiconductor –
              redox electrolyte junction.

present about the real performance of such systems under working
conditions.

Storage Cells

     While regenerative photoelectrochemical cells compete with
solid state photovoltaic devices it appears most attractive to use
such systems immediately for the storage of chemical energy. It was
already been shown in Fig. I.15, how this can be performed. Fig.
IV.10 explains in somewhat more detail how such a system works.

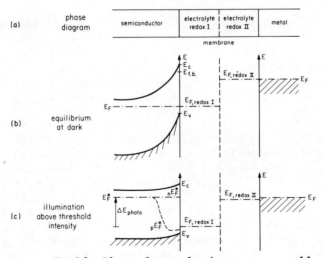

Figure IV.10   Photoelectrolytic storage cell.

This cell can only operate in forward direction if the photovoltage gained in the semiconductor electrode exceeds the open cell voltage. Since the photovoltage is limited by the band gap, a semiconductor used for such a purpose must have a wide enough energy gap. Because the photovoltage depends on illumination intensity, there must also be a critical light intensity which must be surpassed in order to charge this kind of battery.

One can construct a kind of power characteristics of such cells in the same way as was shown for the regenerative cell in Fig. IV.4 by plotting the current voltage curves of the semiconductor and the counter electrode in the same current voltage diagram. This is shown in Fig. IV.11 in a schematic way for two different redox systems which react differently at the counter electrode, one having a very low, the other a large overvoltage. The maximal stored energy is given by the product of the cell voltage and the photocurrent at the

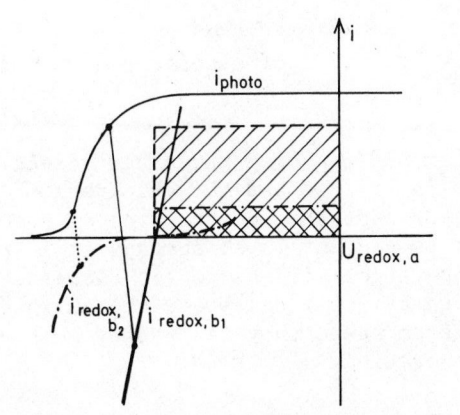

Figure IV.11   Current – voltage characteristics of a photoelectro-chemical storage cell with two redox systems (a,b) and energy storage efficiency.

working point of the cell. Fig. IV.11 illustrates how much this depends on the overvoltage of the redox reactions. The optimal energy storage efficiencies will be obtained if both redox reactions as well that at the semiconductor electrode as that at the counter electrode have very low overvoltages, what also will optimize the losses during discharge of the battery. The latter must practically be done in a different cell with metallic electrodes of high activity for redox reactions.

Fig. IV.12 shows possible electrode arrangements for a device in which redox energy can be stored during the input of sunlight by

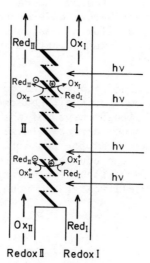

Figure IV.12   Model of a solar redox - battery.

oxidizing one system and reducing the other. Since the capacity of
redox electrolytes is limited and the thickness of the electrolyte
layer will have to be kept thin on the side of the semiconductor
electrode in order to prevent light absorption in the electrolyte,
the best mode of operation will be to let flow the redox electrolytes slowly through the cell and store the charged redox systems
in separate tanks from where they can easily be brought to the
cells for power generation.

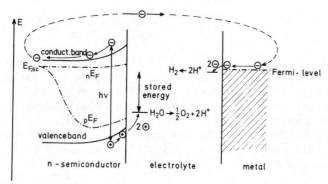

Figure IV.13   Energy scheme of galvanic cell for photoelectrolysis
               of water.

A redox battery with two soluble redox systems needs a separating membrane in order to prevent the loss of energy by mutual electron transfer between the two redox systems. Such a separator can be omitted if the products are very inert and do not interact with the other electrode or are very little soluble and therefore do not reach the other electrode. Most attractive is certainly the decomposition of water where these conditions are fulfilled. Besides this, hydrogen is one of the most versatile sources of energy. Much emphasis has therefore been put on the photodecomposition of water. Fig. IV.13 shows the energy conditions with regard to the width of the band gap and the position of the band edges which are necessary for the performance of the two redox reactions of water. Since energy losses by entropy production are unavoidable, the band gap of the semiconductor must exceed the decomposition voltage of water by some tenth of a volt.

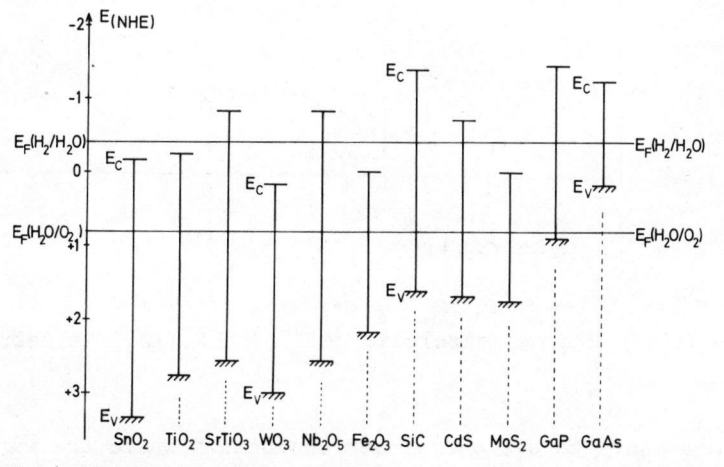

Figure IV.14   Position of band edges of various semiconductors at pH=7 and the fermi energies of water decomposition.

Besides this, the position of the valence band must be below the oxidation potential, the position of the conduction band above the reduction potential of water. Fig. IV.14 gives a survey about the band edge positions of a number of semiconductors in aqueous solution which have a wide enough gap for this purpose. One sees that very few of the semiconductors in this figure however fulfill the other condition. Those which do it like strontium titanate or niobium oxide have a very wide band gap and absorb therefore very little of the sunlight. Cadmium sulfide would have a suitable band edge position. However, it decomposes under illumination. The situation is even worsened by the fact that the redox reactions of the water decomposition are both complex and have to pass intermediates, the formation of which needs a high activation energy. This can only

be overcome by a high catalytic activity of the surface. Unfortunately however, semiconductor surfaces seem to be very poor catalysts for most complex redox reactions. Therefore, the minimal band
gap which will be needed to drive the water decomposition reaction
even if the band edge position is close to optimal, has probably to
exceed 2.2 eV. This is particularly so, if the redox reaction at the
semiconductor is the water oxidation. Since semiconductors with a
wider band gap are normally n-type, it is difficult to see how photodecomposition of water in such cells can be made economically.
Fig. IV.15 shows the situation for strontium titanate where the water photodecomposition can be obtained, however, with a very low
efficiency because the band gap is too wide.

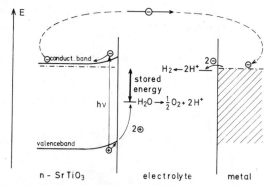

Figure IV.15 Photoelectrolytic cell with SrTiO₃ - electrode.

If one wants to use p-type semiconductors where the water reduction can be obtained by illumination, one suffers from the small
band gaps and has to apply an auxiliary external voltage. Such systems could be used for a photoassisted electrolysis where only part
of the power needed comes from the direct light absorption in the
semiconductor electrode. The additional voltage could be taken from
a photovoltaic cell coupled in series to the electrolysis cell. A
p-type tungsten selenide electrode appears suitable for this purpose
because its conduction band is located somewhat above the hydrogen
potential in acidic solution.

Fig. IV.16 shows photocurrent voltage curves for hydrogen evolution at such electrodes. One sees that the hydrogen evolution even
under illumination begins at these electrodes just at the reversible
hydrogen potential and not at much more anodic potentials where
according to the theoretical expectations a depletion layer should
begin to be formed. The figure shows a theoretical curve for such

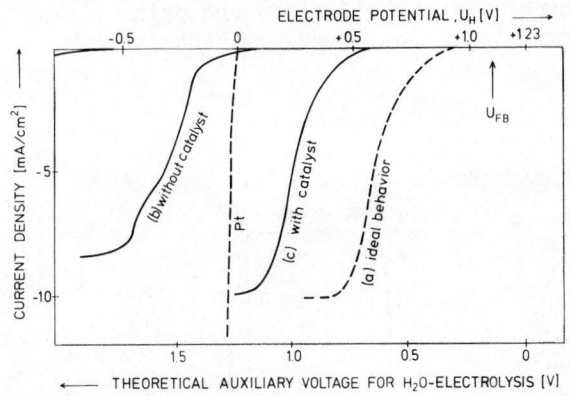

Figure IV.16   Photocurrent voltage curves for H_2-evolution of p-type
WSe$_2$ in 1M H_2SO_4.

ideal behavior. The reason for this very large overvoltage is the
low catalytic activity of the surface for the formation of hydrogen
molecules. The intermediate, the hydrogen atom, is only weakly adsorbed on the surface and therefore its formation needs a large
excess of energy. If one adds to the surface a small amount of a
catalyst, for instance small platinum particles, the photocurrent
begins at much more anodic potentials and this means a gain of energy compared to normal electrolysis. This gain is in this particular
example insufficient for practical use. However, this experiment demonstrates that such kind of photoassisted decomposition of water is
in principle possible. Fig. IV.17 illustrates how the catalytic ac-

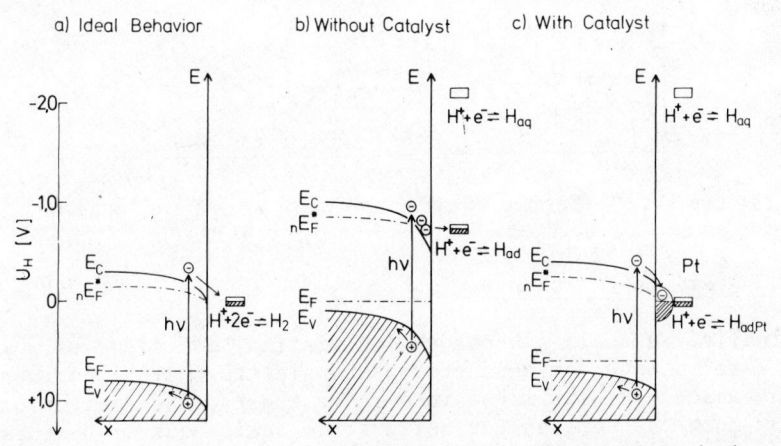

Figure IV.17   Energy relations for photoelectrolytic hydrogen
evolution at p-type semiconductor electrode.

tion of the noble metal on the surface can be explained in terms of energy correlations. Fig. IV.18 gives the principle of photoassisted water decomposition with p-type semiconductors.

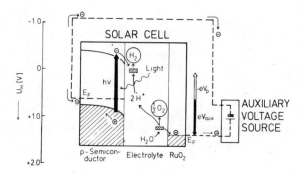

Figure IV.18   Principle of a cell for photoassisted water decomposition.

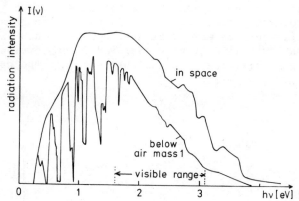

Figure IV.19   Energy distribution of solar radiation.

Energy Conversion Efficiency

Finally, we shall discuss the theoretical efficiencies which can be reached with semiconductor-based electrochemical devices for the conversion of solar light into redox energy. The limitations are very much the same as for solid state semiconductor devices and the derivation of the theoretical conversion efficiencies can be done in a fully analogous way.

Fig. IV.19 shows the spectrum of sunlight in space and on the surface of the earth after having passed one atmosphere. The systems we are discussing are threshold absorbers, what means they absorb only that part of the sun spectrum which has a higher energy than their threshold for absorption. In a semiconductor the threshold is given by the band gap. Light quanta with higher energy generate electron hole pairs with excess translational energy which however is lost very quickly in form of heat. Therefore, the sunlight with quantum energy above the threshold contributes to the energy conversion only the energy of the band gap.

Besides these two kinds of losses in the conversion process there is a considerable entropy production which reduces the free energy left over for useful work. The difference between the free energy and the enthalpy of the system is indicated by the distance between the quasi-Fermi levels in the semiconductor and the band edges (cf Fig. III.1). A further energy loss is suffered by the band bending which provides the driving force for charge separation. As a result the available free energy for electrolysis which is given by the difference between the position of the Fermi level in the bulk and the quasi-Fermi level of the minority carriers at the surface is considerably lower than the band gap energy. This difference decreases somewhat with increasing illumination intensity. If we assume that this loss can be expressed by a constant, we can calculate the conmersion efficiency from the following equation:

$$(1) \quad \eta \approx \frac{(E_{gap}-C) \int_{E_{gap}}^{\infty} I(\nu) \dfrac{dE_\nu}{E_\nu}}{\int_{0}^{\infty} I(\nu)\, dE_\nu}$$

The denominator and the numerator of the above equation are delineated in Fig. IV.20. One sees that the energy of the band gap controls the available conversion efficiency critically. For very small

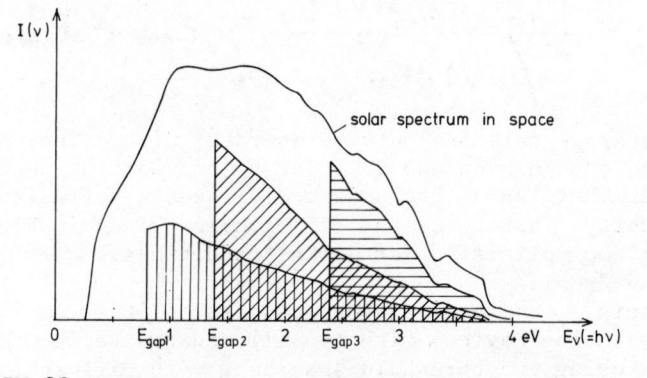

Figure IV.20   Solar energy conversion by an ideal threshold converter.

gaps the losses by the deterioration of the energetic light quanta
are dominating, at large gaps the losses by the limited light absorption. There is obviously an optimum in the energy range between
1 and 2 eV.

Assuming that the losses by entropy production, band bending
and overvoltages in the redox reactions amount to a value of 0.6 eV
one obtains from the efficiency equation above together with the solar spectrum the curve of Fig. IV.21 for the conversion efficiency
in dependence on the band gap.

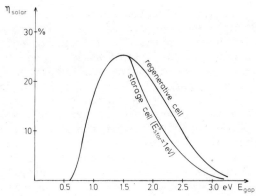

Figure IV.21   Solar energy conversion efficiencies of regenerative cells and storage cells.

Fig. IV.21 also contains the efficiency for storage in a solar
redox battery with an open cell voltage of 1 eV. The latter efficiency can be obtained by replacing in equ. (1) the factor $(E_{gap} - C)$
by the energy of the storage cell $E^o_{storage}$ as shown in the
following equation:

$$(2)\quad \eta_{storage} \approx \frac{E^o_{storage} \int_{E_{gap}}^{\infty} I(v)\, \dfrac{dE_v}{E_v}}{\int_0^{\infty} I(v)\, dE_v} \quad ; \quad E_{gap} > E^o_{storage} + C$$

Clearly, a storage cell can only be operated with a semiconductor
having a wide enough band gap in order to provide the necessary
photovoltage. Charging of the system will need a sufficient illumination intensity. Therefore, the efficiencies given in Fig. IV.21
are probably too optimistic for normal conditions.

In principle, there exist various possibilities to improve the
efficiency by using devices with two illuminated semiconductor electrodes or by using two threshold absorbers with different band gaps
in series. Such attempts for application in electrochemical cells
will be discussed in the series of lectures given by A. Nozik.

In summarizing, one can say that the conversion efficiencies of electrochemical solar cells with semiconductor electrodes are very similar to those for solid state devices. Additional problems arise by the possibility that the electron transfer reactions at the interfaces can be slow. This disadvantage may however be compensated by the larger flexibility in the adjustment of the redox potentials of the electrolytes to the properties of the semiconductors and by the very simple formation of the heterojunction at which the unfavorable effects of interfacial electronic states are less pronounced. The most serious problem of such cells remains the photodecomposition which has to be overcome before such devices can reach practical importance.

References

1. A. J. Bard, in: "Electrode Processes 1979" (Eds. S. Bruckenstein, J. D. E. McIntyre, B. Miller, E. Yeager), Proceedings 80-3, The Electrochemical Society: Princeton 1980, p. 136.

2. A. Fujishima and K. Honda, Nature 238, 37 (1972).

3. A. Fujishima, K. Kohayakawa and K. Honda, J. Electrochem. Soc. 122, 1487 (1975).

4. H. Gerischer, J. Electroanal. Chem. 58, 263 (1975).

5. J. Gobrecht and H. Gerischer, Solar Energy Materials 2, 131 (1979).

6. A. K. Gosh and H. P. Maruska, J. Electrochem. Soc. 124, 1516 (1977).

7. W. Kautek, H. Gerischer and H. Tributsch, Ber. Bunsenges. Phys. Chem. 83, 1000 (1979).

8. H. H. Kung, H. S. Jarrett, A. W. Sleigh and A. Ferretti, J. Appl. Phys. 48, 2463 (1977).

9. J. Manassen, D. Cahen, G. Hodes and A. Sofar, Nature 263, 97 (1976).

10. R. Memming, Electrochim. Acta 25, 77 (1980).

11. A. J. Nozik, Am. Rev. Phys. Chem. 29, 189 (1978).

12. B. A. Parkinson, A. Heller and B. Miller, Appl. Phys. Lett. 33, 521 (1978); J. Electrochem. Soc. 126, 954 (1979).

13. H. Tributsch, Ber. Bunsenges. Phys. Chem. 81, 369 (1977); Ber. Bunsenges. Phys. Chem. 82, 169 (1978).

14. H. Tributsch, J. Electrochem. Soc. 125, 1086 (1978).

15. H. Tributsch, H. Gerischer, C. Clemen and E. Bucher, Ber. Bunsenges. Phys. Chem. 83, 655 (1979).

16. M. S. Wrighton, A. B. Ellis and S. W. Kaiser, Adv. Chem. Ser. 163, 71 (1977).

17. and loc. cit. I, 5, 7, III, 6.

PHOTOELECTROCHEMICAL DEVICES FOR SOLAR ENERGY CONVERSION

A. J. Nozik

Solar Energy Research Institute
1617 Cole Blvd.
Golden, CO 80401

INTRODUCTION

A great deal of interest has developed in recent years in the field of photoelectrochemistry based on photoactive semiconductor electrodes, and especially in the application of these systems to solar energy conversion (1-9). A general classification scheme for the various types of photoelectrochemical cells is presented in Figure 1. The division is first into (a) cells wherein the free energy change in the electrolyte is zero (these are labelled electrochemical photovoltaic cells), and (b) cells wherein the free energy change in the electrolyte is non-zero (these are labelled photoelectrosynthetic cells). In the former cell, only one effective redox couple is present in the electrolyte and the oxidation and reduction reactions at the anode and cathode are inverse to each other. The net photoeffect is the circulation of charge external to the cell, producing an external photovoltage and photocurrent (a liquid junction solar cell); no chemical change occurs in the electrolyte.

In the photoelectrosynthetic cell, two effective redox couples are present in the electrolyte, and a net chemical change occurs upon illumination. If the free energy change of the net electrolyte reaction is positive, optical energy is converted into chemical energy and the process is labelled photoelectrolysis. On the other hand, if the net electrolyte reaction has a negative free energy change, optical energy provides the activation energy for the reaction, and the process is labelled photocatalysis. All three types of photoelectrochemical cells will be discussed here. In Section II a general review of the principles of photoelectrochemical cells will be presented, while in Section III recent developments in

263

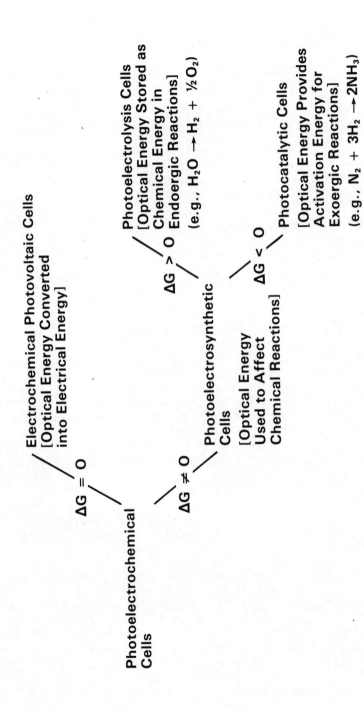

Fig. 1 Classification of photoelectrochemical cells.

various aspects of photoelectrochemical devices will be discussed.

GENERAL DISCUSSION OF PHOTOELECTROCHEMICAL DEVICES

Semiconductor Electrolyte Junctions - Conventional Picture

All phenomena associated with photoelectrochemical systems are based on the formation of a semiconductor-electrolyte junction when an appropriate semiconductor is immersed in an appropriate electrolyte. The junction is characterized by the presence of a space charge layer in the semiconductor adjacent to the interface with the electrolyte. A space charge layer generally develops in a semiconductor upon contact and equilibration with a second phase whenever the initial chemical potential of electrons is different for the two phases. For semiconducotrs, the chemical potential of electrons is given by the Fermi level in the semiconductor. For liquid electrolytes, the chemical potential is determined by the redox potential of the redox couples present in the electrolyte; these redox potentials are also identified with the Fermi level of the electrolyte.

If the initial Fermi level in an n-type semiconductor is above the initial Fermi level in the electrolyte (or any second phase), then equilibration of the two Fermi levels (or chemical potentials) occurs by transfer of electrons from the semiconductor to the electrolyte. This produces a positive space charge layer in the semiconductor (also called a depletion layer since the region is depleted of majority charge carriers). As a result, the conduction and valence band edges are bent upwards, establishing a potential barrier against further electron transfer into the electrolyte (see Figure 2). An inverse but analogous situation occurs with p-type semiconductors having an initial Fermi level below that of the electrolyte.

A charged layer also exists in the electrolyte adjacent to the interface with the solid electrode-the well known Helmholtz layer. This layer consists of charged ions from the electrolyte adsorbed on the solid electrode surface. The width of the Helmholtz layer is generally of the order of a few angstroms. The potential drop across the Helmholtz layer depends upon the specific ionic equilibrium obtaining at the surface.

A very important consequence of the presence of the Helmholtz layer for semiconductor electrodes is that it markedly affects the band bending that develops in the semiconductor when it equilibrates with the electrolyte. Without the Helmholtz layer, the band bending would simply be expected to equal the difference in initial Fermi levels between the two phases (i.e., the difference between

All photoelectrochemical cells are based on the semiconductor-
electrolyte junction; the energy level diagram for this junction
is shown below:

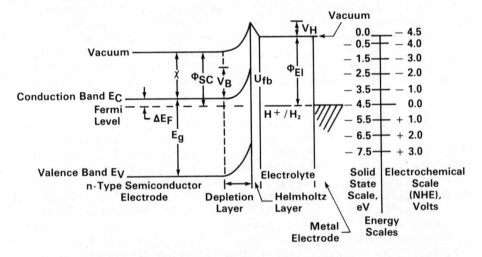

Relationships are shown between the electrolyte redox couple (H^+/H_2), the Helmholtz
layer potential drop (V_H), the semiconductor band gap (E_g), electron affinity (χ), work
function (Φ_{sc}), band bending (V_B), and flat-band potential (U_{fb}). The electrochemical and
solid state energy scales are shown for comparison. Φ_{El} is the electrolyte work function.

Fig. 2   Energy level diagram and energy scales for semiconductor-
electrolyte junctions.

their respective work functions). However, the potential drop across the Helmholtz layer modifies the net band bending as shown in Figure 2.

In Figure 2, the energy scales commonly used in solid state physics and in electrochemistry are shown for comparison. In the former, the zero reference point is vacuum, while in the latter it is the standard redox potential of the hydrogen ion-hydrogen (H^+/H_2) redox couple. It has been shown (10,11) that the effective work function or Fermi level for the standard (H^+/H_2) redox couple at equilibrium is -4.5 eV with respect to vacuum. Hence, by using this scale factor, the energy levels corresponding to any given redox couple can be related to the energy levels of the valence and conduction bands of the semiconductor electrode.

To make the connection between the energy levels of the electrolyte and the semiconductor it is necessary to introduce the flat-band potential, U_{fb}, as a critical parameter characterizing the semiconductor electrode. The flat-band potential is the electrode potential at which the semiconductor bands are flat (zero space charge in the semiconductor); it is measured with respect to a reference electrode, usually either the standard normal H^+/H_2 redox potential (n.h.e.) or the standard calomel electrode (s.c.e.). Hence, the band bending is given by

$$V_b = U_e - U_{fb} \tag{1}$$

where U_e is the electrode potential (Fermi level) of the semiconductor. At equilibrium in the dark, U_e is identical with the potential of the redox couple in the electrolyte.

The effect of the Helmholtz layer on the semiconductor band bending is contained within the flat-band potential. This important parameter is a property both of the semiconductor bulk and the electrolyte, as seen from the following relation:

$$U_{fb}(n.h.e.) = (\chi + \Delta E_F + V_H) - 4.5 = (\phi_{sc} + V_H) - 4.5 \tag{2}$$

where χ is the electron affinity of the semiconductor, ϕ_{sc} is the work function of the semiconductor, ΔE_F is the difference between the Fermi level and majority carrier band edge of the semiconductor, V_H is the potential drop across the Helmholtz layer, and 4.5 is the scale factor relating the H^+/H_2 redox level to vacuum.

The most accurate compilation of experimental flat-band potentials for many semiconductors, measured for both n- and p-type materials at several pH values and accounting for frequency dependent effects, has been presented by Gomes & Cardon (12); the U_{fb} values

are reported to be accurate within 0.05-0.1 V. From the flat-band
potential, and a knowledge of the carrier density, the effective
mass, and the band gap of the semiconductor, the conduction and
valence band edges can be determined with respect to the standard
redox scale. These data are presented in Figure 3 for several semiconductors at pH = 1.0.

Photo-Induced Charge Transfer Reactions

When the semiconductor-electrolyte junction is illuminated with
light, photons having energies greater than the semiconductor band
gap are absorbed and create electron-hole pairs in the semiconductor. Photons absorbed in the depletion layer produce electron-hole
pairs that separate under the influence of the electric field present in the space charge region. Electron-hole pairs produced by
absorption of photons beyond the depletion layer will separate if
the minority carriers can diffuse to the depletion layer before recombination with the majority carriers occurs.

The photoproduction and subsequent separation of electron-hole
pairs in the depletion layer cause the Fermi level in the semiconductor to return toward its original position before the semiconductor-electrolyte junction was established, i.e., under illumination
the semiconductor potential is driven toward its flat-band potential.
Under open circuit conditions between an illuminated semiconductor
electrode and a metal counter electrode, the photovoltage produced
between the electrodes is equal to the difference between the Fermi
level in the semiconductor and the redox potential of the electrolyte. Under close circuit conditions, the Fermi level in the system
is equalized and no photovoltage exists between the two electrodes.
However, a net charge flow does exist. Photogenerated minority
carriers in the semiconductor are swept to the surface where they
are subsequently injected into the electrolyte to drive a redox
reaction. For n-type semiconductors, minority holes are injected
to produce an anodic oxidation reaction, while for p-type semiconductors, minority electrons are injected to produce a cathodic reduction reaction. The photo-generated majority carriers in both
cases are swept toward the semiconductor bulk, where they subsequently leave the semiconductor via an ohmic contact, traverse an external circuit to the counter electrode, and are then injected at the
counter electrode to drive a redox reaction inverse to that occurring at the semiconductor electrode.

Semiconductor Electrode Stability

The photo-generated holes and electrons in semiconductor electrodes are generally characterized by strong oxidizing and reducing
potentials, respectively. Instead of being injected into the elec-

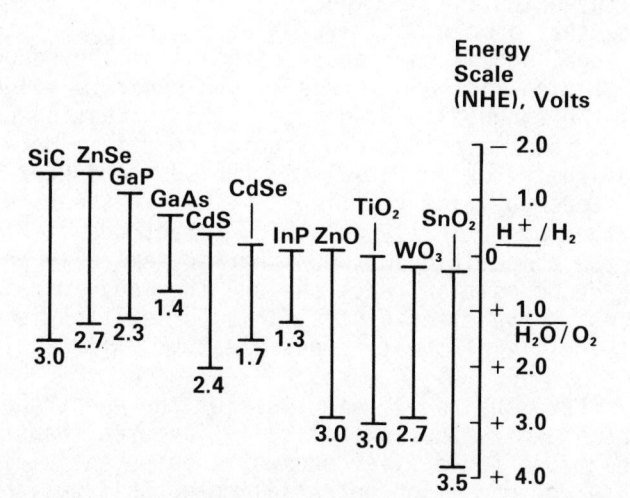

Flat-band potential locates the semiconductor bands with respect to the redox couples of the electrolyte. These relationships are shown for several semiconductors. Those with the valence band above H^+/H_2 operate with zero external bias. For electrochemical photovoltaic cells, the flat-band potential determines the open circuit voltage.

Fig. 3 Positions of conduction and valence band edges of several semiconductors relative to the redox potentials of H^+/H_2 and O_2/H_2O.

trolyte to drive redox reactions, these holes and electrons may
oxidize or reduce the semiconductor itself, and cause decomposition.
This possibility is a serious problem for practical photoelectrochemical devices, since photodecomposition of the electrode leads
to inoperability or to short electrode lifetimes.

A simple model of electrode stability has been presented by
Gerischer (13) and by Bard & Wrighton (14). The redox potential of
the oxidative and reductive decomposition reactions are calculated
and put on an energy level diagram like Figure 2. The relative positions of the decomposition reactions are compared with those of
the semiconductor valence and conduction band edges. Absolute
thermodynamic stability of the electrode is assured if the redox
potential of the oxidative decomposition reaction of the semiconductor lies below (has a more positive value on the s.c.e. scale) the
valence band edge, and if the redox potential of the reduction decomposition reaction lies above (has a more negative value on the
s.c.e. scale) the conduction band edge. This situation does not
exist in any of the semiconductors studied to date. More typically,
one or both of the redox potentials of the semiconductor oxidative
and reductive decomposition reactions lie within the band gap, and
hence become thermodynamically possible. Electrode stability then
depends upon the competition between thermodynamically possible
semiconductor decomposition reactions and thermodynamically possible
redox reactions in the electrolyte. This competition is governed by
the relative kinetics of the two possible types of reactions.

In cases where the redox potentials of the electrode decomposition reactions are more thermodynamically favoured than the electrolyte redox reactions (oxidative decomposition potential more negative, reductive decomposition potential more positive, than the
corresponding electrolyte redox reactions), the products of the
electrolyte redox reactions have sufficient potential to drive the
electrode decomposition reactions. Hence, this situation usually
results in electrode instability, assuming that the electrode decomposition reaction is not kinetically inhibited. This is the
case with ZnO, Cu_2O, and CdS in simple aqueous electrolytes, and
these semiconductors are indeed unstable under these conditions.

It appears that the more thermodynamically favoured redox reactions also become kinetically favoured, so that these reactions predominate. This effect has been used to stabilize semiconductor electrodes by establishing a redox couple in the electrolyte with a redox potential more negative than the oxidative decomposition potential (or more positive than the reductive decomposition potential),
such that this electrolyte redox reaction occurs preferentially compared to the decomposition reaction, and scavenges the photo-generated minority carriers. However, this stabilization technique can
only be used for electrochemical photovoltaic cells, and it is discussed later in further detail.

Electrochemical Photovoltaic Cells

An energy level diagram for electrochemical photovoltaic cells is shown in Figure 4. One predominant redox couple, represented as C+/C, is present in the electrolyte such that C is oxidized to C+ at the anode and C+ is reduced to C at the cathode. The usual cell configuration comprises one semiconductor electrode and one metal electrode; the semiconductor electrode may be n- or p-type although in practice it is usually always n-type. Hence, in this case, photo-generated holes drive an oxidation reaction at the semiconductor surface while photo-generated electrons are driven into the bulk of the semiconductor, exit at a rear ohmic contact, circulate externally through a load, and then are injected at the metal cathode to produce the reduction reaction. No net chemical change is produced in the electrolyte, and electrical power is delivered to the external load through the photovoltage and photocurrent generated by the semiconductor electrode.

The maximum possible open-circuit photovoltage is equal to the band bending, which is the difference between the flat-band potential of the semiconductor and the redox potential of the predominant redox couple in the electrolyte. The photocurrent depends upon the semiconductor band gap and the quantum efficiency. The relation between the maximum short circuit photocurrent and the band gap has been well established for sunlight (15).

The factors determining the maximum theoretical efficiency for electrochemical photovoltaic cells are completely analogous to those for solid state photovoltaic cells. It is necessary to optimize the product of the external photovoltage and photocurrent, and this is achieved with a semiconductor band gap of about 1.3-1.4 eV. The general principles of the operation of electrochemical photovoltaic cells were presented by Gerischer (10).

For p-n photovoltaic devices, the optimum band gap yields upper limit conversion efficiencies of about 25%. For solid state Schottky photovoltaic cells, the calculated maximum efficiencies are lower for the same semiconductor materials; this is because the potential barrier heights (band bending) at the semiconductor-metal junctions are low compared with the semiconductor band gap. The upper limit efficiency for solid state Schottky cells is about 10-12%.

In this regard, electrochemical photovoltaic cells may have a major potential advantage, because in well-behaved electrodes the band bending may be increased by adjusting the redox potential of the electrolyte. For n-type electrodes a redox couple with a more positive redox potential could be introduced into the electrolyte, while for p-type electrodes more negative redox potentials are required.

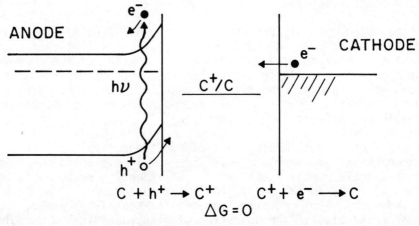

Fig. 4   Energy diagram for electrochemical photovoltaic cells.

Other important advantages of liquid junction photovoltaic cells compared with solid state cells are the ease with which the liquid junction potential barrier can be established (the semicon-cutor electrode is simply immersed in the electrolyte), and the fact that polycrystalline semiconductor films can apparently be used without a drastic decrease in efficiency, compared with single cry-stal semiconductor electrodes. The latter effect is probably due to the intimate and perfect contact of the liquid electrolyte with the crystallite grains, in contrast to the imperfect contact bet-ween two solid, polycrystalline phases. The maximization of the band bending by adjustment of the redox potential of the electro-lyte is limited by considerations of electrode stability. If the redox potential of the redox couple in the electrolyte is made too positive with n-type electrodes, or too negative with p-type elec-trodes, then the electrode decomposition reaction may become more thermodynamically and/or kinetically favoured. On the other hand, electrodes that are unstable in a given electrolyte may be stabil-ized by adding to the electrolyte a redox couple having a redox potential more thermodynamically and/or kinetically favoured com-pared with the decomposition potential. Thus, electrochemical photovoltaic cells have much greater flexibility with respect to possible solutions to the electrode stability problem than photo-electrolysis cells. For the latter, the redox couples in the elec-trolyte are fixed by the desired net chemical reaction in the cell, and their redox potentials cannot be adjusted with respect to the potentials of the electrode decomposition reactions. For the former, the redox potentials of the electrolyte redox couples may be adjust-ed to stabilize the electrode within the constraint of maintaining sufficient band bending, and hence open circuit voltage.

A discussion of recent research directions and results related to the problem of stabilizing semiconductor electrodes against photoelectrochemical corrosion is presented in Section III.

Photoelectrosynthetic Cells

Photoelectrolysis Cells. In photoelectrolysis cells, the anodic reaction has a more positive redox potential than the catho-dic reaction so that the overall cell reaction has a positive free energy change.

Two types of photoelectrolysis cells can be distinguished. In the first type, one electrode is a semiconductor and the second a metal. In the second type, one electrode is an n-type semiconduc-tor and the second a p-type semiconductor. The energetics of the former cell is represented in Figure 5 for the photoelectrolysis of water into H_2 and O_2 by using n-type electrodes (analogous analy-ses apply to p-type electrodes). Since there are two redox couples in the electrolyte, the initial Fermi level in the electrolyte can

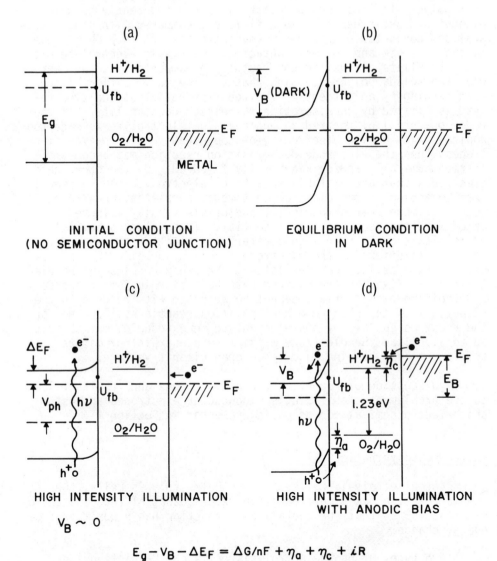

$$E_g - V_B - \Delta E_F = \Delta G/nF + \eta_a + \eta_c + iR$$

Fig. 5 Energy diagrams for semiconductor-metal photoelectrosyn-
        thetic cells.

be anywhere between them depending upon the initial relative concentrations of H_2 and O_2 in the cell. In Figure 5 the initial Fermi level in the electrolyte is arbitrarily drawn just above the O_2-H_2O redox level. Upon equilibration in the dark (Figure 5b) the Fermi level in the semiconductor equilibrates with the electrolyte Fermi level producing a band bending, V_b in accordance with equation (1).

Under illumination (Figure 5c), the Fermi level in the semiconductor rises toward U_{fb} producing a photovoltage, V_{ph}. This voltage can be measured between the two electrodes with an external potentiometer. However, the value of V_{ph} depends upon the initial metal electrode potential, which depends upon the initial relative concentrations of H_2 and O_2: V_{ph} can vary from zero to some finite value. Except under very special circumstances (initial metal electrode potential and the valence band edge of the semiconductor electrode both at the O_2/H_2 redox potential), V_{ph} is not the potential energy available for photoelectrolysis.

With the two electrodes shortened together, the maximum Fermi level possible in the cell is the flat-band potential. In Figure 5, U_{fb} is below the H^+/H_2 potential. Hence, even with illumination intensity sufficient to completely flatten the semiconductor bands, H_2 could not be evolved at the counter electrode because the Fermi level is below the H^+/H_2 potential. In order to raise the Fermi level in the metal counter electrode above the H^+/H_2 potential an external anodic bias, E_b, must be applied, as shown in Figure 5d. This bias also provides the overvoltage at the metal cathode, η_c, required to sustain the current flow, and it increases the band bending in the semiconductor to maintain the required charge separation rate.

The situation depicted in Figure 5 is the one that describes most of the n-type semiconductors studied to date. For these semiconductors, an external bias is required to generate H_2 and O_2 in a semiconductor-metal photoelectrolysis cell; the further U_{fb} lies below H^+/H_2, the greater the bias. The bias can be applied either by an external voltage source or by immersing the anode in base and the cathode in acid (the two compartments being separated by a membrane).

Several semiconductors such as $SrTiO_3$, Nb_2O_5 and ZrO_2 have U_{fb} above the H^+/H_2 potential: therefore, no external bias is required to generate H_2 and O_2 in a Schottky type cell. This has been confirmed for $SrTiO_3$ and $KTaO_3$. Unfortunately, these oxides all have large band gaps (3.4-3.5 eV), which result in essentially zero solar absorptivity, and hence, they are ineffective in systems for solar energy conversion.

For purposes of discussion of Figure 5, the difference between the O_2/H_2O redox potential and the valence band edge at the interface is defined as the intrinsic overpotential of the semiconductor anode, η_a. This overpotential is not the usual overpotential or overvoltage of conventional electrochemistry since it is current independent and is determined only by the band gap, the flat-band potential, and the redox potential of the electrolyte donor state.

A simple identity can be derived from the energy diagram of Figure 5d and it is given at the bottom of Figure 5. This identity can be considered to represent an energy balance describing the distribution of the energy produced by the absorption of a photon by the semiconductor. In the absence of a bias, the input energy of an absorbed photon is equal to the semiconductor band gap. Energy loss terms in the semiconductor resulting from the movement of electrons and holes from their point of creation to the respective electrolyte interfaces are the band bending (V_b), and the difference between the conduction band edge in the bulk and the Fermi level (ΔE_F). The net electron-hole pair potential available at the electrode interfaces is thus $E_g - V_b - \Delta E_F$. In the electrolyte, a portion of this potential energy is recovered as the free energy ($\Delta G/nF$) of the net endoergic reaction in the electrolyte. The rest of the potential energy is lost through the irreversible, entropy-producing terms η_a, η_c, and iR (ohmic heating).

For the case where both electrodes are n and p-type semiconductors, the most interesting situation occurs when the two electrodes are made of different semiconductors. If the electron affinity of the n-type electrode is greater than that of the p-type electrode, then the available electron-hole potential for driving chemical reactions in the electrolyte is enhanced when both electrodes are illuminated. The energetics of this system is shown in Figure 6.

As seen from the energy balance presented at the bottom of Figure 6, the net electron-hole pair potential available at the electrode surface is the sum of the semiconductor band gaps minus the difference in their flat-band potentials and minus the Fermi level terms, ΔE_F, for each semiconductor. The sum of the band bending values produced at each semiconductor is equal to the difference between their flat-band potentials. The amount of potential enhancement is maximized if the difference in flat-band potentials is minimized. The minimum difference in flat-band potentials is determined by the minimum band bending required in each semiconductor electrode to produce efficient charge carrier separation. The total band bending present in the two electrodes is independent of light intensity. However, the distribution of the total band bending between the electrodes is dependent upon light intensity and the carrier densities of each electrode.

p-n CELL

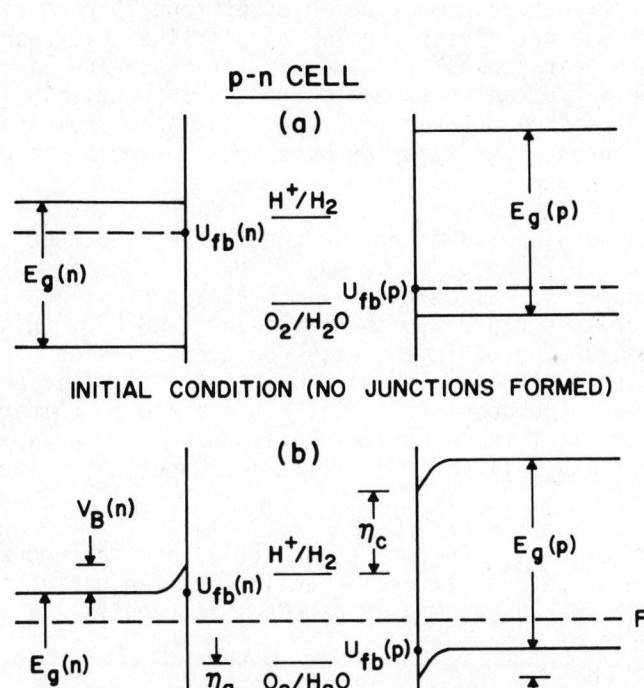

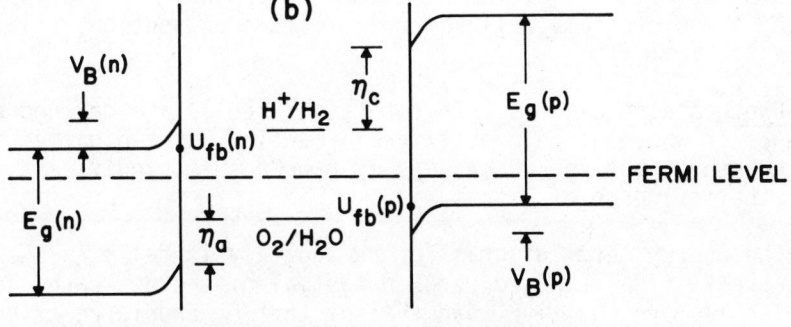

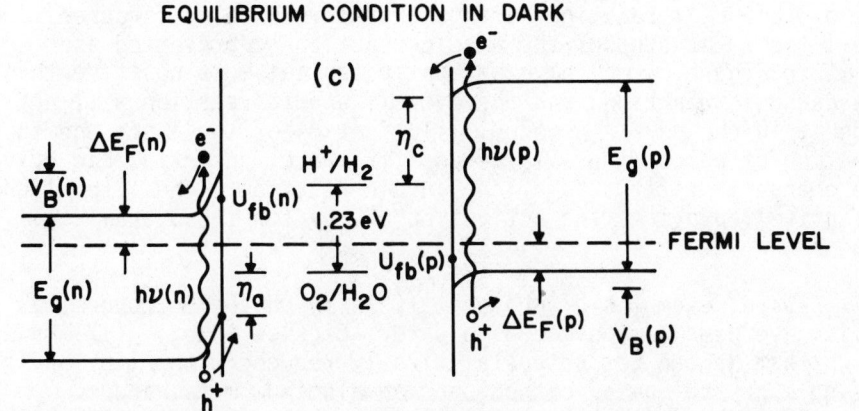

ZERO BIAS, HIGH INTENSITY ILLUMINATION, H_2 AND O_2 EVOLUTION

$$E_g(n) + E_g(p) - \left[U_{fb}(p) - U_{fb}(n) \right] + \Delta E_F(n) - \Delta E_F(p) = \Delta G/nF + \eta_a(n) + \eta_c(p) + iR$$

Fig. 6 Energy diagram for double semiconductor electrode (p/n) photoelectrosynthetic cells.

In the cell, two photons must be absorbed (one in each electrode) to produce one net electron-hole pair for the cell reaction. This electron-hole pair consists of the minority hole and minority electron from the n-type and p-type electrodes, respectively, and it has a potential energy greater than that available from the absorption of one photon. The majority electrons and holes recombine at the ohmic contacts.

An important possible advantage of the double electrode cell is that for a given cell reaction it may allow the use of smaller band gap semiconductors. Since the maximum photocurrent available from sunlight increases rapidly with decreasing band gap, this could produce higher conversion efficiencies. For the splitting of water into H_2 and O_2 with sunlight, the maximum theoretical efficiency has been estimated to be about 25% (16). This is to be compared to the value of about 16% for the maximum efficiency of water splitting by a photovoltaic cell in series with a conventional electrolysis cell.

Photocatalytic Cells. Photocatalytic cells are defined as those cells wherein the overall cell reaction has a negative free energy change, but the reaction only proceeds at significant rates under the influence of light.

The energy level diagram for photocatalytic cells is shown in Figure 7. The standard redox potential of the net cathodic reaction is more positive than that of the net anodic reaction, and the overall cell reaction is thus thermodynamically favoured. However, the first step of the anodic reaction (represented as $C^- + h^+ \rightarrow I$ in Figure 7) may have a redox potential more positive than the cathodic reaction, and the overall anodic reaction will not proceed in the dark. However, illumination of the anode and the creation of holes with a large positive redox potential can drive the energetic first step of the anodic reaction. Thus, the light effectively provides the activation energy for the overall cell reaction.

Several examples of photocatalysis in photoelectrochemical cells have been demonstrated to date. Dickson & Nozik (17) have shown that N_2 can be photocatalytically reduced with Al by using a single crystal p-GaP cathode and an aluminum metal anode. Krauetler & Bard (18), using platinized n-TiO2 powders, have shown that carboxylic acids can be oxidized to CO_2 and alkanes and that CN^- can be oxidized to OCN^-.

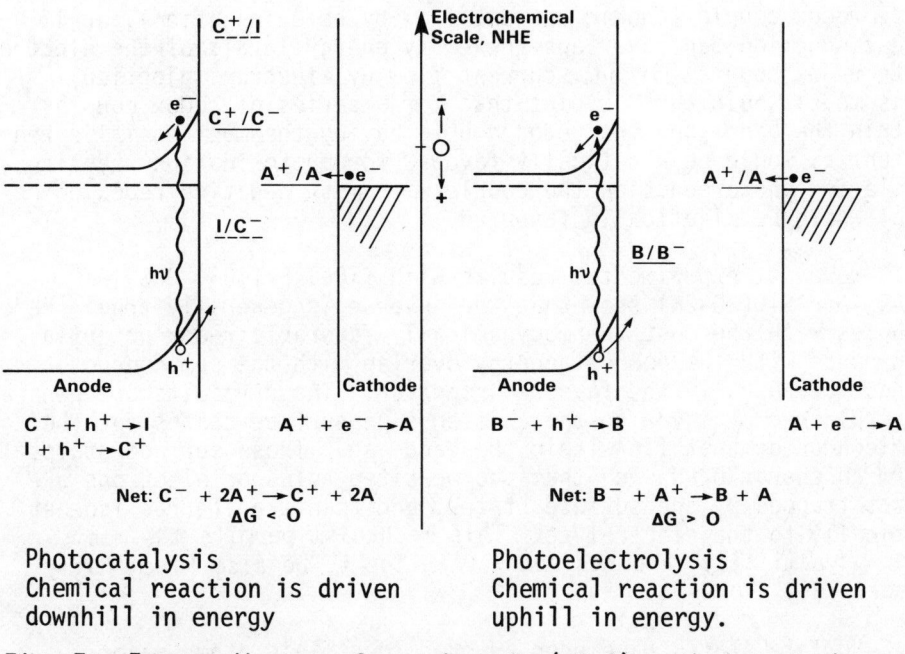

Fig. 7. Energy diagrams for endoergic ($\Delta G > 0$) and exoergic ($\Delta G < 0$) photoelectrosynthetic cells.

RECENT DEVELOPMENTS AND SPECIAL TOPICS IN PHOTOELECTROCHEMICAL
ENERGY CONVERSION

General Considerations

Effects and Importance of Surface States. In the usual model
(10, 19) of charge transfer across semiconductor-electrolyte interfaces, it is assumed that this process occurs isoenergetically.
That is, as shown in Figure 8 electrons can only be exchanged between the semiconductor and the electrolyte if the energy levels of
the valence band overlap the occupied energy levels of the electrolyte redox couple (anodic current flow by hole injection), or if
the conduction band overlaps the empty energy levels of the electrolyte redox couple (cathodic current flow by electron injection).
This model would thus predict that for a series of redox couples
within the band gap, the redox couple having the most positive redox
potential would be kinetically favored for photoxidation reactions,
while for photoreduction the couple with most negative redox potential would be kinetically favoured.

However, experimental results with TiO_2, $SrTiO_3$, CdS, GaP,
GaAs, and Si (20-26) show that the inverse is generally true. Redox
couples with the most thermodynamically favorable redox potential,
and hence with the poorest energy overlap with the semiconductor
bands, are usually the favoured reaction. These results are generally explained by invoking the existence of surface states for the
semiconductor that lie within the band gap. These surface states
form an energy band such that photoexcited holes or electrons are
first trapped by the surface states, and then transferred isoenergetically to the electrolyte. This mechanism permits the most
thermodynamically favoured redox reaction to be also kinetically
favorable.

Surface states have been invoked to explain other major features of the behavior of semiconductor electrodes in photoelectrochemical cells. One such important effect is the large shift from
the flat-band potential (producing a large band bending) generally
required before the onset of the photocurrent. These shifts have
been explained (27) by electron-hole recombination processes at surface states. In Figure 9, an experimental curve for O_2 evolution
is shown compared to the theoretical curve in the absence of surface
recombination. Much larger shifts, in the order of 0.5 to 0.7 volts,
are seen for p-type semiconductors such as p-GaP and p-GaAs.

Another important role for surface states has been proposed to explain how semiconductor band edges can become unpinned at the semiconductor-electrolyte interface. This important effect is discussed
in detail below in the next section.

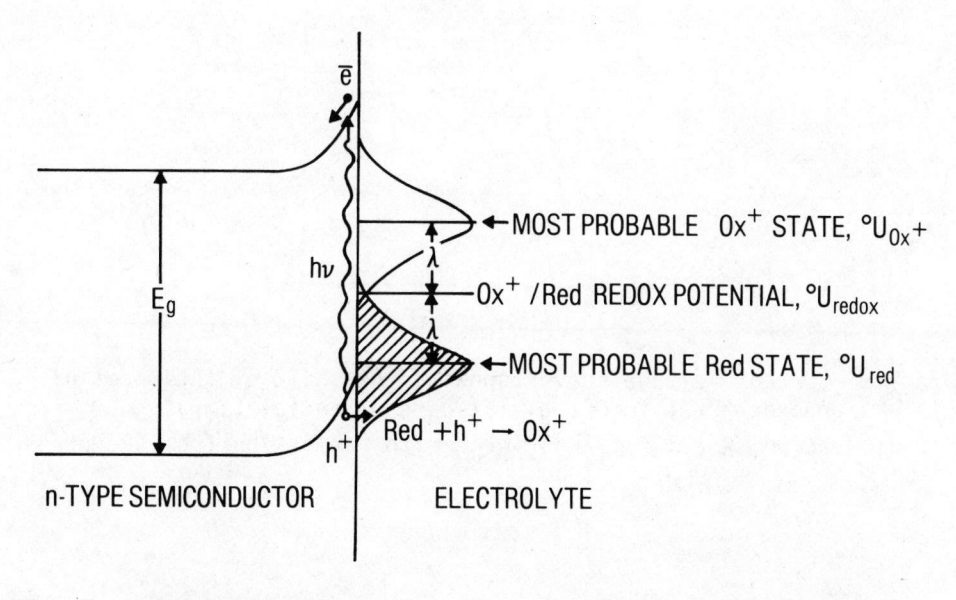

Fig. 8 Isoenergetic charge transfer of hole from n-type semicon-
        ductor to occupied state of electrolyte redox couple.
        Redox energy levels have a distribution due to solvent-
        ion interactions.

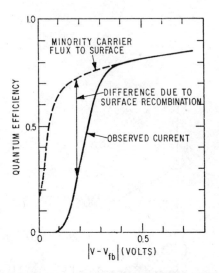

Fig. 9 Current-voltage curves showing the shift of the onset of
photocurrent from the flat-band potential (V_{fb}).
"(from Ref. 27)" -Reprinted with permission from CRC press.

Unpinned Band Edges. In the conventional model of semiconductor-electrolyte junctions (19) it is assumed that changes in applied potential appear across the semiconductor space charge layer and that the drop across the Helmholtz layer is constant. This means that the positions of the semiconductor band edges are fixed or pinned with respect to the redox energy levels in the electrolyte. This leads further to the restriction that only redox couples lying within the semiconductor band gap can undergo photoinduced redox chemistry. Photogenerated holes or electrons (in the absence of hot carrier effects) come from the valence or conduction band edges, and redox couples outside the band gap are therefore inacessable. The possibility of hot carrier effects is discussed in a later section.

However, recent experiments (28) with Si, GaAs, and MX_2 (M = Mo, W, X = S, Se, Te) indicate that the band edges can become unpinned, and that applied potentials can shift the semiconductor band positions with respect to the electrolyte redox levels. The effect here is to be able to oxidize or reduce redox couples that lie outside the band gap as determined from values of the flat-band potential obtained in the dark. Several explanations for the unpinning of the bands have been proposed.

One explanation applies to small band gap semiconductors where inversion can occur (29). When the band bending in the semiconductor is sufficiently large such that the Fermi level at the surface lies closer to the minority carrier band than to the majority carrier band, then the surface becomes inverted and a large charge density develops in the semiconductor space charge layer   (see Figure 10). This inversion makes the capacitance of the semiconductor space charge layer comparable to that of the Helmholtz layer such that potential charges applied to the electrode appear across the Helmholtz layer and effectively unpin the band edges.

Experiments showing this effect were performed with single crystal (111) p-Si electrodes with a resistivity of about 5.5 ohm cm in non-aqueous electrolytes consisting of absolute methanol containing tetramethylammonium chloride or acetonitrile containing tetraethyl ammonium perchlorate. The flat-band potentials of p-Si in the two electrolytes were determined from Mott-Schottky plots (in the dark) in the depletion range of the p-Si electrode, from open-circuit photopotential measurements, and from the values of electrode potential at which anodic photocurrent is first observed in n-type Si electrodes. These three methods all yielded consistent flat-band potential values for p-Si of +0.05V (vs SCE) ± .05V in both methanol and acetonitrile. These values, combined with the band gap of 1.12eV for p-Si, places the conduction band edge in methanol and acetonitrile at -0.85V (vs SCE). The redox couples chosen for the two electrolytes were 1,3 dimethoxy 4 nitrobenzene (E_0 = -1.0V vs SCE) for methanol, and 1 nitronaphthalene (E_0 = -1.08), 1,2 dichloro 4-nitro-

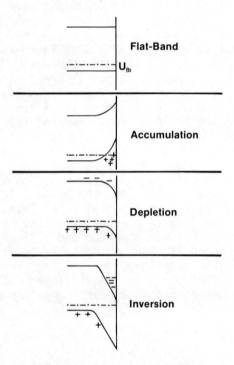

Fig. 10   Inversion in p-type semiconductors.

benzene (E_0 = -0.95), and anthraquinone (E_0 = -0.95) for acetonitrile. These redox couples lie from 0.1V to 0.24V above the conduction band edge of p-Si, and hence, in the conventional model could
not be photoreduced by p-Si.

Photocurrent-voltage data showed that all the supra-band-edge
redox species listed above are reduced upon illumination of the p-Si
electrode; no appreciable reduction was achieved in the dark. Typical results for p-Si with methanol and acetonitrile are shown in
Figure 11. Capacitance measurements as a function of electrode potential, ac signal frequency, and light intensity are shown in Figures 12 and 13.

It is instructive to compare these data with the ideal curves
obtained for a metal-insulator-semiconductor (MIS) device (30). In
the latter case, the space-charge density and the capacitance decrease in the depletion region and reach a minimum value at the onset
of the inversion region. As the inversion region develops, the
space-charge density increases rapidly. The effect of this increased charge density on the measured capacitance depends upon the frequency of the applied ac signal. At high frequency (>100 Hz for
Si/SiO_2) the electron concentration cannot follow the ac signal, and
the measured capacitance is flat and minimized in the inversion
region. However, at low frequency (<100 Hz) the electrons in the
space-charge layer can follow the ac signal and the capacitance increases rapidly with the increased degree of inversion. The effect
of illumination in these experiments is to increase the measured
capacitance in the inversion region at the high frequencies such
that the low frequency behavior is produced (30).

The capacitance data presented in Figure 12 and 13 generally
exhibit the behavior described above for MIS (metal/SiO_2/p-Si) devices. In acetonitrile (Figure 12), the capacitance decreases in
the depletion region, and in the dark remains at a flat minimum in
the inversion region at ac signal frequencies above 100 Hz. Under
illumination, the capacitance increases sharply with increased light
intensity and shows a rapid increase in the inversion region. At
frequencies below 100 Hz, the capacitance also increases in the inversion region in the dark as expected for MIS behavior. Similar
results are obtained with methanol (Figure 13) except that a sharp
decrease in capacitance is also generally observed well into the
inversion region at all frequencies. This drop is associated with
reduction of trace amounts of water in the electrolyte which leads
to large dark current flow.

Although the p-Si sample is etched in HF solution before each
run, it is also exposed to air. Hence, a thin oxide layer (10-20 Å)
exists on the surface. Therefore, the similar behavior of the capacitance of the p-Si electrode in non-aqueous electrolyte to that of

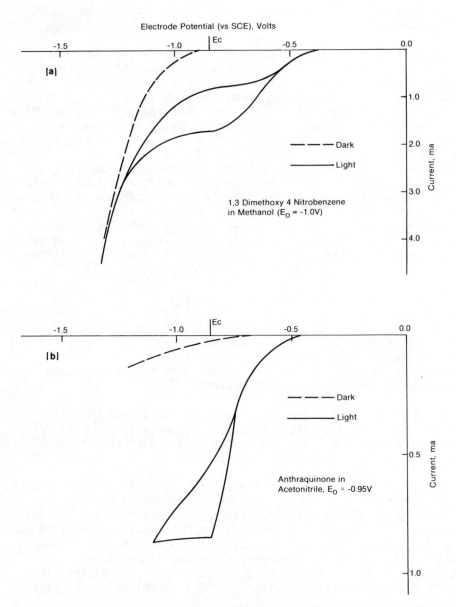

Fig. 11   Current voltage curves for p-Si with methanol and aceto-
          nitrile containing redox couples above the conduction band
          edge as determined from flat-band potential measurements.

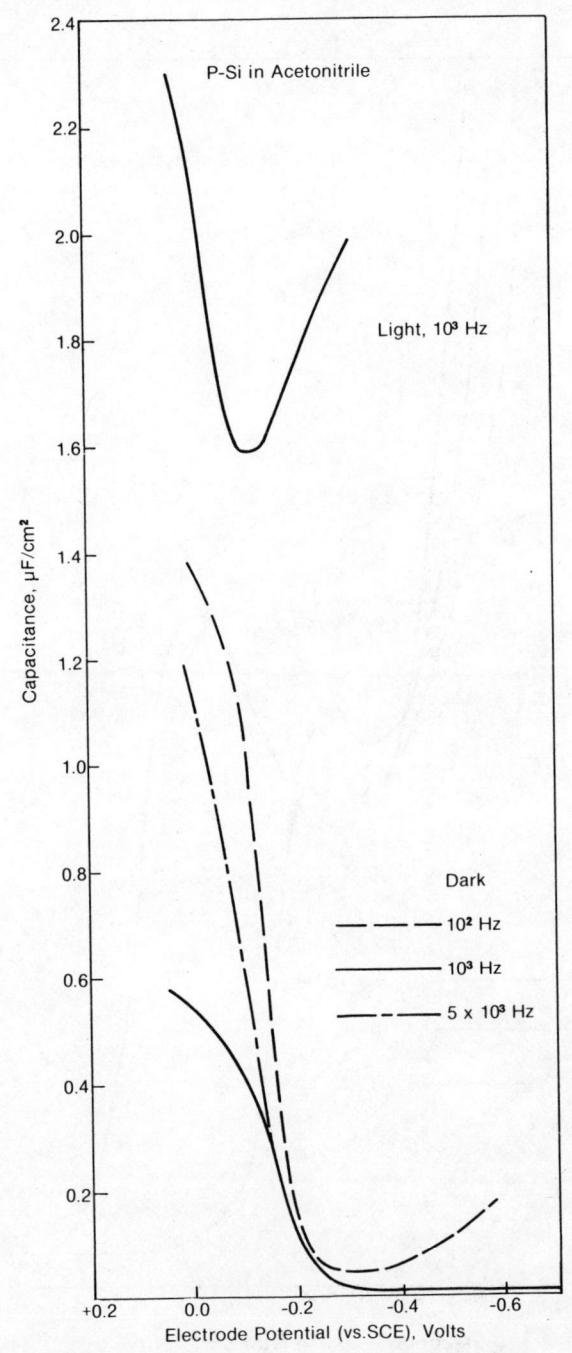

Fig. 12 Capacitance - voltage curves for p-Si in acetonitrile as
         a function of frequency and light intensity showing be-
         havior expected for MOS devices in inversion.

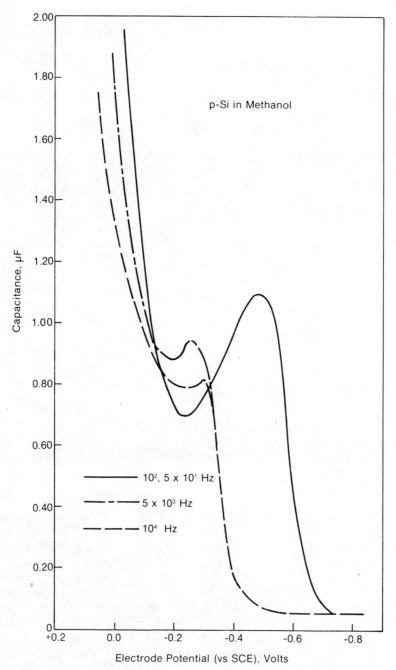

Fig. 13 Capacitance voltage curves for p-Si in methanol as a function of frequency.

an MIS device is not surprising. The effect of the oxide layer is to inhibit current flow and facilitate the maintenance of a large charge density in the semiconductor space charge layer.

As seen in Figure 12, the effect of light on the system is to further increase the capacitance in the inversion layer. This, of course, enhances the unpinning effect as the capacitance of the space-charge-layer approaches or exceeds that of the Helmholtz layer.

Another model that has been used to explain band edge unpinning is based on the concept of Fermi level pinning as seen in solid state Schottky barriers (31). In this model, it is believed that surface states pin the Fermi level and produce band bending that is independent of the redox potential of the electrolyte. Photoelectrochemical experiments with $MoSe_2$, GaAs and Si (28, 32) show that: (a) a number of A^+/A systems with widely different redox potentials produce the same photovoltage for a given semiconductor; (b) certain A^+/A systems produce photovoltages even when the redox potentials are outside the band gap; and (c) photoeffects can be found for redox couples with redox potentials that span a range greater than the semiconductor band gap. If the Fermi level is pinned by surface states, then the system is equivalent to a Schottky barrier in series contract with electrolyte, such that the band edges move with applied potential while the band bending is constant.

Fermi level pinning would produce junction effects that are completely independent of the redox species in the electrolyte since the junction would be controlled strictly by semiconductor surface and bulk properties. However, experimental results (28) show that the junction properties are only independent of the electrolyte redox species for certain critical values of the redox potential. This means that the Fermi level would not always be pinned, and this raises some questions concerning the possible mechanism for the Fermi level pinning model.

Hot Carriers. Another assumption of the conventional model (10, 19) is that the energy of injected photogenerated carriers is given by the position of the minority carrier band edge at the semiconductor-electrolyte interface. That is, carriers are accelerated to the surface by the electric field in the semiconductor space charge layer, but they lose energy (as heat) in the process via carrier - phonon collisions. As a result, the carriers are in thermal equilibrium with the lattice before injection.

In a modification of this model (32), photogenerated minority carriers which have not undergone full intraband relaxation may also be injected into the electrolyte; this process is called hot carrier injection (32) (see Figure 14). This process can occur if the thermalization time (τ_{Th}) of the photogenerated carriers in the semicon-

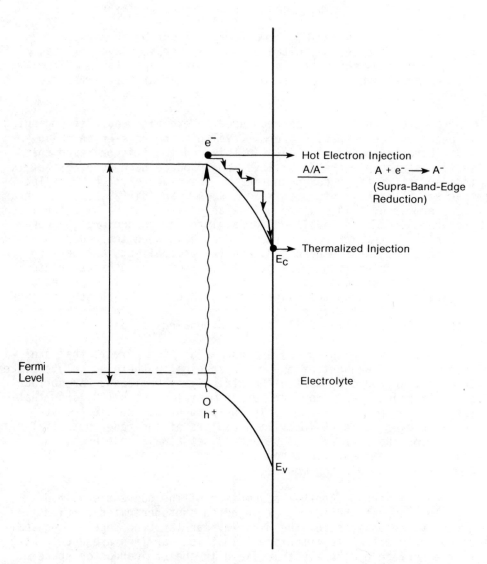

Fig. 14 Hot carrier injection process in p-type semiconductor
         showing electron injection above the conduction band
         edge at the interface.

ductor space charger layer is greater than both the charge transfer
or tunneling time of these carriers into the electrolyte and the
effective relaxation time of the injected carriers in the electrolyte. These various characteristic times are shown in Figures 15
and 16 for hot hole injection from n-type semiconductors and hot
electron injection from p-type semiconductors, respectively.

The calculations for the various characteristic times have
been estimated using classical and quantum mechanical models (33).
In one particular model, the aqueous electrolyte is treated as a
large gap semiconductor (details are given in Ref. 34) and photo-
excited carriers find themselves in the potential well created by
the position-dependent potential in the semiconductor-electrolyte
barrier. This well has characteristic quantized levels and carriers
can be injected from these quantized levels into the electrolyte.
Photogenerated holes may either cascade up the quantized levels,
i.e., thermalize, or they may tunnel through the surface barrier
into the electrolyte. For a p-type semiconductor whose work function is greater than that of a contiguous electrolyte a similar
potential well will exist for conduction electrons.

On the electrolyte side, the ion to which the electron is
transferred is represented by a square well. The width of the well
is taken to be of molecular dimensions, which is estimated to be
about 2 A. The depth of the well is determined for each eigenstate
by the condition that the energy of each eigenstate for the well is
the same energy as an eigenstate in the semiconductor. The latter
condition insures isoenergetic tunneling. In the real system this
condition is satisfied by the existence of broad extrinsic molecular states in the electrolyte produced by electron-phonon interactions. The well is considered to be a distance h from the surface
(see Fig. 17). On the semiconductor side two approximations are
used. First, the parabolic potential of the depletion region is
approximated as a linear potential which results in a triangular
well. This permits simplifications in the determination of wave
functions for the quantized levels. Secondly, the effective mass
m^* of the charge carriers is introduced to reflect the effect of
lattice periodicity in the semiconductor. This is a well-known
approach and is justified in the present case by the fact that
the most localized quantized states span at least seven to eight
atomic layers. However, the effective mass approximation is not
applied in the representation of states in the square well used to
simulate the molecular ion in the electrolyte.

The basic quantum mechanical problem to be solved is to find
the eigenstates for the potential well depicted in Fig. 17. From
the nature of these solutions the properties of the system are deduced in terms of the following characteristic times: τ_{Th}, the time
for thermal relaxation of an electron in an excited state in the

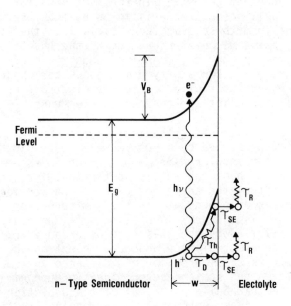

Fig. 15   Hot hole injection from n-type semiconductors showing the
characteristic time constants for various competing de-
excitation pathways.

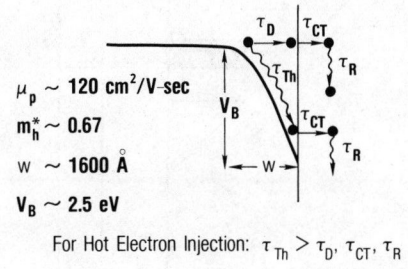

For Hot Electron Injection: $\tau_{Th} > \tau_D,\ \tau_{CT},\ \tau_R$

Fig. 16   Hot electron injection from p-type semiconductors showing
the characteristic time constants for competing de-exci-
tation pathways.

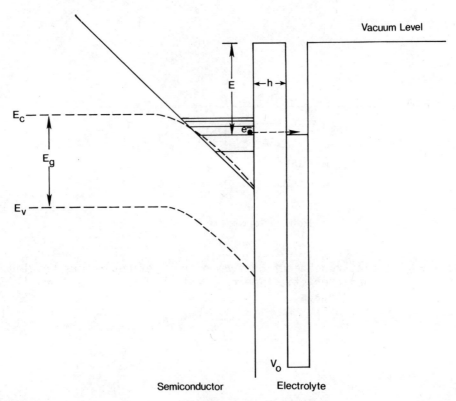

Fig. 17   Tunneling model for charge transfer across semiconductor-
electrolyte interfaces.

depletion layer; τ_t, the time required for the carrier to be transferred to the ion in the electrolyte, and τ_r, the time required for
the ionic energy level to relax by reorientation of solvent dipolar
species surrounding the ion such that reverse tunneling from the
electrolyte to the semiconductor is prevented.

These calculations are described in detail in reference 32;
the results are comparable to those based on classical and simpler
quantum mechanical models.

From these analyses and calculations several general criteria
for obtaining hot carrier injection at semiconductor-electrolyte
junctions are evident. The overall criterion is that both the
tunneling time of the photogenerated minority carriers and the effective relaxation time of the electrolyte are faster than the
thermalization time of these carriers in the semiconductor.

Strong electronic-vibrational interaction in the electrolyte
renders the tunneling process from the semiconductor irreversible,
obviating oscillations which increase residence time in the semiconductor and thus intraband thermalization therein. This irrever-
sibility has its basic origin in the electronic particle tunneling
from the semiconductor where the electron-phonon interaction is weak
to a strongly-coupled state of the electrolyte where the electronic-
vibrational interaction is strong.

Hot carrier injection is favored in semiconductor electrodes
which have a low effective mass for the minority carrier. This is
because this low effective mass will produce more widely-spaced
quantized levels in the depletion layer which then result in long
intraband (now interlevel) thermalization times. Low minority
carrier effective mass also means that the part of the photon energy
which exceeds the band gap will preferentially go to the minority
carrier; this will produce hot carriers in the depletion region by
virtue of absorption of photons with energy greater than the band
gap. In this case, hot carrier injection is more favored in direct
band gap semiconductors.

Heavy doping will favor hot carrier injection because the depletion layer thickness will be reduced. This effect leads to enhanced quantization in the depletion layer, and hence, longer thermalization times.

The occurrence of hot carrier injection in photoelectrochemical
reactions would be very significant for the following reasons: (1)
the nature of the permitted photoinduced reactions at semiconductor
electrodes could be controlled either by the nature of the anode
reaction or by an external bias; (2) the photogenerated carriers
would not be in thermal equilibrium in their respective bands so

that quasi-thermodynamic arguments, such as the use of the quasi-Fermi level to describe the energetics of photoelectrochemical reactions would not be valid; (3) the influence of surface states would be restricted to the class of states originating from the chemical interaction of the electrolyte with the semiconductor surface; and (4) the maximum theoretical conversion efficiency for photoelectrochemical energy conversion may be much greater compared with the case of thermalized injection.

Experiments have not yet been reported that conclusively prove the existence of hot carriers. However, some recent experiments (35) on the photoenhanced reduction of N_2 in a photoelectrochemical cell may provide qualitative evidence for a hot electron injection process. The system studied is a photoelectrochemical cell which contains a p-GaP cathode and an Al-metal anode immersed in a non-aqueous electrolyte of titanium tetraisopropoxide and $AlCl_3$ dissolved in glyme (1,2-di-methoxyethane). When N_2 is passed through the electrolyte and the p-GaP electrode is illuminated with band gap light, the N_2 is reduced and is recovered as NH_3; Al is consumed in the process and acts as the reducing agent. Although the reduction of N_2 to NH_3 with Al is thermodynamically favored ($\Delta G < 0$), the cell reaction does not procedd in the dark. The activation energy for the process is provided by light absorbed in the p-GaP electrode; hence, this system is an example of photocatalysis in a photoelectrochemical cell. The cell has been successfully operated in both flow and static modes; in the former, N_2 is continuously bubbled through the electrolyte. Experiments using $^{15}N_2$ have also been carried out and $^{15}NH_3$ has been identified from FTIR spectra.

The cell and electrolyte used in this work are closely related to those used by Van Tamelen and co-workers (36) to demonstrate normal electrolyte fixation of N_2. In those previous experiments, an external voltage source was used with either two Pt electrodes or with an Al anode and a nichrome cathode to fix N_2. In the photoelectrochemical system, no external voltage source is required to achieve N_2 fixation; the activation energy for the reaction is provided by light alone.

In typical flow runs, the reduced nitrogen yields, expressed as moles NH_4^+ per mole of Ti ion, varied between about 2%-5%; this corresponded to reaction rates of about 10^{-4} mol NH_3/hr/cm^2 electrode. Blank runs, in which either N_2 was replaced by Ar, p-GaP was replaced by PT, or no light was used, produced significant yields of NH_4^+.

The chemical processes occurring in the cell are equivalent to those for the pure electrolytic case as described by Van Tamelen et al. Titanium (IV) isopropoxide is first reduced to a state wherein molecular N_2 can be bound; this is evidenced by the development of an intense blue-black color which is attributed to a Ti(II) complex. The reduced titanium-molecular nitrogen complex is then reduced further to produce a reduced nitrogen-reduced titanium complex. Finally, NH_3 is produced through protonation of the reduced nitrogen-reduced titanium complex. The overall reaction in the cell can be represented as:

$$N_2 + 2Al + 6H^+ \xrightarrow{h\nu} 2NH_3 + 2Al^{3+}, \Delta G/e^- = -1.72 \text{ eV} \qquad (1)$$

Although Reaction 1 is favored in the dark, it does not proceed because of the high activation energy of intermediate steps.

The experimental result that may provide evidence for hot electron injection from the p-GaP cathode is that substitution of a H_2/Pt anode for the Al anode did not result in N_2 reduction at the p-Gap cathode. For thermalized injection the energy of the injected electrons (and, hence, the nature of the allowed cathodic reaction) is independent of the nature of the anodic reaction (and, hence, independent of V_B) since the energy level of E_c is pinned at the semiconductor-electrolyte interface. That is, the cathodic reaction occurring at the p-GaP electrode should be independent of the identity of the anode if the photogenerated electrons in the p-GaP cathode were thermalized in the depletion layer before being injected into the electrolyte. Therefore, the observation that N_2 is only reduced with an Al anode and not with a H_2/Pt anode indicates that the cathodic reaction is dependent upon V_B, and that a hot electron injection process may be involved. However, this result is not unambiguous because, as described in Section III, the bands could become unpinned under the experimental conditions, and this could explain the observed photoreduction behavior. Further experimental and theoretical work is required to firmly establish the importance of hot carrier injection in photoelectrochemical systems.

Surface Modification. As discussed in Section II, electrode decomposition reactions that are thermodynamically favored compared to electrolyte redox reactions are also generally kinetically favored. This undesirable situation can be alleviated through surface modification. It has been shown (37) that covalent attachment of redox species to semiconductor surfaces can effect the kinetics of charge transfer across semiconductor-electrolyte junctions such that the less thermodynamically favored reactions predominate over thermodynamically favored reactions. For example (37), illumination of n-type Si modified with a ferrocene derivative will not result in the oxidation of Si to SiO_2 but rather will result in the oxidation of $Fe(CN)_6^{4-}$ to $Fe(CN)_6^{3-}$ in the electrolyte. The chemically attached ferrocene species acts as a hole mediator for n-Si, and channels the photogenerated hole to the $Fe(CN)_6^{4-}/Fe(CN)_6^{3-}$ redox reaction rather than to the Si/SiO_2 reaction (see Figure 18). This effect can be very useful not only in stabilizing electrodes in

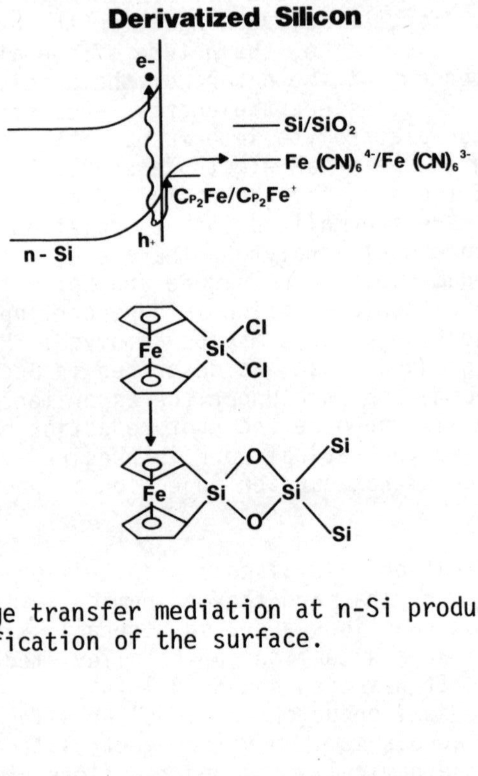

Fig. 18   Charge transfer mediation at n-Si produced by chemical modification of the surface.

electrochemical photovoltaic cells, but also in photoelectrosynthetic cells. For example, in Figure 19, derivatization of the semiconductor AB with R-D/D$^+$ would result in photogenerated charge
being channeled through the D/D$^+$ couple to the H_2O/O_2 reaction,
rather than to the oxidation of AB to B and A$^+$; the stepwide and
net reactions for this sequence are indicated in the figure. Such
charge mediation effects are analogous to the function of the Mn-
complex in Photosystem II in photosynthesis; in that case photo-generated holes are also channeled to the H_2O oxidation reaction, rather
than to the oxidation of the chlorophyll molecule itself.

Electrochemical Photovoltaic Cells

Reduced Surface and Grain Boundary Recombination. The quantum
efficiency of electrochemical photovoltaic (EP) cells is controlled
by electron-hole recombination processes at the surfaces of single
crystal electrodes, and at grain boundaries in thin, polycrystalline film electrodes. Recent work (38) has shown that the recombination velocities at n-GaAs electrodes in EP cells can be markedly
reduced by chemisorption of Ru^{3+} on the surface and/or grain boundaries. Studies (38) of single crystal and polycrystalline n-GaAs
electrodes show that in the former, the power conversion efficiency
is increased from 8.8% to 12%, while in the latter the increase is
from 1.2% to 4.8%. The electrolyte in both cases is a 1 M KOH solution containing 0.8 M K_2Se and 0.1 M K_2Se_2.

The fact that the Ru^{3+} is confined to the surface and does not
diffuse into the bulk (even at 300°C) was conclusively confirmed
(38) by Rutherford backscattering studies. The number of Ru^{3+} ions
on the surface corresponds to less than a monolayer. These studies
indicate that the surface recombination velocity has been decreased
from 10^6 cm/s to 3 x 10^4 cm/s following the Ru^{3+} treatment. The
dramatic improvement in cell efficiency persists for several weeks.
In addition to Ru^{3+}, other ions, such as Pb^{2+}, Ir^{4+}, and Rh^{3+} also
decrease recombination velocities on n-GaAs. The best conversion
efficiency obtained for a polycrystalline n-GaAs film (9 μm grain
size) was 7.8% following the chemisorption of both Ru^{3+} and Pb^{2+} on
the surface (38). It is interesting to note that while chemisorbed
Ru^{3+} reduces surface and grain boundary recombination on n-GaAs, it
has no effect on p-GaAs.

A simple model to explain these results has been proposed (38)
that invokes splitting of surface or grain boundary states resulting from a strong interaction of these surface states with chemisorbed species. The split states lie outside the band gap, and
hence are inactive as recombination centers. Ions which do not

Derivatized Electrodes
(for stability)

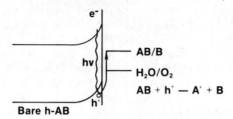

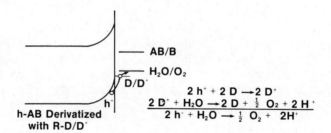

Fig. 19   Chemical modification of an n-type semiconductor surface
          to yield H_2O oxidation rather than self-oxidation of the
          electrode.

strongly chemisorb (such as Bi^{3+}), do not split the surface states sufficiently and do not effect recombination velocities.

Similar passivation effects of atomic hydrogen on grain boundaries in polycrystalline silicon have also been reported (39). However, the explanation here is that the surface or grain boundary states are filled upon interaction with covalently bonding atoms, and these filled states then become inactive as recombination centers. Further work is required to establish the correct model.

Non-Aqueous Electrolytes. One approach to increasing the stability of electrochemical photovoltaic cells is to use non-aqueous electrolytes. Work on this approach (40) is divided between organic-based electrolytes (methanol, ethanol, N,N-dimethyl-formamide, acetonitrile, propylene carbonate, ethylene glycol, tetrahydrofuran, nitromethane, benzonitrile), and room temperature molten salts ($AlCl_3$-butyl pyridinium chloride). These studies are relatively new and final conclusions concerning the relative merits of aqueous vs. non-aqueous electrolytes have not yet been made. To date, the most efficient and stable EP cells are based on aqueous electrolytes.

Storage Systems. As is well known, a major problem with photovoltaic cells is the lack of a viable storage system; galvanic battery storage is available but it is expensive and bulky. One potential advantage of EP cells is that the cells can be built with storage capability; this storage can be insitu or externally connected through a fuel cell -- redox flow system. The former approach can involve either the presence of a third storage electrode (41) in the cell, or it can be based on a solid electrolyte system (42). In the latter approach oxidized and reduced chemical species are first produced and then circulated through storage tanks to fuel cells. These three approaches are all being investigated (40), and it is too early to tell whether any of them will be technically and economically viable. An important question under analysis is whether insitu storage is inherently superior to exsitu storage.

General Status and Prognosis for Electrochemical Photovoltaic Cells. The main thrusts of research activities in EP Cells are in thin film electrodes, new layered compounds, non-aqueous electrodes, and storage systems. Work on thin film electrodes is concentrating on electrodeposited, thermally evaporated, or chemical-vapor-deposited thin films of CdSe, CdTe, GaAs, and $CdSe_x Te_{1-x}$ (40). Research on new layered compounds is primarily devoted to $MoSe_2$ (Eg = 1.4 eV) and WSe_2 (Eg = 1.6 eV). The most efficient (12%) cell produced to date is Ru^{3+}-treated single crystal n-GaAs in 0.8 M Se^{2-}/0.1 M Se_2^{2-}/1 M OH^-; the polycrystalline version of this cell (formed by CVD) has an efficiency of 7.3% (43). Polycrystalline thin films (formed by electrodeposition) of $n-CdSe_{1-x}Te_x$ in polysulfide electrolyte have efficiences of 8% (44). Single crystals of n-CdTe and n-CdSe in polyselenide or polysulfide electrolytes show efficiencies of 8.4%. The relatively high efficiency of the polycrystalline thin films is impressive, and nearly comparable to the 9.1% efficiency obtained for thin film CdS/Cu_2S and 9.4% efficiency for $CdS/CuInSe_2$ (45). The new layered compounds show 8.5% efficiency for WSe_2 and 7.5% efficiency for $MoSe_2$ as single crystal electrodes in aqueous electrolytes (45). Considering the short time period under which EP cells have been under active development ($\sim$3 years), progress has been remarkable. However, much additional work is required before it will become apparent how well EP cells can compete with solid state photovoltaic cells.

Photoelectrosynthetic Cells.

Derivatized Electrodes. As discussed in the above section, chemical modification of the semiconductor surface can result in control of the kinetics and pathways of charge transfer across the semiconductor-electrolyte interface. For photoelectrosynthetic cells, this offers the possibility of favoring H_2O oxidation over electrode self-oxidation for n-type electrodes. For p-type electrodes, where photo-decomposition (via a reduction process) is not as serious a problem, chemical modification can be used to enhance the catalytic properties of the electrode. This has been demonstrated (46) for p-type Si in aqueous electrolyte. Normally it is difficult to evolve H_2 on p-type electrodes, and a large negative deviation from the flat-band potential is required before H_2 evolution is achieved. However, by binding a paraquat (PQ^{2+}) derivative to the surface, the photogenerated electron is first efficiently captured by the PQ^{2+} species attached to the surface, and the reduced PQ^+ species then reduced H^+ to H_2. The overall scheme if shown in Figure 20, and the net effect is that H_2 is evolved at a much less negative potential than required for the electrode without surface modification.

Photo-Oxidation and Photo-Reduction on the Same Surface and in Particulate System. Photoelectrochemical reactions generally occur in a heterogeneous structure. That is, an n-type semiconductor electrode is connected either to a metal counter-electrode or to a p-type semiconductor. If these connections are made within a monolithic structure (i.e., no separated electrodes and no external wires), the devices are called photochemical diodes (47) (see Figure 21). The heterogeneous structure is required to produce the most favorable energy level scheme for the efficient separation of electrons and holes. Thus, in Figure 21, the hole

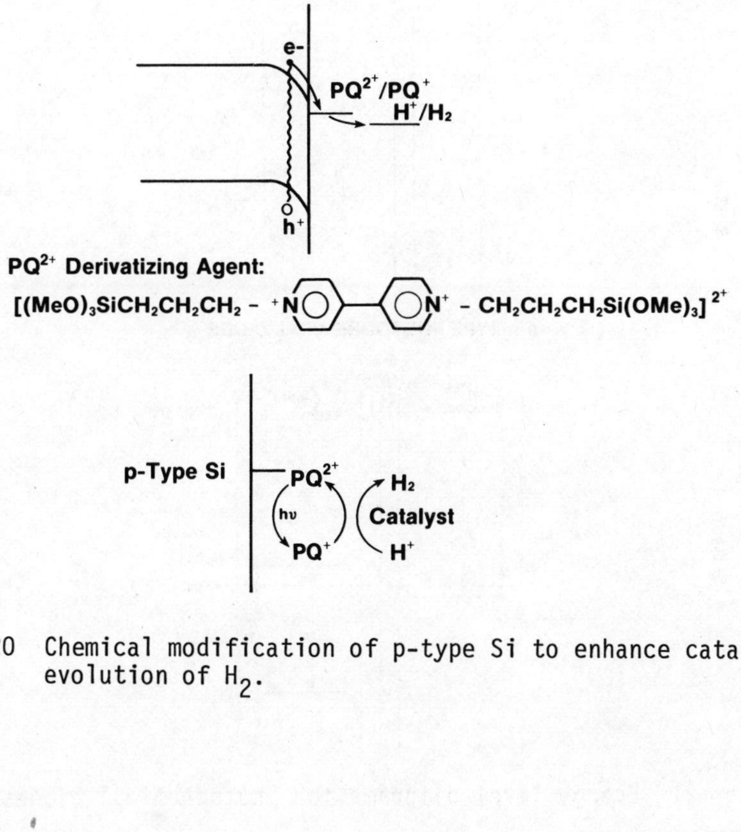
e-
PQ2+/PQ+
H+/H2
h+
PQ2+ Derivatizing Agent:
[(MeO)3SiCH2CH2CH2 - +N N+ - CH2CH2CH2Si(OMe)3] 2+
p-Type Si
PQ2+
H2
hυ
Catalyst
PQ+
H+

Fig. 20 Chemical modification of p-type Si to enhance catalytic evolution of H_2.

SCHOTTKY-TYPE PHOTOCHEMICAL DIODE

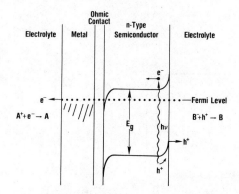

p-n TYPE PHOTOCHEMICAL DIODE

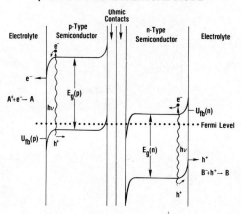

Fig. 21   Energy level diagrams for photochemical diodes.

is driven toward the n-type semiconductor surface by the positive space charge layer, while the electron is driven toward the semiconductor bulk and out the metallic-coated backside. For doubly illuminated p/n photochemical diodes, holes are driven to the surface of the n-type semiconductor and electrons are driven to the surface of the p-type semiconductor.

If a homogeneous semiconductor structure is in contact with the electrolyte, then the band bending at the liquid interface would bend in the same direction everywhere. This would create a potential well for electrons in n-type semiconductors and for holes in p-type semiconductors. The existence of these wells would inhibit majority carrier charge transfer from the semiconductor to the electrolyte depending upon the width and height of the potential well.

Recent work by Somorjai and co-workers (48) on single crystal $SrTiO_3$ shows that photoinduced H_2O splitting can occur without platinization of one surface if the water layer in contact with the illuminated surface has a very high OH^- concentration. The rate of H_2O splitting is highest if a photochemical diode is formed by attaching a platinum layer to the back of the crystal. However, a smaller but finite rate of H_2O splitting can also be produced without a platinum layer. Even if the $SrTiO_3$ crystal is not reduced (to make it conductive), and the bands are consequently flat, a finite rate of H_2O splitting can be observed. The energetics of these various situations are depicted in Figure 22. The photochemical diode configuration provides the best scheme for electron and hole injection into the liquid. The non-platinized, but reduced, $SrTiO_3$ permits electron injection either by tunnelling through the barrier (Figure 22) or by photoexcitation over the barrier; the rates of these processes are slower than the former case. Finally, electron and hole injection from unreduced and non-platinized $SrTiO_3$ (flat-bands) would depend upon the relative rates of bulk diffusion versus recombination.

The dramatic effect of the OH^- concentration on the rate of H_2O splitting is not understood (48).

In addition to the work on relatively large single crystals, work is also being done on fine particles. These particles have been studied both as photochemical diodes and as homogeneous semiconductor powders. N_2 has been photoreduced to NH_3 using photochemicals diodes made of p-GaP/Al (35, 49), and TiO_2 powders (50). CO_2 has been photoreduced using powders of p-GaP, TiO_2, WO_3, and Fe_2O_3 (51).

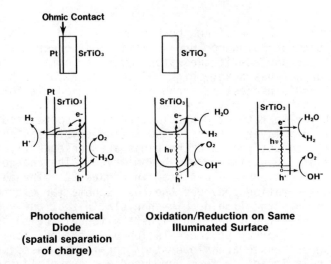

Fig. 22   Energy level diagrams for homogeneous, non-platinized
          reduced (middle) and unreduced (right) n-SrTiO$_3$ particles
          and photochemical diodes (left).

Dye Sensitization. A great deal of work is in progress (52) on improving the visible response of large band gap, but relatively stable, semiconductor electrodes by dye sensitization techniques. The general mechanism for this approach is shown in Figure 23. A dye (A) with strong absorption properties in the visible is bonded to the semiconductor surface. Upon excitation to A^*, an electron is transferred to the semiconductor conduction band leaving A^+ at the surface. The electron in the semiconductor subsequently reduces H^+ to H_2 at a metal cathode, while A^+ oxidizes H_2O to O_2 to regenerate A. As seen in Figure 23, the net overall reaction is the photolysis of H_2O into H_2 and O_2.

The problems associated with dye sensitization are: (1) the dyes are generally organic species that have relatively poor long term stability and do not survive the repeated oxidation-reduction cycles; (2) the dye layer on the electrode must be very thin to permit charge transfer from the excited dye to the electrode, but this requirement often is in conflict with the dye layer also exhibiting intense optical absorption; and (3) certain dye-semiconductor electrode combinations show very low quantum efficiencies. With regard to the last problem, reported quantum efficiencies for electron transfer to semiconductor electrodes from dye molecules range from 4×10^{-3} (53) for the rose bengal-TiO_2 system, to 1.0 for the tris (4,7-dimethyl-1, 10-phenanthroline) ruthenium (II) system (54).

To date, no system has been found that exhibits total conversion efficiencies greater than about 1% coupled with long-range stability.

Layered Compounds and Other New Materials. Another active area of research in photoelectrosynthesis involves the study of new semiconductor electrode materials. Layered compounds are particularly interesting since the top of the valence band is comprised of metal d_{z^2} orbitals rather than anion 4p orbitals (55). This means that an optical transition near the band gap energy involves only metal to metal transitions (the conduction band is also made of d-like metal orbitals) and does not disrupt a metal-anion bond. This type of optical transition is believed to be less susceptible to photocorrosion than the usual transitions involving anion-like orbitals in the valence band.

Results with layered chalcogenides (56-60) such as MoS_2 (Eg= 1.8 eV), $MoSe_2$ (Eg = 1.4 eV), $MoTe_2$ (Eg = 1.0 eV), WSe_2 (Eg = 1.6 eV), and WS_2 (Eg = 2.0 eV) show that enhanced resistance to photocorrosion is indeed observed. However, photocorrosion readily occurs at the edges of the layered materials (11 to C-axis), and this presents problems for the planar face also since atomic-size step dislocations in the surface become photocorrosion sites. Much intense research in layered compounds is proceeding at the present

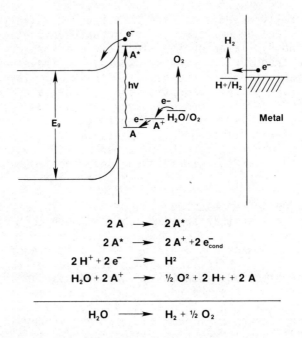

$$2\,A \longrightarrow 2\,A^*$$

$$2\,A^* \longrightarrow 2\,A^+ + 2\,e^-_{cond}$$

$$2\,H^+ + 2\,e^- \longrightarrow H^2$$

$$H_2O + 2\,A^+ \longrightarrow \tfrac{1}{2}\,O^2 + 2\,H+ + 2\,A$$

$$H_2O \longrightarrow H_2 + \tfrac{1}{2}\,O_2$$

Fig. 23 Energy level scheme for dye sensitization of n-type semiconductors.

time because of their very favorable band gaps and their potential for long-term stability.

General Status and Prognosis for Photoelectrosynthesis. The photoelectrosynthetic approach to solar energy conversion is very appealing and exciting and deserves continued intense attention. The potential advantages of photoelectrosynthesis over photovoltaics coupled to dark electrosynthesis are in higher net conversion efficiency, better engineering designs for solar reactors, and unique catalytic effects possible with modified semiconductor electrode surfaces.

The main problem in the field is the lack of semiconductor materials that exhibit high conversion efficiency and long term stability. However, one very promising approach involves the chemical modification or derivatization of the semiconductor surface to enhance photo-stability and catalytic activity. This approach is interesting in that it represents a common intersection with the current directions of photochemical approaches to solar energy conversion (61). In the latter, systems are being studied that involve semiconductor particles as catalysts to help drive the photochemical redox reactions. These semiconductors act as electron pools to facilitate the redox chemistry and catalyze the desired reactions.

It appears that the optimum system may involve semiconductors (to absorb the light and separate the photo-induced charge carriers), in the form of fine particles (for optimum engineering design), but modified on the surface with chemical compounds (to enhance stability, catalytic activity, and perhaps also optical absorptivity).

ACKNOWLEDGEMENT

This work was performed under the auspices of the Office of Basic Energy Sciences, U.S. Department of Energy under Contract EG-77-C-01-4042.

REFERENCES

1. A.J. Nozik, Ann. Rev. Phys. Chem., 29, 189 (1978).
2. H. Gerischer, In "Solar Power and Fuels", J.R. Bolton, Ed;
    Academic Press, N.Y. 1977.
3. L.A. Harris and R.H. Wilson, Ann. Rev. Mater. Sci. 8, 99
    (1978).
4. A.J. Bard, Science 207, 139 (1980).
5. M. Wrighton, Acc. Chem. Res. 12, 303 (1979).
6. M. Tomkiewicz and H. Fay, Appl. Phys. 18, 1 (1979).
7. "Interfacial Photoprocesses: Energy Conversion and Synthesis",
    Advances in Chemistry Series 184, ACS 1980.
8. H. Gerischer, "Topics in Applied Physics", Vol. 31, Solar
    Energy Conversion, B.O. Seraphin, Ed., Springer-Verlag,
    Berlin (1979).
9. M.A. Butler and D.S. Ginley, J. Mat. Sci. 15, 1 (1980).
10. H. Gerischer, Electroanal. Chem. Interfac. Electrochem. 58,
    263 (1975).
11. H. Gerischer, In "Physical Chemistry; An Advanced Treatise",
    H. Eyring, Ed., Vol. 9A, Academic Press (1970).
12. W.P. Gomes and F. Cardon, Proc. Conf. on Electrochem. and
    Physics of Semiconductor-Liquid Interfaces Under Illumina-
    tion (A. Heller, Ed.), Princeton: Electrochemical Society
    (1977).
13. H. Gerischer, J. Electroanal. Chem. 82, 133 (1977).
14. A.J. Bard and M.S. Wrighton, J. Electrochem. Soc. 124, 1706
    (1977).
15. M. Wolf, Proc. I.R.E. 48, 1246 (1960).
16. A.J. Nozik, Proc. Conf. on Electrochem. and Phys. of Semicon-
    ductor-liquid Interfaces Under Illumination (A. Heller, Ed.)
    Princeton: Electrochemical Society (1977).
17. C.R. Dickson and A.J. Nozik, J. Am. Chem. Soc. 100, 8007
    (1978).
18. B. Kraeutler, and A.J. Bard, J. Am. Chem. Soc. 100, 2239
    (1978).
19. H. Gerischer, In "Adv. in Electrochem. and Electrochem. Eng.
    P. Delchay, Ed., Interscience, N.Y. (1961).
20. E.C. Dutoit and F. Cardon, Ber. Bunsenges. Phys. Chem. 80,
    1285 (1976).
21. S.N. Frank and A.J. Bard, J. Am. Chem. Soc. 99, 4667 (1977).
22. T. Inoue, T. Watanabe, H. Fujishima, and K. Honda, Chem. Lett.
    1073 (1977).
23. A. Fujishma, T. Inoue, K. Honda, J. Am. Chem. Soc. 101, 5582
    (1979).
24. R. Memming, J. Electrochem. Soc. 125, 117 (1978).
25. R.H. Wilson, J. Electrochem. Soc. 126, 1187 (1979).
26. H. Tamura, H. Yoneyama, T. Kobayashi, in "Photoeffects at Semi-
    Conductor-Electrolyte Interfaces," A.J. Nozik, Ed., ACS
    Symposium Series, Houston ACS Meeting (1980) in press:
    W.P. Gomes, F.Van Overmeire, F.Vanden Kerchove,F.Cardon,ibid.

27. R.H. Wilson, Electron Transfer Processes at Semiconductor-
     Electrolyte Interfaces", CRC Critical Reviews of Solid
     State and Materials Science (1980) in press.
28. A.J. Bard, A.B. Bocarsly, F.F. Fan, E.G. Walton, and M.S.
     Wrighton, J. Am. Chem. Soc. 102, 3671 (1980); A.B. Bocarsly,
     D.C. Bookbinder, R.N. Dominey, N.S. Lewis, and M.S. Wrighton,
     102, 3683 (1980).
29. J.A. Turner, J. Manassen, and A.J. Nozik, Appl. Phys. Letts.
     37, 489 (1930).
30. S.M. Sze, "Physics of Semiconductor Devices", Chap. 9, Wiley,
     N.Y. (1969).
31. C.A. Mead, Solid State Elec. 9, 1023 (1966).
32. A.J. Nozik, D.S. Boudreaux, R.R. Chance, and F. Williams, Adv.
     Chem. Series No. 184, ACS, Washington 1980.
33. D.S. Boudreaux, F. Williams and A.J. Nozik, J. Appl. Phys. 51,
     2158 (1980).
34. F. Williams, S. Varma, S.J. Hillenius, J. Chem. Phys. 64, 1549
     (1976).
35. C.R. Dickson and A.J. Nozik, J. Am. Chem. Soc. 100, 8007 (1978).
36. E.E. Van Tamelen, Acc. Chem. Res. 3, 363 (1970).
37. M.S. Wrighton, R.G. Austin, A.B. Bocarsly, J. Bolts, O. Haas,
     K.D. Legg, L. Nadjo, and M. Palazzotto, J. Am. Chem. Soc.
     100, 1602 (1978); J. Bolts and M. Wrighton, ibid 100, 5257
     (1978).
38. A. Heller, B. Parkinson, and B. Miller, Appl. Phys. Letts. 33,
     521 (1978); W.D. Johnston, H.J. Leamy, B.A. Parkinson, and
     B. Miller, J. Electrochem. Soc. 127, 90 (1980).
39. C.H. Seager and D.S. Ginley, Appl. Phys. Letts. 34, 337(1979).
40. Abstracts of Second Electrochemical Photovoltaic Cell Contrac-
     tor's Meeting, Solar Energy Research Institute, Golden,
     CO, Jan. 1980.
41. G. Hodes, J. Manassen, D. Cahen, Nature 261, 403 (1976).
42. P.G.P. Ang and A.F. Sammells, in "Photoeffects at Semiconduc-
     tor-Electrolyte Interfaces", A.J. Nozik, Ed., ACS Symposium
     Series, Houston ACS Meeting (1980), in press.
43. A. Heller, ibid; A. Heller, H.J. Lewerenz, and B. Miller, Ber.
     Bunsengesellschaft, Phys. Chem. (1980), in press.
44. G. Hodes, Nature 285, 29 (1980).
45. S.K. Deb, private communication.
46. M. Wrighton, Fourth DOE Solar Photochemistry Conference, Rad-
     iation Laboratory, University of Notre Dame, Indiana, June
     (1980).
47. A.J. Nozik, Appl. Phys. Letts. 30, 567 (1977).
48. F.T. Wagner and G.A. Somorjai, in press.
49. M. Koizumi, H.Yoneyama, H. Tamura, J. Amer. Chem. Soc. (1980),
     in press.
50. G.N. Schauzer and T.D. Guth, J. Amer. Chem. Soc. 99, 7189
     (1977).
51. T. Inoue, A. Fujishima, S. Konishi, and K. Honda, Nature 277,
     637 (1979); M. Halman, Nature 275, 155 (1978).

52. A.Hamnett, M.P. Dare-Edwards, R.D. Wright, K.R. Seddon, and
     J.B. Goodenough, J. Phys. Chem. 83, 3280 (1979);
     H. Tsubomura, M. Matsumura, Y. Nomura, and T. Amamiya,
     Nature 261, 5559 (1976); H. Gerischer, in "Topics in
     Current Chemistry", A. Davison, Ed., Springer, N.Y. (1976),
     Vol. 31; R. Memming and H. Tributsch, J. Phys. Chem. 75,
     562 (1971).

53. M.T. Spitler and M. Calvin, J. Chem. Phys. 66, 4294 (1977).

54. W.D.K. Clark and N. Sutin, J. Amer. Chem. Soc. 99, 14 (1977).

55. H. Tributsch, Ber. Bunsenges. Phys. Chem 81, 361 (1977); ibid,
     82, 169 (1978); ibid, J. Electrochem. Soc., 125, 1086 (1978);
     ibid, Solar Energy Materials 1, 705 (1979).

56. J. Gobrecht, H. Gerischer, and H. Tributsch, Ber. Bunsenges.
     Phys. Chem. 82, 1331 (1978).

57. B.A. Parkinson, T.E. Furtak, D. Canfield, K. Kam, and G. Kline,
     in press.

58. J. Gobrecht, H. Gerischer, and H. Tributsch, J. Electrochem.
     Soc. 125, 2085 (1978).

59. H. Gerischer, J. Gobrecht, and J. Turner, Ber. Bunsenges.
     Phys. Chem. 84, 596 (1980).

60. T. Kawai, H. Tributsch, and T. Sakata, Chem. Phys. Letts. 69,
     336 (1980).

61. Proceedings of Third International Conference on Photochemical
     Conversion and Storage of Solar Energy, Univ. of Colorado,
     Boulder, CO, Aug. (1980), in press.

THE IRON THIONINE PHOTOGALVANIC CELL

W. John Albery

Department of Chemistry
Imperial College
London, S W 7

INTRODUCTION

A typical thin layer photogalvanic cell is illustrated in figure 1. In this device for solar energy
conversion the sunlight enters the cell through a transparent electrode and is trapped in the electrolyte. [1,2]
Photochemical reactions, usually involving electron transfer, lead to high energy products. One of these products
diffuses to the illuminated electrode and reacts there,
while the other product diffuses across the cell to the
dark electrode to be converted back into original reactant.

The Reaction Scheme

The simplest reaction scheme, involving two redox
couples [3,4] A, B and Y, Z, is as follows:-

$$A \xrightarrow{h\nu} A^*$$

$$A^* + Z \longrightarrow B + Y$$

$$(B + Y \xrightarrow{k} A + Z)$$

313

Illuminated
   Electrode $B \pm e \longrightarrow A$

Dark
   Electrode $Y \mp e \longrightarrow Z$

An example of this system [5] is :-

A	$Ru\ (bpy)_3^{2+}$
B	$Ru\ (bpy)_3^{3+}$
Y	Fe (II)
Z	Fe (III)

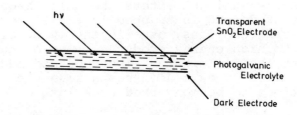

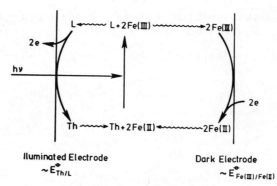

Fig. 1. A photogalvanic cell and a schematic representation of the iron thionine system.

For an efficient cell it is obvious that one must avoid the thermal back reaction of B and Y described by the rate constant k. Hence one of the problems with this type of device is that the homogeneous kinetics may destroy the energetic species B and Y. In my second lecture we shall explore this constraint in more detail. Another crucial requirement is that the illuminated electrode should be selective. If the illuminated electrode is not selective between B and Y then we will have :-

$$B \; \overset{+}{\underset{-}{}} \; e \; \longrightarrow \; A$$

and

$$Y \; \overset{-}{\underset{+}{}} \; e \; \longrightarrow \; Z$$

$$\overline{\qquad\qquad\qquad\qquad\qquad\qquad\qquad}$$

$$B \; + \; Y \; \longrightarrow \; A \; + \; Z$$

The beastly electrode is merely an efficient catalyst for the back reaction. In my third lecture we shall deal with the problems of electrode selectivity and the thionine coated electrode in particular. My fourth lecture will be concerned with the efficiency of photogalvanic cells and the progress (or lack of it) made to date. In this lecture I propose to outline the theoretical analysis which leads to the recipe for the ideal photogalvanic cell.

The Differential Equation

The transport and kinetics of the energetic species B in the electrolyte are described [3] by the following two differential equations :-

$$D \; \frac{\partial^2 b}{\partial x^2} \; + \; I \varepsilon a \; - \; k \, b \, y \; = \; 0 \qquad\qquad (1)$$

and

$$\frac{\partial I}{\partial x} \; = \; - I \varepsilon a \qquad\qquad (2)$$

where lower case letters have been used for concentrations and I is the irradiance. The equation in I
describes the penetration of the light into the cell,
while eqn (1) describes the steady state concentration
profile of B arising from diffusion (D) photogeneration and the thermal back reaction. In most
systems it is B, the product from the dye, that reacts
on the illuminated electrode. Its concentration will be
smaller than that of Y, the other energetic product,
because extra Y has to be added in order to carry the
current across the cell to the dark electrode without
causing concentration polarisation.[6] So we can assume
that y >> b. With this assumption the differential
equations can be solved with the boundary conditions that
arise from the electrode kinetics.

The Characteristic Lengths

     The solutions that arise can be understood in terms
of a set of characteristic lengths. (For a steady state
system where the concentrations vary with distance,
characteristic lengths play the same part as characteristic
times in ordinary reaction kinetics.) The lengths are
summarised in Table 1 and illustrated in figure 2.

TABLE 1

Characteristic Lengths

The Cell Length X_1	the distance across the cell
The Absorption Length $X_\varepsilon = (\varepsilon a)^{-1}$	the distance in which I falls by e^{-1}, if the light is absorbed in a Beer Lambert profile.
The Kinetic Length $X_k = (D / ky)^{\frac{1}{2}}$	the distance B diffuses before being destroyed by reaction with Y.
The Generation Length $X_g = (D / \phi I_o \varepsilon)^{\frac{1}{2}}$	the distance A diffuses before being excited by the incoming irradiance I_o.

The Kinetic Length

     For an efficient cell it is vital that the species B
should be photgenerated within a distance X_k of the illuminated electrode. Molecules of B generated outside this
distance are likely to react with Y before they reach the
electrode, since the time taken to diffuse X_k, given by
$X_k^2 D^{-1}$, is larger than the lifetime of B given by $(ky)^{-1}$.
The distance over which the photogeneration takes place is
that in which the light is absorbed, given by X_ε. Hence

for an efficient cell it is vital that the following condition holds : $X_\varepsilon < (0.10)X_k$ (3)

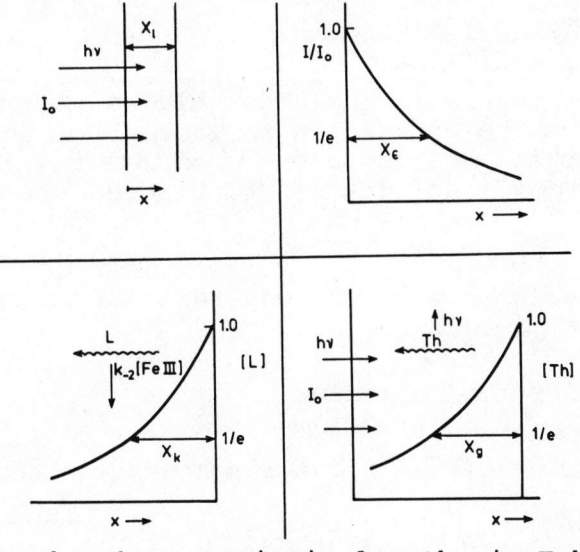

Fig. 2. The characteristic lengths in Table 1.

Bleaching and the Generation Length

Next we turn to X_g and the question of bleaching.
For an efficient cell the dye close to the illuminated
electrode must not be bleached by the incoming solar
radiation.[4] If bleaching occurs, very little radiation
is absorbed close to the electrode and hence little
power can be generated. In many practical systems
bleaching does not occur because the back reaction of B
with Y regenerates the dye A. However this recipe
for avoiding bleaching cannot lead to an efficient cell
because energy is being wasted by the back reaction.
Therefore we need to avoid bleaching with (hopefully) no
back reaction between B and Y. There must be
sufficient A to mop up all the incoming photons. The
analysis shows that the required condition is :-

$$X_\varepsilon < (0 \cdot 10) X_g \qquad (4)$$

When this condition holds, the light is absorbed in a
narrow layer; the lifetime of A in the photon flux is
long enough for the reservoir of A in the illuminated
part of the solution to be able to diffuse in and out of
the narrow illuminated layer. If condition (4) does
not hold and if there is no back reaction, then the
solution becomes bleached and the light may penetrate

all the way across the cell to the "dark" electrode. This
type of cell has been advocated by Lichtin [7], but very
little light is absorbed at each electrode interface.

The Recipe for Success

Now although condition (4) is irksome, it has an important consequence in that the value of X_g is almost God
given.[2] From the definition in Table 1, using the values
for I of the solar irradiance (4×10^{-8} mol cm^{-2} s^{-1}), for
D of 10^{-5} cm^2 s^{-1} and for ε of a good dark dye
we find

$$X_g/cm = 10^{-3} \tag{5}$$

This tells us straightaway that for an efficient cell
we must have

$$X_\varepsilon/cm = 10^{-4} \tag{6}$$

and from the typical value of ε

$$a / mol\ dm^{-3} = 10^{-1} \tag{7}$$

Given the value of X_ε we can also deduce the minimum
value of X_k from eqn (3)

$$X_k/cm > 10^{-3} \tag{8}$$

Now the concentration of Y is fixed by the fact
that Y must carry the current from the illuminated region
to the dark electrode without causing concentration polarisation at the dark electrode. Therefore the concentration of Y depends on X_1 the thickness of the cell. We
take $X_1 = 10^{-2}$cm and this gives, using the flux of an
efficient cell,

$$y / mol\ dm^{-3} \sim 10^{-2} \tag{9}$$

From eqn (8) we then find the crucial limitation on the
rate constant of the back reaction :-

$$k / dm^3\ mol^{-1}\ s^{-1} < 10^3 \tag{10}$$

Finally the same considerations apply to the concentration of Z as to the concentration of Y in that the flux
through the cell must not deplete Z in the vicinity of the
illuminated electrode or else the capture of A^* will be
inefficient. From the point of view of the voltage
generated by the cell, it is desirable to have z as small
as possible. So, like Y, we find

$$z / mol\ dm^{-3} \sim 10^{-2} \tag{11}$$

However we must enter an important proviso, and that is
that this concentration of Z must also be sufficient to
capture A^* efficiently. For the photochemical scheme :-

$$A^* \xrightarrow{k_T} A$$

$$A^* + Z \xrightarrow{k_q} B + Y$$

we require, as well as eqn (11), that

$$z > k_T / k_q \tag{12}$$

For many systems a concentration of 10^{-2} mol dm^{-3} is sufficient to satisfy eqn (12).

Therefore in eqn (5) to (12) we have the recipe for a successful photogalvanic cell as regards the homogeneous part of the system. The concentration profiles for the ideal cell are illustrated in figure 3.

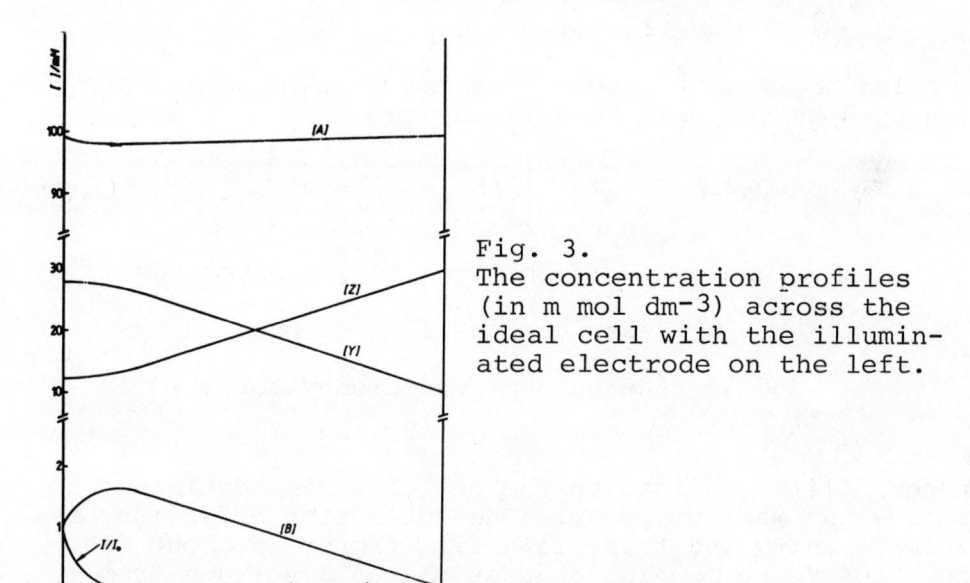

Fig. 3.
The concentration profiles (in m mol dm-3) across the ideal cell with the illuminated electrode on the left.

The Electrode Kinetics

At the illuminated electrode, as discussed above, we need the A,B couple to be reversible so that B is easily converted to A, but we also require that, as far as possible, the conversion of Y to Z is blocked. The electrode will then be close to the standard electrode potential of the A, B couple $E^{\ominus}_{A,B}$. Very little B reaches the dark electrode because the illuminated electrode is close to the region where B is generated (10^{-4} cm) as opposed to the dark electrode being $\sim 10^{-2}$ cm away. This means that the dark electrode has merely to convert the photogenerated Y back to Z at a potential close to the standard electrode potential of the Y, Z couple, $E_{Y,Z}^{\ominus}$. In fact if there is little concentration polarisation, and if the Y,Z couple is reversible, the dark electrode potential will shift very little when the cell is illuminated. The change in voltage on illumination occurs at the illuminated electrode from the manyfold increase in the concentration of B.

Current Voltage Characteristics

To describe the current it is helpful to define the collection efficiency N where N compares the flux of electrons out to the flux of photons in :-

$$N = \frac{\text{flux of electrons}}{\text{flux of photons}} \tag{13}$$

Detailed analysis [4] shows that the maximum power, W_m, developed by the cell is given by :-

$$W_m \simeq 0.8 \, A \, F \, \phi \, I_o \left[|\Delta E^{\ominus}| - 0.04 \right] \tag{14}$$

where A is the area of the electrode.

I_o describes the flux of photons

and ϕ the quantum efficiency for the

generation of B and Y

In eqn (14) the load on the cell has been optimised to achieve the maximum power. The collection efficiency is close to unity and there is a fill factor of about 0.8. The 0.04 V term which reduces the cell voltage from $\Delta E^{\ominus}$ arises from the fact that the concentration of B is less than that of A at the illuminated electrode and so there is a Nernst term. If one expresses this power output as a power conversion efficiency then the theoretical ideal cell could achieve efficiencies as high as 20%. However this requires that all the different conditions in the recipe for success should be met. Can this be achieved ?

HOMOGENEOUS KINETICS

The Iron Thionine System

In lecture 1 we discussed the ideal theoretical A, B, Y, Z system. In this lecture we examine, in more detail, the most successful candidate for a photogalvanic cell, the iron thionine system.[8, 9] Thionine, Th, Semi-thionine, S$^{\bullet}$, and Leucothionine, L, are :-

Th

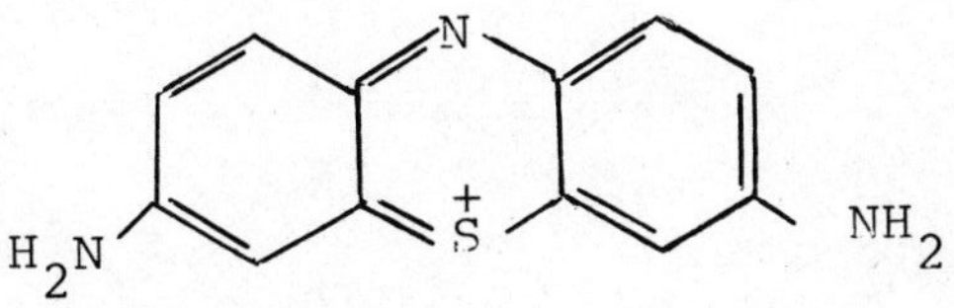

S$^{\bullet}$

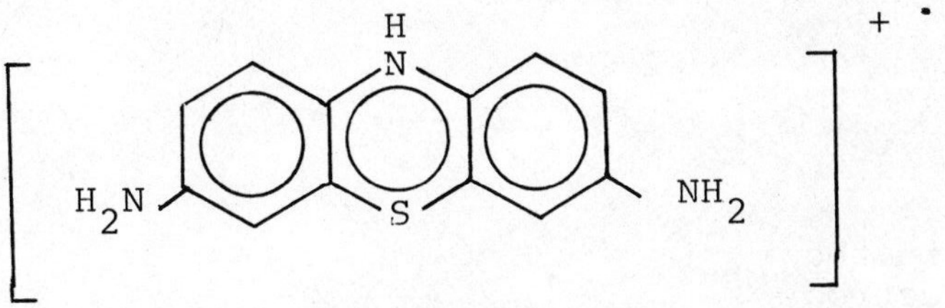

L

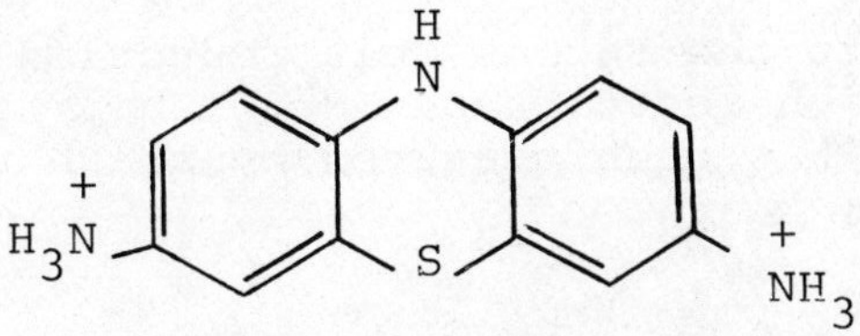

Thionine is a deep purple dye with an absorption band well placed to trap solar radiation; leucothionine is colourless and together they constitute a two electron redox couple :-

$$Th + 2e + 3H^+ \rightleftharpoons L$$

The Reaction Scheme

The combination of a one electron couple with a two electron couple is more complicated than the A, B, Y, Z system. The reaction scheme (ignoring protons) is :-

$$Th \longrightarrow Th^*$$

$$Th^* + Fe(II) \longrightarrow S^\bullet + Fe(III)$$

$$S^\bullet + S^\bullet \xrightarrow{k_3} Th + L$$

$$(L + Fe(III) \xrightarrow{k_{-2}} S^\bullet + Fe(II))$$

$$(S^\bullet + Fe(III) \xrightarrow[k_{-1}]{} Th + Fe(II))$$

Illuminated
    Electrode $$L \longrightarrow Th + 2e$$

Dark
    Electrode $$2 Fe(III) + 2e \longrightarrow 2 Fe(II)$$

The key to the successful production of L is the fast dismutation reaction of $S^\bullet$. The rate constant for this reaction has been measured by flash photolysis[10] to be

$$k_3 / dm^{-3} mol^{-1} s^{-1} = 7 \times 10^9.$$

The mechanism can be written more simply as:-

$$Th \; \underset{k_{-1}[Fe(III)]}{\overset{h\nu}{\rightleftarrows}} \; S^\bullet \; \underset{k_{-2}[Fe(III)]}{\overset{}{\rightleftarrows}} \; L$$

This mechanism has been established by the early work of Hatchard and Parker [8] by detailed work at the Royal Institution[10] and by analysis of the dependence of $[L]$ on variables such as irradiance, $[Fe (III)]$ and $[Fe (II)]$. [11, 12] A crucial feature of the mechanism is that L is made through the fast reaction generated by k_3 but decomposes through k_{-2}. Hence it is possible to trap L with a high energy transition state at k_{-2} but still make it through the low energy transition state 3.

Quantum Efficiencies

The overall quantum efficiency for the production of L, ϕ, is broken down in two contributions ϕ_1 the quantum efficiency for producing $S^\bullet$ and ϕ_2 the fraction of $S^\bullet$ that makes L :-

$$\phi = \phi_1 \ \phi_2$$

The maximum value of ϕ_2 is $\frac{1}{2}$ and it might be thought that this would lead to the cell being half as efficient as the corresponding A B Y Z cell. However this factor is exactly compensated by the electrochemical oxidation of L which requires two electrons rather than one.

For this system the back reaction rate constant for the kinetic length, X_k, is k_{-2}. For the overall quantum efficiency we need to know both ϕ_1 and ϕ_2. Analysis of the steady state concentration of $S^\bullet$ gives

$$\frac{1}{\phi_2} = 2 + \frac{k_{-1} [Fe (III)]}{(\phi_2 k_3 f)^{\frac{1}{2}}} \tag{15}$$

where f is the flux $(mol \ cm^{-3} \ s^{-1})$ for the generation of $S^\bullet$. The bimolecular nature of the dismutation reaction means that ϕ_2 depends on the flux for a solar cell

$$f \simeq \phi_1 \ I_o / X_\epsilon \tag{16}$$

Hence the last term in eqn (15) , which we wish to be negligible so that $\phi_2 = \frac{1}{2}$, is proportional to $X_\epsilon^{\frac{1}{2}}$. This is a further advantage of having a system with a small X_ϵ. If all the $S^\bullet$ are made in a narrow layer

they have a better chance of finding each other rather
then Fe (III). Using typical values and eqn (15) we
find that we require

$$(k_{-1}^2 / k_3) / dm^3 \ mol^{-1} \ s^{-1} \ < \ 30 \qquad (17)$$

The Parameters

In order to evaluate a system for a solar cell we
therefore need to know the parameters listed in Table 2

TABLE 2

The Kinetic Parameters

Stern Volmer		Ideal Value
Constant	$(k_q/k_T)/dm^3 \ mol^{-1}$	$> 10^3$
Quantum	ϕ_1	1.0
Efficiencies	ϕ_2 or $(k_{-1}^2/k_3)/dm^3 \ mol^{-1} \ s^{-1}$	< 30
Thermal Back Reaction	$k_{-2} /dm^3 \ mol^{-1} \ s^{-1}$	$< 10^3$

Rotating Transparent Disc Electrodes

The traditional technique used to investigate these
systems has been flash photolysis. We have developed a
complimentary technique, the rotating transparent disc
electrode, 13, 14, 15, 16 which is perhaps more appropriate for these investigations. This is because the
transparent electrode mimics the illuminated electrode
in the cell but with the system better defined.

The apparatus is illustrated in figure 4. The transparent disc electrode consists of the end of a quartz rod
which is coated with a thin layer of $Sn \ O_2$ to make the
electrode. Light from an ordinary slide projector is
focussed on to the top of the rod. The electrode is
rotated at speeds between 1 and 50 Hz. This rotation
imposes a known pattern of flow on the solution, and this

stirring motion is responsible for bringing reactants
to the electrode and removing the products of both the
electrochemical and photochemical reactions. Because
the pattern of flow can be calculated the rate of transport of reactants and products controlled by the rotation speed can also be calculated. This means that
the rotation speed is a valuable experimental variable.

In particular the thickness of the stagnant layer,
X_D, adjacent to the electrode can be calculated from the
Levich equation [17, 18] :-

$$X_D = 0.64 \, D^{1/3} \, \nu^{1/6} \, W^{-1/2} \tag{18}$$

where ν is the kinematic viscosity and W is the rotation speed in Hz. Within the stagnant layer transport is by diffusion, as it would be in a photogalvanic
cell. Outside the stagnant layer the solution is well
stirred.

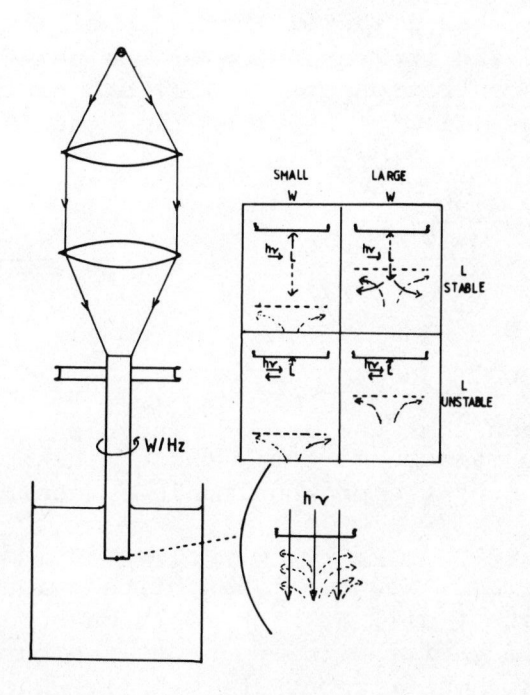

Fig. 4. The Rotating Transparent Disc Electrode. The
         dotted arrows show the imposed hydrodynamics.
         The four insets show how the fate of L depends
         on W the rotation speed and k_{-2}.

In investigating photogalvanic systems we perform two types of experiment. In the first type of experiment we do not add any Fe (III) to the solution. Under these conditions the concentration of photo-generated Fe (III) is so small that the back reaction of L with Fe (III) is negligible. The potential of the electrode is set so that all the photogenerated L which reaches the electrode is destroyed. The photo-current then varies linearly with $W^{-\frac{1}{2}}$. The current is lower at higher rotation speeds because to reach the electrode L must be photogenerated inside the stagnant layer; most of the L generated outside the layer is swept away into the solution. Hence the slower the rotation speed the thicker the stagnant layer (eqn (18)) and the higher the current. When the photocurrent varies with $W^{-\frac{1}{2}}$ we can therefore conclude that the back re-action is negligible. Then from the size of the photo-current we can measure the quantum efficiency for the production of S^{*}, ϕ_1. This is because with no added Fe (III) ϕ_2 has its limiting value of $\frac{1}{2}$. Further-more by studying the variation of the photocurrent, i_p, with the concentration of Fe (II) and plotting i_p^{-1} against $[Fe (II)]^{-1}$ we can find the Stern Volmer constant :-

$$\frac{(i_p)_\infty}{i_p} = 1 + \frac{k_T}{k_q [Fe (II)]} \qquad (19)$$

It can be seen that the use of electrochemical detection allows us to measure directly the efficiency of the for-ward reaction of the photogalvanic system.

In the second type of experiment we add Fe (III) to the solution. Now two types of behaviour may be found. First of all, at high rotation speeds the photo-current still varies with $W^{-\frac{1}{2}}$ but the currents are smaller than those without Fe (III). The rotation speed variation shows that L is still stable on the time scale of the experiment. So the diminished currents must arise from the effect of Fe (III) on S^{*} in reducing ϕ_2 below $\frac{1}{2}$. Hence these experiments measure ϕ_2. At lower rotation speeds the photocurrent becomes independent of rotation speed. Now the stagnant layer, X_D, is thicker than the kinetic length, X_k. This means that a photostationary state is established inside the stagnant layer where the main fate of L is to be destroyed by reaction with Fe (III). From the size of the current we can now measure the rate constant for the reaction of Fe (III) with L. In Table 3 we summarise these different techniques.

TABLE 3

Techniques for Measuring Kinetic Parameters

	Transparent Rotating Disc Electrode	Other Techniques
k_q/k_T	No added Fe (III) Gradient of $(i_p W^{\frac{1}{2}})^{-1}$ with Fe(II)$^{-1}$	Flash Photolysis
ϕ_1	No added Fe (III) Intercept of $(i_p W^{\frac{1}{2}})^{-1}$ with Fe(II)$^{-1}$	Flash Photolysis
ϕ_2	Added Fe (III) Gradient of i_p with $W^{-\frac{1}{2}}$ at high W	Flash Photolysis
k_{-2}	Added Fe (III) Size of i_p at low W	Flash Photolysis Stopped Flow.

The Thionine System

The problem with thionine as a constituent of a photogalvanic cell is that, although its rate constant k_{-2} has a satisfactory value of $500 \ dm^3 \ mol^{-1} \ s^{-1}$, its solubility is far too low to achieve X_ϵ of 10^{-4} cm. Thus it is impossible, using thionine, to achieve the necessary condition of eqn (3) $(X_\epsilon < 0 \cdot 10 \ X_k)$. Lichtin and co workers [19, 20] have tried to overcome these problems by using acetonitrile as a solvent. Even so, the solubility is not sufficient. Furthermore, there is another problem in that thionine forms dimers at concentrations of $\sim 10^{-3} \ mol \ dm^{-3}$. These dimers are inactive from the point of view of the photo redox system; the energy is lost by internal quenching in the dimer. We therefore decided that the right approach must be to synthesise modified thiazine dyes by attaching sulphonate groups to the thiazine nucleus.

The Synthesis of Modified Thiazine Dyes

We have developed Bernthsen's synthesis[20] of methylene blue so that we can couple two different monomers to form sulphonated thiazine dyes as shown in scheme 1.

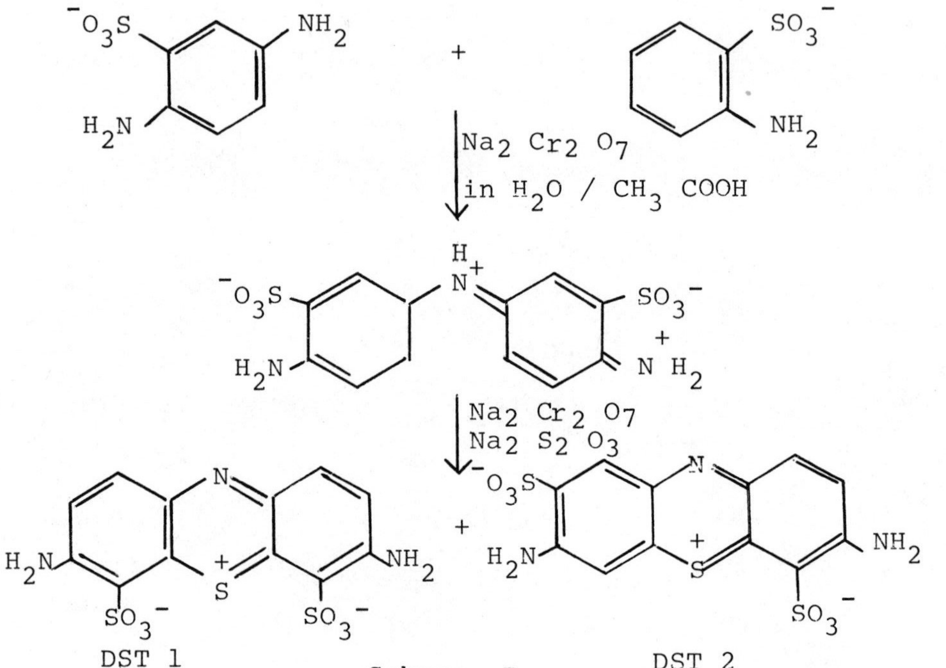

Scheme I

By a modification of the route we were able to make a
third isomer   DST4 :-

DST4

Starting with   N   methyl substituted monomers we have
recently used Scheme   I   to produce two   N, N, dimethyl
isomers   DMST 1   and   DMST 2, in which each NH_2 is
NHMe.   The location of the sulphonate groups in these
isomers has yet to be determined.

The Properties of the Modified Dyes

     All of these isomers are considerably more soluble
in water than thionine and we estimate that concentrations
up to 3×10^{-2} mol dm^{-3} can be achieved.   Furthermore,
there is no evidence of dimerisation.   In Table 4 we have
collected together values for the different parameters
required to assess the suitability of the dyes for use in
a photogalvanic cell.

     As regards the absorption of light the disulphonated
thionines are all similar to thionine.   The same holds
true of the Stern Volmer constant; concentrations of
Fe (II)   of about 10^{-2} mol dm^{-3} are sufficient to sca-
venge Th^*.   The values of ϕ_1 cluster around   ½   and
this is an unavoidable loss.   Work at the Royal Institution[22]   shows that for thionine this factor arises
from inefficient intersystem crossings from singlet to
triplet.   The values of k_1/k_3 are reasonably satisfactory and mean that ϕ_2 will be close to its limiting
value of   ½, especially if X_ϵ is as small as 10^{-4} cm.
The values of k_{-2} are all close to the critical value
of 10^{-3} dm^3 mol^{-1} s^{-1} except for   DST4.   The larger
value for this compound is the reason why we have not
measured its other parameters.   Introduction of the sulphonate groups shifts $E^{\ominus}$ to more negative values.

This is desirable since it increases the voltage of the cell. However this desirable shift is accompanied by an undesirable increase in k_{-2}. From the limited data we can construct the "linear free energy relation", shown in figure 5.   The implications of this linear free energy relation on the efficiency of the cell will be discussed in the fourth lecture.

TABLE   4

Properties of Disulphonated Thionines at $25^{\circ}C$

in 0.05 M H_2SO_4

	Thionine	DST1	DST2	DST4	DMST1	DMST2
$\dfrac{\lambda max}{nm}$ [a]	599	585	580	593	615	626
$\dfrac{\varepsilon}{cm^2\mu mol^{-1}}$ [b]	129	161	115	–	181	184
$\dfrac{k_T/k_q}{mmol\ dm^{-3}}$ [c]	4.0	1.0	1.0	–	3.3	–
ϕ_1 [d]	0.55	0.56	0.44	–	0.2	0.65
$\dfrac{k^2_{-1}/k_3}{dm^3mol^{-1}s^{-1}}$ [e]	8	27	50	–	–	–
$\dfrac{k_{-2}}{dm^3mol^{-1}ms^{-1}}$ [f]	0.47	1.0	2.2	10	0.5	2.5
$\dfrac{E_{\frac{1}{2}}}{mV}$ [g]	208	188	177	145	163	164

Notes

a) λmax is the wavelength of maximum absorption.

b) ε is the extinction coefficient at the maximum absorption (in natural ln units).

c) k_T / k_q is the Stern Volmer term and gives the critical concentration (eqn (1)) of Fe (II) to trap Th*.

d) ϕ_1 is the quantum efficiency for production of S$^\bullet$ (assuming there is sufficient Fe (II)).

e) ϕ_2 can be calculated from these values using eqn (15); there are complications for DMST1 and DMST2 but the values of ϕ_2 are similar to those for the other isomers.

f) k_{-2} is the rate constant for L + Fe (III).

g) $E_{\frac{1}{2}}$ is the half wave potential for the reduction of Th, measured with respect to a saturated calomel electrode; since the systems are reversible, $E_{\frac{1}{2}} \simeq E^{\ominus}$.

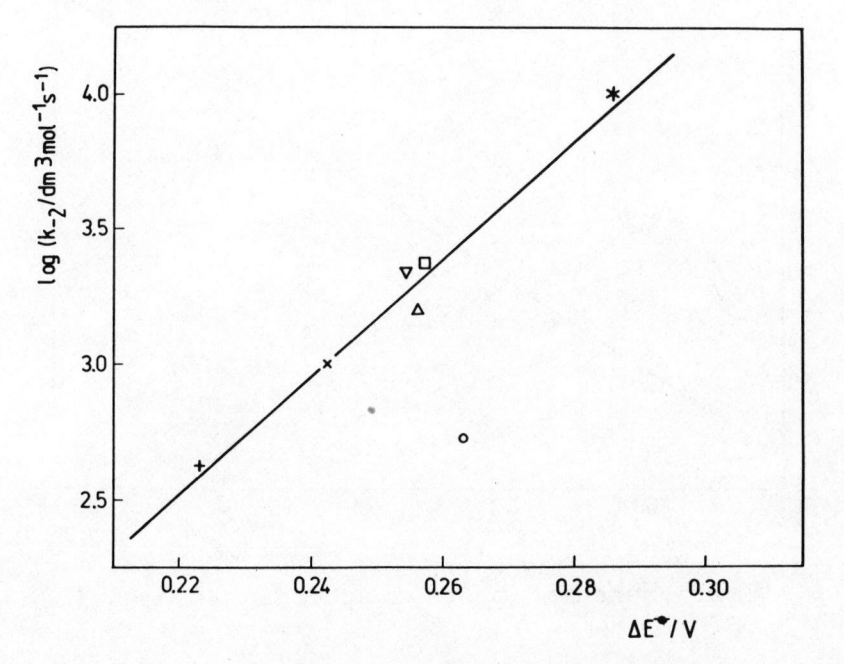

Fig. 5.   Linear Free Energy Relation for Thiazine dyes.

Self Quenching

A problem with increasing the concentration of the dye to 10^{-1} M to reduce X_ε to 10^{-4} cm is that there may be self quenching of Th^* by Th. Unfortunately preliminary results on DST2 show that indeed this is the case. Figure (6) shows plots of the observed $(\phi_1^{-1})_{obs}$ against DST2 according to the following equation

$$(\phi_1^{-1})_{obs} = \phi_1^{-1} \left[1 + \frac{k_T + k_{Th} [DST2]}{k_q [Fe(II)]} \right] \qquad (20)$$

From the gradient we find that k_{Th} is close to the diffusion controlled limit :-

$$k_{Th} / dm^3 \ mol^{-1} \ s^{-1} = 10^9 \qquad (21)$$

This is perhaps not surprising since the dimer is shown to undergo efficient internal quenching.

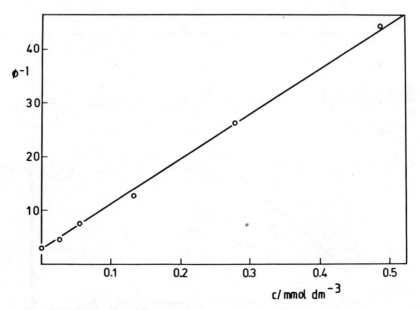

Fig. 6.   Self quenching plotted according to eqn (20) for DST2.

In order to further increase the solubility of the
dye we are now synthesising tetra sulphonated derivates.
We hope that the increased charge may, besides, decrease
the self quenching. If this is not the case then more
Fe (II) will have to be added in order to trap Th*; this
will lead to a reduction in the cell voltage through the
Nernst term.

Summary of Progress to Date

To sum up disulphonation has produced isomers which
are nearly soluble enough, which do not dimerise and
which have the required kinetic characteristics. However
to increase the voltage from the cell it would be desirable to find electron releasing substituents that
shift $E^{\ominus}$ to more negative values but have little effect
on k_{-2}. Self quenching may be prevented by increasing
the number of sulphonate groups.

ELECTRODE SELECTIVITY

Introduction

We now turn from the processes occuring in the
bulk of the solution to the kinetics of the electrode
reactions. As discussed in the first lecture it is
vital that the illuminated electrode should discriminate
between the photogenerated products, in general B and Y,
or in particular L and Fe (III). If the electrode is not
selective then it merely catalyses the back reaction.
With a semi-conductor electrode the separation of photogenerated holes and electrons is achieved with the help
of the field in the space charge layer. The counter-
part of the combination of B and Y on an unselective
electrode is the surface recombination of holes and
electrons. Thus in the photogalvanic cell the selective
electrode both separates B and Y and prevents their recombination. That is why it is a vital part of the cell.

The Problem

Since for any system B and Y are energetic species,
and furthermore, to be formed, they have to undergo
rapid electron transfer reactions one will need a rather
special sort of electrode to prevent the recombination.
Semi-conductor electrodes (n type SnO_2) have been used
with limited success for the iron thionine cell. [23]
However we found that it was possible to make a thionine
coated modified electrode that had the right kinetic

properties.   It is the ease with which this selective
electrode can be made, that makes the iron thionine
system the most attractive candidate for a photogalvanic
cell.

The Manufacture of the Thionine Coated Electrode

Up to twenty layers of thionine can be irreversibly
coated on to an electrode made of Pt, Au or SnO_2 by
holding the electrode at about $1 \cdot 1$ V for several minutes
in a thionine solution.[24, 25, 26]   The coating process
can be followed using a ring-disc electrode.   The ring
electrode, which is downstream of the disc, monitors the
consumption of thionine as it is plated on to the disc
[25].   From the integrated ring current we can calculate
the amount of thionine that was consumed in the coating
process.

Properties of the Thionine Coated Electrode

The electrode can then be removed, washed, dried,
waved about in the air and put into background electrolyte.   One can then observe cyclic voltammograms such
as those in figure 7, which result from the reduction
and oxidation of the coated thionine.

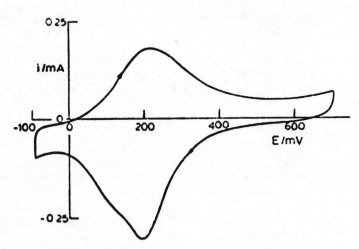

Fig. 7.   Typical cyclic voltammogram of the thionine
           coated electrode in background electrolyte.

From the area we can find the amount of coated thionine
and this is in good agreement with that calculated from
the ring-disc plating experiment.   The proportions of
reduced and oxidised thionine can also be measured using
the optical absorption of the coated thionine.   The
absorbance of the film on a quartz plate is measured as
a function of potential.   The system obeys the Nernst
equation except that the slope is about ¼ of that
expected for a two electron system.   This sluggish response to changes in potential can also be linked with
the broad cyclic voltammograms in figure (7) and perhaps
with the broadened absorption spectrum for the coated
thionine.   It may be that there is a Gaussian distribution of coated species, each with a somewhat different $E^{\ominus}$.

Spectra taken by XPES confirm that thionine is
coated on the electrode; the coat is so thick that the
Pt signal cannot be seen.

Electrode Kinetics

Turning to the electrode kinetics current voltage
curves for the reduction of thionine and of Fe (III) on
clean and on coated electrodes are shown in figure 8.

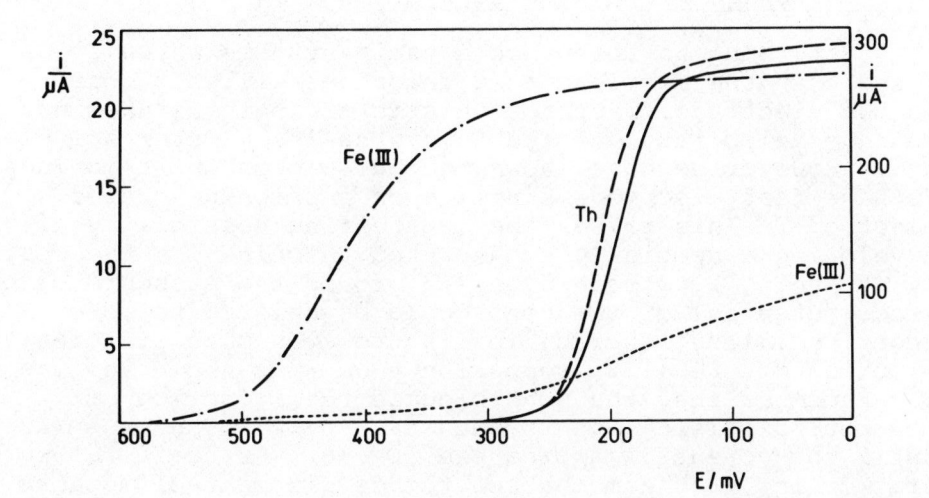

Fig. 8.   Current voltage curves for the reduction of
          Fe (III) (−·−·) and (---) and of Th (− − −) and
          (—) on clean and Th coated Pt respectively.

It can be seen that the coating process hardly affects
the electrode kinetics of the thionine reduction, while
the current voltage curve for the Fe (III) reduction
is shifted and much reduced. This is exactly the type
of electrode that we need for an efficient photogalvanic
cell.

    The selective electrode kinetics of the thionine
coated electrode extends to other systems. For instance,
the coat hardly affects the kinetics of organic systems
such as quinone / hydroquinone, but other inorganic
systems such as the reduction of Ru $(bpy)_3^{3+}$ and Ce (IV)
are blocked.[25, 26] For inorganic systems it appears
that the ions are reduced by direct reaction with the
leucothionine present in the coat rather than with the
metal itself. This explains why the current voltage
curves are shifted close to $E^{\ominus}$ for thionine / leucothionine in the coat. Because of its more rapid
kinetics Ru $(bpy)_3^{3+}$ requires much less L than Fe(III);
hence the former is reduced efficiently a hundred mV or
so more positive than $E^{\ominus}_{coat}$ while Fe (III) is reduced
inefficiently at a potential that is several hundred
millivolts more negative than $E^{\ominus}_{coat}$. We presume that
the organic compounds are able to adsorb on to the
coated electrode and hence react more efficiently than
the aquated inorganic ions.

Application to Photogalvanic Systems

    Returning to the photogalvanic systems we have
found that the disulphonated thionines will not coat
on the electrode. Perhaps this is not surprising since
we have deliberately tried to synthesise a water soluble
dye. However we have shown that disulphonated thionines
do have fast electrode kinetics on a thionine coated
electrode. This raises the interesting possibility of
developing a specially substituted thionine for the coat.
By attaching electron releasing groups the standard electrode potential of the coat could be shifted to more
negative values, thereby leading to even more efficient
blocking of Fe (III) reduction than that shown in figure
8. Remember that the coat blocked the reduction of
powerful oxidising agents such as Ru (III) and Ce (IV)
until the potential approached $E^{\ominus}$ of the coat. So we
are working at the moment on the development of methoxy
substituted thionines to provide the selective paint that
we need for the electrode. The fact that thionine coated
electrodes can be made so easily and that they have the
right selective properties are two of the most important
reasons why the iron thionine system is the best prospect

for an efficient photogalvanic cell.

THE EFFICIENCIES OF PHOTOGALVANIC CELLS

Introduction

In the previous lecture we saw how the use of thionine coated electrode will give us the necessary electrode selectivity to separate L and Fe (III) and prevent their recombination. We consider that this problem can therefore be solved. We shall now explore how the efficiency depends on the behaviour of the homogeneous part of the system. We remember that it is always advantageous to reduce X_ε and absorb the light as close as possible to the electrode. We shall therefore present our results as a function of X_ε. Other variables that can be changed in the cell are :-

1) $\left[\text{Fe (II)}\right]$

2) $\left[\text{Fe (III)}\right]$

3) $E^{\ominus}$ of thiazine dye

4) k_{-2} of thiazine dye

5)   pH.

We now discuss these in turn.

The Concentration of Fe (II)

The concentration of Fe (II) must be large enough to capture the triplet state of Th and to prevent flourescence or self quenching. On the other hand increasing the concentration of Fe (II) reduces the voltage of the cell through the Nernst concentration term at the dark electrode. From the results for DST2 self quenching can only be prevented if $\left[\text{Fe (II)}\right]$ is some hundred times greater than the concentration of the thiazine dye. Dye concentrations of 10^{-2} to 10^{-1} mol dm^{-3} are required to obtain the ideal low value of X_ε of 10^{-4} cm. So that this requirement may lead to an unfavourable Nernst term of 0·1 V.

The Concentration of Fe (III)

There is also an optimum concentration of Fe (III). There must be sufficient Fe (III) to prevent concentration polarisation at the dark electrode. The required condition is

$$[Fe~(III)] > \phi_1 ~N~I_o~X_1~/~D \qquad\qquad (22)$$

where N is the collection efficiency describing the fraction of photogenerated L that is converted into current on the electrode. We have written this equation with X_1 but more strictly the distance should be the shorter of either X_1 or the thickness of the natural diffusion layer. In our calculations we take $X_1 = 10^{-2}$ cm.

While eqn (22) requires [Fe (III)] to be greater than a critical value, the larger the concentration of [Fe (III)] , the greater the rate of the back reaction of L with Fe (III) and the smaller the value of N. N is given by

$$N = X_k~/~(X_k + X_\varepsilon) \qquad\qquad (23)$$

From eqn (22) and (23) by differentiation we find that the optimum value of [Fe (III)]$_m$ is given by

$$[Fe~(III)]_m \simeq \frac{(\phi_1~I_o~X_1~/~D)}{1 + (\phi_1~I_o~X_1~X_\varepsilon^2 k_{-2}/D^2)^{1/3}} \qquad\qquad (24)$$

The presence of Fe(III) may also affect the fate of $S^{\cdot}$ but in nearly all cases of practical interest the more severe restriction arises from the reaction of Fe (III) with L. The voltage developed by the cell depends on the Nernst term containing Fe (III)- the larger the [Fe (III)] the more voltage. However because of the logarithmic nature of the Nernst term and because [Fe (III)]$_m$ does not vary greatly this effect is relatively small. Substitution of eqn (24) in eqn (23) gives an approximate expression for the collection efficiency when [Fe (III)] has been optimised :-

$$N \simeq \left[1 + (\phi_1~I_o~X_1~X_\varepsilon^2 k_{-2}/~D^2)^{1/3}\right]^{-1} \qquad\qquad (25)$$

Using the typical values of

$$\phi_1 = 0.5$$

$$I_0 / mol\ cm^{-2}\ s^{-1} = 4 \times 10^{-8} \qquad (26)$$

$$X_1 / cm = 10^{-2}$$

$$D / cm^2\ s^{-1} = 6 \times 10^{-6}$$

we display in figure (9) contour diagrams showing
how N and [Fe (III)]$_m$ vary with X_ε and k_{-2}.
Because of the cube root in eqn (24) and (25) neither
N nor [Fe (III)]$_m$ vary strongly with k_{-2}. This means
that while it is relatively easy to obtain N = 0·1
it is very much more difficult to obtain N = 0·9.
The last steps are much more difficult than the first.
We also indicate the improvement in N for the disulphonated thionines compared to thionine, and we include
a point for thionine in acetonitrile which was used by
Lichtin in his cell.

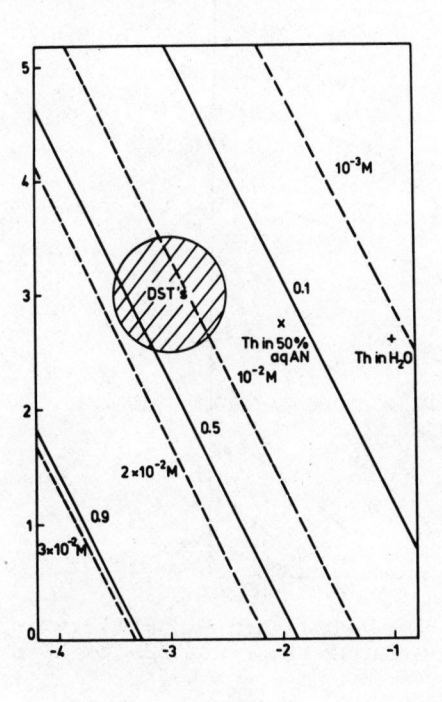

Fig. 9. Results from
eqn (24) and (25)
for N (solid lines)
and [Fe(III)]$_m$
(broken lines) as a
function of log
$(k_{-2}/dm^3\ mol^{-1}\ s^{-1})$
on the y axis and
log (X_ε/cm) on the
x axis. The pos-
itions of Th and
the DSTs are shown.

The variation of Power with $\Delta E^{\ominus}$ and k_{-2}

While the current depends mainly on efficient collection, the power developed by the cell also depends on the voltage described by $\Delta E^{\ominus}$. We have shown that the fill factor only varies between 0.6 and 0.8 and hence we can use $N\Delta E^{\ominus}$ as an estimate of the power produced by the cell. Clearly we would like $\Delta E^{\ominus}$ as large as possible but, alas, increasing $\Delta E^{\ominus}$ nearly always increases k_{-2} as well. For a series of compounds it is quite common for $\log k_{-2}$ and $\Delta E^{\ominus}$ to be related by a "Linear Free Energy Relation". We have already seen some data for thionine plotted in this way in figure 5. These data are replotted together with some results [27] for the rival Ruthenium Iron system in figure 10. Now for each series of compounds there will be an optimum compromise between $\Delta E^{\ominus}$ and k_{-2}. We write the LFER as

$$\ln k_{-2} = \ln (k_{-2})_o + \alpha |\Delta E^{\ominus}| F /RT$$

Then from eqn (26) we obtain

$$\frac{1}{\Delta E^{\ominus}N} = \frac{1}{\Delta E^{\ominus}} \left[1 + B^{1/3} \exp (P) \right] \qquad (27)$$

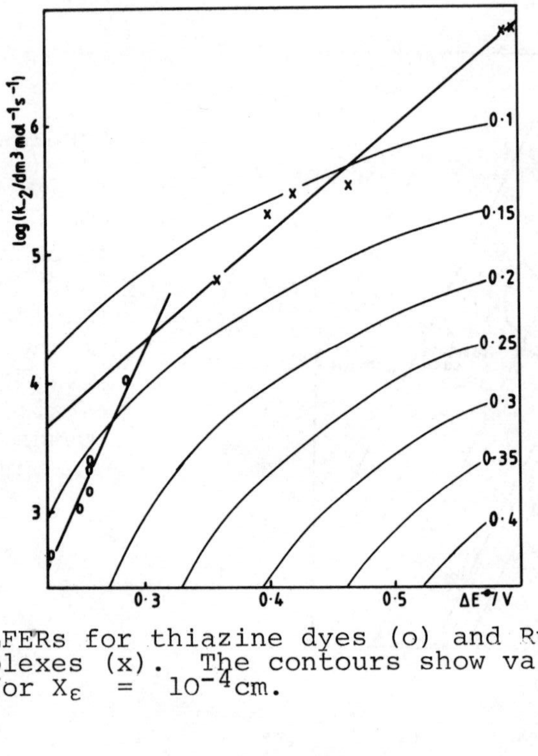

Fig. 10. LFERs for thiazine dyes (o) and Ru (II) complexes (x). The contours show values of $N\Delta E^{\ominus}_m$ for $X_\varepsilon = 10^{-4}$cm.

where
$$B = \frac{\phi_1 I_o X_1 X_\varepsilon^2 (k_{-2})_o}{D^2} \tag{28}$$

and
$$P = \alpha F \Delta E^{\ominus} / 3RT \tag{29}$$

Differentiation gives the condition for the optimum compound in the series that obey the LFER :-

$$B^{1/3} = \frac{\exp(-P)}{P - 1} \tag{30}$$

The value of the maximum power $(\Delta E^{\ominus} N)_m$ and collection efficiency N_m are given by

$$(\Delta E^{\ominus} N)_m = (\Delta E^{\ominus})_m - 3RT / \alpha F \tag{31}$$

and
$$N_m = 1 - P^{-1} \tag{32}$$

Taking the values used above for figure 9 and the appropriate values for the LFER, in figure 11 we show how $\Delta E_m^{\ominus}$, N and $N\Delta E_m^{\ominus}$ vary with X_ε. These results were obtained from eqn (27) to (32) for different values of P. From figure 11 it can be seen that the reduction of X_ε towards 10^{-4}cm is essential to increase $\Delta E_m^{\ominus}$, to obtain a collection efficiency close to unity and thereby any significant power conversion. The existing dyes have values of $\Delta E^{\ominus}$ in the right range but they still need to be made even more soluble. We can now return to figure 10 and using eqn (25) plot contours of $\Delta E^{\ominus} N$ for the target value of $X_\varepsilon = 10^{-4}$cm. This gives us the map for our future research.

The ideal values of $\Delta E_m^{\ominus}$ and N are 1.1 V and 1 respectively. With these values and a putative cell that absorbed all the solar radiation of energy greater than 1.8 eV we have calculated [2] that the sunlight engineering

efficiency would be 20%. In using the value of I_0 in
eqn (26) we have assumed that the dye absorbs only $\frac{1}{4}$
of the required photons. The value of 0.2 for $N\Delta E_m^{\ominus}$
for X_ε = 10^{-4}cm means that a further factor of 5
is lost in the operation of the cell. We can predict
therefore that the overall efficiency of the cell under
these conditions could only be 1%. This disappointing
result arises partly from the steep slope of the LFER
in figure 5 where α = 1.18. This is an atypical
gradient and it may be that more compounds like DMST1
will be found which lie below the line.

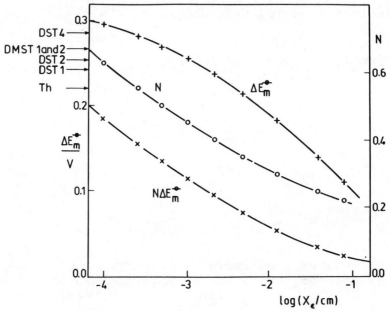

Fig. 11. Results for optimum $\Delta E^{\ominus}_m$, N and $N\Delta E^{\ominus}_m$ for
thiazine dyes as a function of X_ε.

Turning to the rival Ruthenium system we can carry
out the same analysis using the LFER in figure 10 the
results are presented in figure 12. Again it is
essential to reduce X_ε to 10^{-4}cm, but even so $N\Delta E_m^{\ominus}$
is $\frac{1}{2}$ as good as the Thiazine optimum. The optimum
compound lies outside the range of existing compounds in
a region where it can be seen in figure 10 that the
thiazine dyes have lower values of k_{-2} for the same
$\Delta E_m^{\ominus}$ and better values of $N \Delta E_m^{\ominus}$. The slope of the
Ruthenium LFER is 0.49. This is a sensible value and it
is therefore much less likely that there will be ex-
ceptions. Furthermore there is as yet no selective

electrode for Ruthenium / Iron and Ruthenium is more expensive than thionine. For all these reasons we consider that there are better prospects for the iron thiazine system than for the iron ruthenium system.

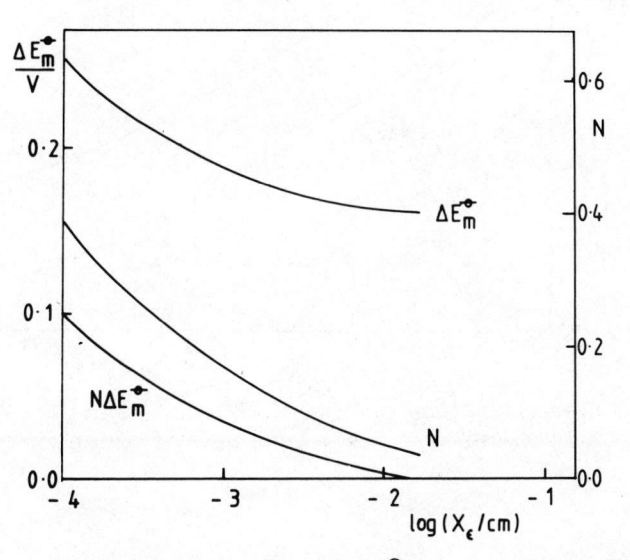

Fig. 12.   Results for optimum $\Delta E^{\ominus}_m$, N and N$\Delta E^{\ominus}_m$ for iron ruthenium system.

Variation with pH

Another variable that can be used to alter $\Delta E_m^{\ominus}$ and k_{-2} is pH. We have shown [28] that

$$\Delta E^{\ominus} \simeq \text{constant} + 1.5 \text{ pH}$$

$$\text{and} \quad \log k_{-2} \simeq \text{constant} + \text{pH}.$$

giving another LFER with $\alpha = 2/3$. This therefore has a smaller slope than the LFER produced by substitution. Figure 13 shows the results of the analysis for varying k_{-2} and $\Delta E_m^{\ominus}$ with pH. For $X_\epsilon = 10^{-4}$cm about twice as much power could be produced at pH of 4 to 5. So this variable may well be worth further investigation.

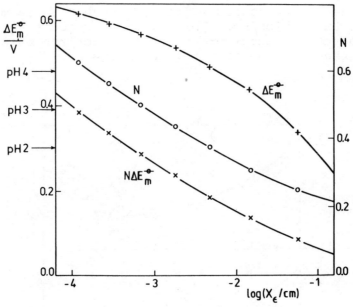

Fig. 13.   Results for optimum $\Delta E^{\ominus}_m$, N and N$\Delta E^{\ominus}_m$ for a
            thiazine dye as a result of changing the pH.

Final Summary

We now summarise in Table 5 the main factors that
lead to loss of efficiency in an iron thiazine cell
with X_ε = 10^{-4} cm.   It is clear that to obtain worth-
while efficiencies it will be vital to use sensitizers
or a cocktail of dyes to absorb more of the solar
radiation.   Interesting results have already been obtained by Lichtin[29],[30].   Even so the best photogalvanic
cell is always likely to be less efficient than a semiconductor cell.   However it may be cheaper and this
type of cell will always have one great advantage - it
is easier to change the electrolyte than the electrode.

Table 5

Factors affecting efficiencies

	a	Best possible b
Adsorption of Radiation	$\frac{1}{8}$	$\frac{1}{2}$
Quantum Efficiency ϕ_1	$\frac{1}{2}$	$\frac{1}{2}$ c
Voltage Factor	$\frac{1}{5}$	$\frac{1}{5}$
	1%	5%

Notes a) A single thiazine dye only absorbs $\sim \frac{1}{8}$ of the total insolation.

b) An ideal cocktail should absorb $\sim \frac{1}{2}$ of the total insolation.

c) The ideal result would be 1 but results for thiazine dyes show that $\phi_1 \simeq \frac{1}{2}$.

d) The ideal cell should produce 1.8 V. The results in figures 11 and 13 show that voltages greater than 0.4 V are unlikely to be achieved.

Acknowledgements

I am grateful to the following: -

to Drs Souto Bachiller and Wilson for synthesising the new thiazine dyes, to Dr Foulds, Mr Bartlett and Mr Davies for the kinetic characterisation and to Dr Hillman, Mr Colby and Mr Boutelle for their work on electrode selectivity. I am particularly grateful to Dr Foulds for lively discussion and help in preparing these lectures.

References

1. M.D. Archer and W.J. Albery, Affinidad.
    346 : 257 (1977).

2. W.J. Albery and M.D. Archer, Nature.
    270 : 399 (1977).

3. W.J. Albery and M.D. Archer, J. Electroanal.
    Chem. 86 : 1 (1978).

4. W.J. Albery and M.D. Archer, J. Electroanal.
    Chem. 86 : 19 (1978).

5. C.T. Lin and N. Sutin, J. Phys. Chem.
    80 : 97 (1976).

6. W.J. Albery and A.W. Foulds, J. Photochem.
    10 : 41 (1979).

7. N.N. Lichtin - Chapter 5 - in "Solar Power
    and Fuels", J.R.R. Bolton, ed. Academic
    Press. New York. (1977).

8. C.G. Hatchard and C.A. Parker, Trans.Faraday Soc.
    54 : 1093 (1961).

9. D.E. Hall, W.D.K. Clark, J.A.Eckert, N.N. Lichtin,
    and P.D. Wildes, Ceram. Bull. 56 : 408
    (1977).

10. M.I.C. Ferreira and A. Harriman, Faraday Trans.I.
    73 : 1085 (1977).

11. W.J. Albery, W.R. Bowen and M.D. Archer,
    J.Photochem. 11 : 15 (1979).

12. W.J. Albery, W.R.Bowen, M.D.Archer and
    M.I.C. Ferreira, J. Photochem. 11 : 27
    (1979).

13. W.J. Albery, M.D. Archer and R.G. Egdell
    J. Electroanal. Chem. 82 : 199 (1977).

14. W.J. Albery, W.R. Bowen, F.S. Fisher and
    A.D. Turner, J. Electroanal. Chem. 107 : 1
    (1980).

15. W.J. Albery, W.R. Bowen, F.S. Fisher and
    A.D. Turner, J. Electroanal. Chem. 107 : 11
    (1980).

16. W.J. Albery, P.N. Bartlett, W.R. Bowen,
    F.S. Fisher and A.W. Foulds, J. Electroanal.
    Chem. 107 : 23 (1980).

17. V.G. Levich, "Physicochemical Hydrodynamics", Prentice Hall, New Jersey. (1962) p 68.

18. W.J. Albery, "Electrode Kinetics", Clarendon Press, Oxford (1975), p 53.

19. P.D. Wildes, N.N. Lichtin and M.Z. Hoffman, J. Am. Chem. Soc. 97 : 2288 (1975).

20. A. Bernthsen, Chem. Ber. 17 : 2854 (1884).

21. W.J. Albery, P.N. Bartlett, A.W. Foulds, F.A. Souto-Bachiller and R. Whiteside, J.C.S. Perkin II, submitted for publication.

22. A. Harriman, Private Communication.

23. D.E. Hall, P.D. Wildes and N.N. Lichtin, J. Electrochem. Soc. 125 : 1365 (1978).

24. W.J. Albery, W.R. Bowen, F.S. Fisher, A.W. Foulds, K.J. Hall, A.R. Hillman, R.G. Egdell and A.F. Orchard, J. Electroanal.Chem. 107 : 37 (1980).

25. W.J. Albery, A.W. Foulds, K.J. Hall and A.R. Hillman, J.Electrochem.Soc. 127 : 654 (1980).

26. W.J. Albery, A.W. Foulds, K.J. Hall, A.R. Hillman, R.G. Egdell and A.F. Orchard, Nature. 282 : 793 (1979).

27. C.T. Lin, W. Böttcher, M. Chou, C. Creutz and N. Sutin, J. Am. Chem. Soc. 98 : 6536 (1976).

28. W.J. Albery, P.N. Bartlett, J.P. Davies, A.W. Foulds, A.R. Hillman and F. Souto-Bachiller, Disc. Faraday Soc. 70 - to be published.

29. N.N. Lichtin, P.D. Wildes, T.L. Osif and D.E. Hall, Adv. Chem. Ser. 173 : 296 (1979).

30. P.D. Wildes, D.R. Hobart and N.N. Lichtin, Solar Energy 19 : 567 (1977).

CHARGE SEPARATION AND REDOX CATALYSIS IN

SOLAR ENERGY CONVERSION PROCESSES

K. Kalyanasundaram and M. Grätzel

Institut de Chimie Physique
Ecole Polytechnique Fédérale de Lausanne
Ecublens, 1015 Lausanne
Switzerland

1. INTRODUCTION

Photochemical conversion and storage of solar energy is receiving increased attention[1.1,1.2] all over the world, and there has been some distinct progress in the area of photogeneration of fuels in homogeneous solutions. This review paper is an attempt to review the "state of art" in this area, indicate the various approaches and discuss their success, scope and limitations. The types of photochemical reactions that have been investigated in homogeneous solutions can be broadly classified as seen in (Fig. 1).

(i) Photoredox reactions leading to production of fuels,
     e.g. a) photodissociation of water into H_2 and O_2

$$H_2O \longrightarrow H_2 + \frac{1}{2} O_2 \qquad (1)$$

     b) photoreduction of CO_2, N_2, etc.

$$CO_2 + H_2O \longrightarrow HCOOH + \frac{1}{2} O_2 \qquad (2)$$

$$N_2 + 3 H_2O \longrightarrow 2 NH_3 + \frac{3}{2} O_2 \qquad (3)$$

(ii)   Molecular energy storage reactions such as photoisomera-
        tions. Here the light energy is stored in transformed
        'metastable' products and energy subsequently released
        as heat with the aid of suitable catalysts

$$A \xrightarrow{h\nu} B \qquad\qquad (4)$$

$$e.g.\ norbornadiene \underset{\Delta}{\overset{h\nu}{\rightleftarrows}} quadricyclene \quad (5)$$

(iii)   Homolytic bond fission reactions where the visible light
        is used to carry out an endergonic bond fission reaction

$$A \xrightarrow{h\nu} B + C \qquad\qquad (6)$$

$$e.g.\quad FeBr^{2+} \xrightarrow{h\nu} Fe^{2+} + \tfrac{1}{2} Br_2 \qquad (7)$$

Fig. 1 Types of Photochemical Reactions Employed for Light-Energy
        Conversion.

        Attractive among the various schemes that have been proposed
are those which are intimately related to the green plant photosynthesis. All of the type (i) reactions give energy-rich fuels
such as H_2, HCOOH, alcohols and NH_3 and are carried out efficiently
in plants. The important point to note is that, in very simple
terms, all these processes are endergonic photoredox reactions.
In this review we would like to focus our attention from this view-
point and try to assess the progress that has been made in recent
years. Two major sources of gross inefficiency in homogeneous
systems based on the redox reactions are the very efficient reverse
electron transfer between the redox products and/or the destructive
side reactions of the free radical intermediates. Plants overcome
these problems by carrying out reactions in organised molecular en-
vironment and through the use of various enzymes to mediate these
complex multi-electron transfers. So here our emphasis will be to
explore how redox reactions carried out in organised molecular media
such as functional micelles can lead to better charge separation
between the redox products and also review the recent success in
the development of suitable enzyme analogs labelled 'redox catalysts' to mediate multi-electron transfers. Various aspects of
in vitro photosynthesis has been addressed recently by several
authors[1.3-1.16] and the discussion here will be complimentary to
the materials covered in these reviews.

Table 1. Some Endergonic Fuel-Generation Reactions Involving N_2, CO_2, and H_2O

Reaction	ΔH^O (kj/mole)	ΔG^O (kj/mole)	n	ΔE^O (V)	λ_{max} (nm) thresh.	
					One photo-system	Two photo-systems
$H_2O(l) \rightarrow H_2(g) + \frac{1}{2}O_2(g)$	286	237	2	1.23	611	877
$CO_2(g) \rightarrow CO(g) + \frac{1}{2}O_2(g)$	283	257	2	1.33	581	845
$CO_2(g) + H_2O(l) \rightarrow HCOOH(l) + \frac{1}{2}O_2(g)$	270	286	2	1.48	543	804
$CO_2(g) + H_2O(l) \rightarrow HCHO(g) + O_2(g)$	563	522	4	1.35	576	840
$CO_2(g) + 2 H_2O(l) \rightarrow CH_3OH(l) + \frac{3}{2} O_2(g)$	727	703	6	1.21	616	881
$CO_2(g) + 2 H_2O(l) \rightarrow CH_4(g) + 2 O_2(g)$	890	818	8	1.06	667	932
$N_2(g) + 3 H_2O(l) \rightarrow 2 NH_3(g) + \frac{3}{2} O_2(g)$	765	678	6	1.17	629	895
$CO_2(g) + H_2O(l) \rightarrow \frac{1}{6} C_6H_{12}O_6(s) + O_2(g)$	467	480	4	1.24	607	872

Table I adapted from a recent review paper of Bolton[1.10] summarises the energetics of various reduction schemes. It includes n, the number of electrons to be transferred in an electrochemical reaction, E, the redox potential gap, ΔH^O and ΔG^O, the standard enthalpy and free energy changes, λ_{max} (nm), the threshold in the visible photon energy which is required to achieve the given reaction in one or two photosystems (with 0.6eV energy loss for each electron transfer). Thermodynamic limitations on the conversion of light energy to chemical energy in reactions of this type have been discussed recently by Bolton,[1.10] Almgren[1.12] and also by Ross.[1.13-1.14]

The possibility of generating fuels H_2 and O_2 via photodissociation of water with sunlight has been addressed over the years by several authors. We would like to briefly mention the four instructive chemical cycles proposed by Balzani.[1.15] As shown in Figure 2, these schemes involve simple electron-transfer reactions of possibly metal ions, with decreasing threshold energy required to drive the reaction.

Cycle	Threshold λ

C1 $X + H_2O \xrightarrow{h\nu} X^+ + H + OH^-$

$$X^+ + \tfrac{1}{2}H_2O \rightarrow X + H^+ + \tfrac{1}{4}O_2$$

$$\frac{\qquad\qquad\qquad\qquad\qquad\qquad}{\tfrac{1}{2}H_2O \xrightarrow{h\nu} H + \tfrac{1}{4}O_2}$$
     332 nm

C2 $Y + H_2O \xrightarrow{h\nu} Y + H^+ + OH$

$$Y + H_2O \rightarrow Y + OH^- + \tfrac{1}{2}H_2$$

$$\frac{\qquad\qquad\qquad\qquad\qquad\qquad}{H_2O \xrightarrow{h\nu} \tfrac{1}{2}H_2 + OH}$$
     367 nm

C3 $Z + H_2O \xrightarrow{h\nu} ZO + H_2$

$$ZO \rightarrow Z + \tfrac{1}{2}O_2$$

$$\frac{\qquad\qquad\qquad\qquad\qquad\qquad}{H_2O \xrightarrow{h\nu} H_2 + \tfrac{1}{2}O_2}$$
     420 nm

or $cis\text{-}ML_4H_2^{2+} \xrightarrow{h\nu} ML_4^{2+} + H_2$

$$ML_4^{2+} + H_2O \rightarrow ML_4 + 2H^+ + \tfrac{1}{2}O_2$$

$$\frac{ML_4 + 2H^+ \rightarrow cis\text{-}ML_4H_2^{2+}}{H_2O \xrightarrow{h\nu} H_2 + \tfrac{1}{2}O_2}$$

C4 $ML_n + H^+ \xrightarrow{h\nu} (ML_n - H)^+$

$$(ML_n - H)^+ \rightarrow \tfrac{1}{2}(ML_n)_2^{2+} + \tfrac{1}{2}H_2$$

$$\frac{\tfrac{1}{2}(ML_n)_2^{2+} + \tfrac{1}{2}H_2O \rightarrow ML_n + H^+ + \tfrac{1}{4}O_2}{\tfrac{1}{2}H_2O \rightarrow \tfrac{1}{2}H_2 + \tfrac{1}{4}O_2}$$
     841 nm

Fig. 2   Classification of Methods for Photochemical Decomposition of Water (after Balzani et al.,[1.15])

Bolton has recently proposed an additional cycle (C5) involving two photochemical systems working, in series, with a

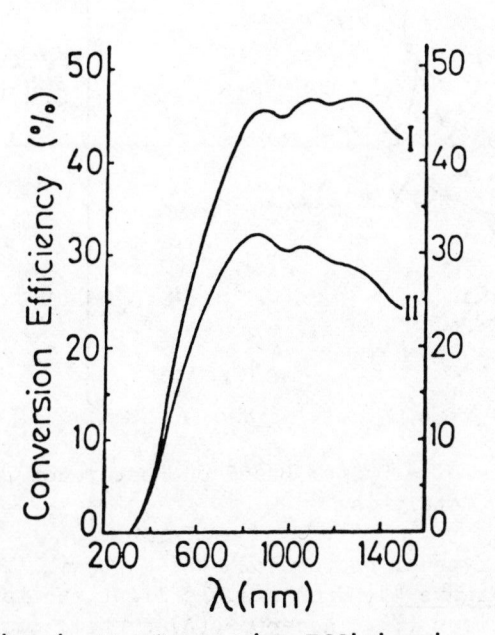

$$\text{(8)}$$

proton-conducting membrane connecting the two. The threshold energies required for each cycle depend on the energy losses involved in the light energy conversion step. Important to realise in the design of photoredox systems for solar light energy conversion is that the available solar radiation is strongly wave length dependent.

Fig. 3 Variation in the Conversion Efficiencies with wavelength for Photochemical Devices with a Threshold. Curve I is a plot of the fraction of incident solar power (in percentages) that is available at various threshold wavelengths and curve II is a plot of thermodynamic conversion efficiencies under optimal rates of energy conversion (for AM 1.2 solar radiation).

Figure 3 presents conversion efficiency in solar energy conversion systems at various threshold wavelengths (Curve I, a plot of fraction of solar power available at E_{thr}, Curve II, thermodynamic limits of conversion efficiency under optimal rate of conversion). According to Curve II, a solar energy conversion system with a 840 nm threshold can store about 32% of solar light and this value decreases drastically with decrease in the threshold wavelengths.

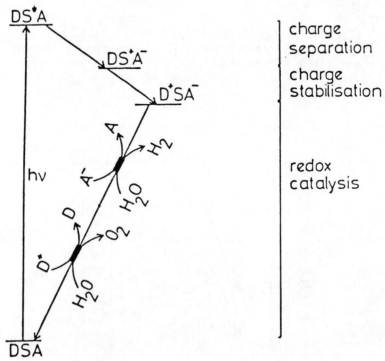

Fig. 4   Schematics of a System Based on Photoredox Reactions
           and Redox Catalysis

The simplest type of photoredox system we will be concerned with is shown in Figure 4. Upon visible light excitation of a donor (D) - sensitizer (S) - acceptor (A) system, one produces an excited state of the sensitizer which undergoes a photoredox reaction either with D or A or both.

$$S \xrightarrow{\;h\nu\;} S^* \qquad\qquad\qquad\qquad (9)$$

$$S^* + D \rightleftarrows S^- + D^+ \text{ (reductive quenching)} \qquad (10)$$

$$S^* + A \rightleftarrows S^+ + A^- \text{ (oxidative quenching)} \qquad (11)$$

The primary redox products are then stabilized either through judi-
cious choice of D/S/A relays or through the use of multiphase systems.
Generation of fuels H_2, O_2 from S^+ (or D^+) and A^- (or S^-) subsequently
is achieved with the aid of redox catalysts.

$$(S^+) \qquad 2D^+ + H_2O \xrightarrow{\text{cat.}} 2H^+ + \frac{1}{2} O_2 + 2D \tag{12}$$

$$(S^-) \qquad A^- + H_2O \xrightarrow{\text{cat.}} A + \frac{1}{2} H_2 + OH^- \tag{13}$$

With respect to the sensitizer relay, involvement of reaction (10)
constitutes a reductive cycle and that of reaction (11) an oxidative.
Note that a given sensitizer can also function as an electron-accep-
tor or donor. The overall success of a scheme like this very much
depends on the efficiency in each of the three principal areas:

(i) Suitable choice of donor-sensitizer-acceptor relays for
efficient light-induced charge transfer. The sensitizer
should have good absorption spectral features with respect
to the solar spectrum. Also the excited state should be
long-lived, have high quantum yields and also good redox
properties. The donor-acceptor relays must satisfy the
thermodynamic requirements for reactions (12) and (13),
viz, the redox potentials $E_o(D/D^+) > 1.23$ V and $E_o(A/A^-)$
< 0.0 V. Furthermore the relays must be stable to extensive redox cycles.

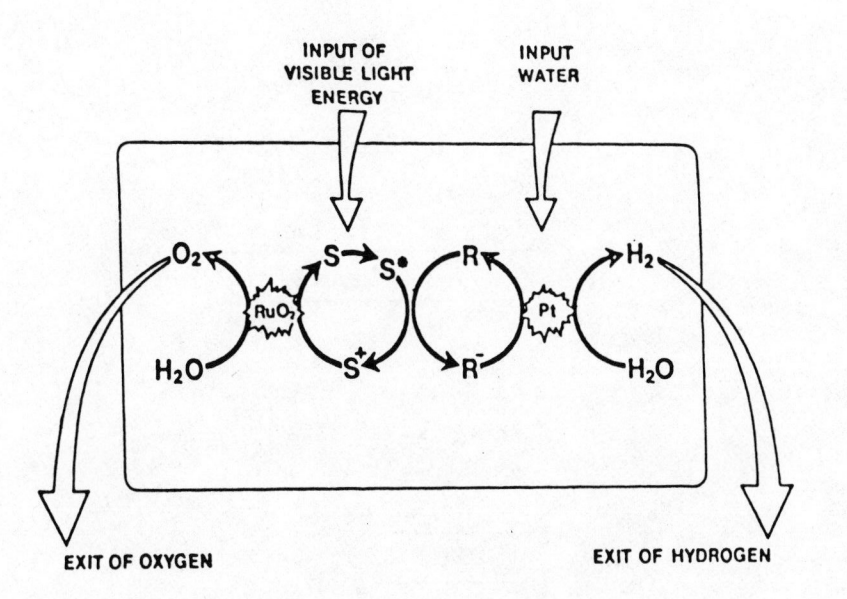

Fig. 5 Coupled Redox Catalysis in the Photodissociation of Water.

(ii) Stabilization of the redox intermediates through the use of multiphase systems. Various multiphase systems such as micelles, vesicles, microemulsions, monolayers and polyelectrolytes have been explored in this context.

(iii) Redox catalysis for fuel generation from the redox products. In the subsequent sections we review the progress in each of these areas. An ultimate goal of these schemes is complete sustained photodissociation of water of the type shown in Figure 5. The two half-cell reactions (O_2, H_2 production) can be carried out in an all-in-one homogeneous system or in photoelectrochemical cells. In the last section we will explore this latter concept further.

Most of the recent design of the half-cycles for the H_2 or O_2 production are based on the dye sensitized redox reactions. Two such cycles (designated as C1' and C2' cycles following Balzani's general scheme) for O_2 and H_2 production are shown in Figure 6. In C1' cycle, a visible light absorbing sensitizer S (a metal complex for example) is photooxidised to a higher oxidation state which in turn oxidises water in the dark. In addition to O_2, the reducing equivalents of H_2O are stored as a stable reduced acceptor QH or QH_2. In the C2' cycle, the sensitizer S (can be same as in   C2'

$$C1' : \quad S^{n+} + 2Q \xrightarrow{h\nu} S^{(n+2)} + 2Q^-$$

$$S^{(n+2)+} + H_2O \longrightarrow S^{n+} + \tfrac{1}{2} O_2 + 2H^+$$

$$2H^+ + 2Q^- \longrightarrow (QH_2 + Q) \text{ or } 2 QH$$

$$\text{Nett:} \quad H_2O + 2Q \xrightarrow{h\nu} (QH_2 + Q) + \tfrac{1}{2}O_2 \quad (\text{or } 2 \ QH)$$

$$C2' : \quad S + A \xrightarrow{h\nu} S^+ + A^-$$

$$S^+ + QH \longrightarrow S + QH^+$$

$$QH^+ \longrightarrow Q + H^+$$

$$A^- + H^+ \longrightarrow \tfrac{1}{2} H_2 + A$$

$$\text{Nett:} \quad QH_2 \xrightarrow{h\nu} Q + \tfrac{1}{2} H_2$$

Fig. 6   Photoredox Cycles for Decomposition of Water

cycle or different) shuttles the electron from a low energy donor QH to an acceptor A for eventual formation of H_2. An idealised complete cycle is one which employs the acceptor QH or QH_2 generated in the Cl' cycle as the electron donor for the C2' cycle. In a complete system the two cycles can be linked as part of a two-compartment all with a semi-permeable membrane (cycle C5 of Bolton) or in a flow-cell. Cl' and C2' cycles then roughly mimic photosystems II and I of photosynthesis respectively. In contrast to the Cl and C2 cycles of Balzani, these do not involve production of undesirable free radicals H and OH.

2. DESIGN OF PHOTOREDOX REACTIONS FOR PHOTODISSOCIATION OF WATER

2.1 Photodecomposition of Water in Homogeneous Solutions

In this section we will focus our attention on studies aimed at the photodecomposition of water in homogeneous solutions. In the absence of detailed knowledge on how to design a complete water-splitting device, all recent attempts have been in the design of half cells where either H_2 or O_2 alone is produced from water. Obviously such half-cells will involve sacrificial electron donors (for H_2) and acceptors (for O_2) which are either oxidised or reduced irreversibly. In spite of this shortcoming, studies of this type have been extremely educative and also informative in the design of photoredox systems for cyclic water decomposition. First we review the known sacrificial systems for photoproduction of H_2 from water and then some recent studies on thermal and light-induced O_2 production from water. The success of these systems are also partly due to the utilisation of suitable redox catalysts but these are discussed in the subsequent section. The H_2 and O_2 production from water are multi-electron transfer processes

$$2A^- + 2H_2O \longrightarrow 2A + H_2 \uparrow + 2OH^- \text{ (2e}^- \text{ reduction)} \qquad (14)$$

$$4D^+ \quad 2H_2O \longrightarrow 4A^+ + 4D + O_2 \uparrow \text{ (4}^{e-} \text{ oxidation)} \qquad (15)$$

and as will be seen later, the efficiencies of these reactions in the absence of the redox catalysts have been poor.

2.2 Photoproduction of H_2 from Water

Of all the redox reactions mentioned in section 1, hydrogen gas production from water by two-electron reduction of water (protons) is by far the simplest

$$2H^+ + 2e^- \longrightarrow H_2 \qquad (16)$$

Currently there are three main ways of photoproduction of H_2 from water:

    a)   photolysis of simple transition metal ions in solution
    b)   photolysis of metal hydrides
    c)   dye sensitized redox reactions (especially of the type C2 mentioned earlier).

Table II: H_2 Evolution Via Low-valent Transition Metal Ion Photolysis

Metal Ion	$E_0(HA^-)$	Medium	$\phi_{H2}(\lambda)$	Reference
Cu^+	0.158	1M HCl	0.70 (274)	(2.1 - 2.3)
Eu^{2+}	-0.43	1M HClO$_4$	0.20 (254)	(2.3 - 2.6)
Fe^{2+}	0.75	2M H$_2$SO$_4$	0.40 (254)	(2.7)
Cr^{2+}	-0.41	H$_2$SO$_4$	0.15 (254)	(2.8)
Ce^{3+}	1.44	1M HClO$_4$	0.001 (254)	(2.9)
V^{3+}	-0.255	H$_2$O-C$_2$H$_5$OH	0.23 (280)	(2.10)
Co^{2+}				(2.11)
Ti^{3+}/Cu^+	0.12	1M HCl	-	(2.12)
$Mo^{2+}{}_{aq}$	-	1M CF$_3$SC$_3$H	0.035 (254)	(2.13)
$Mo_2(SO_4)_4^{4-}$	-	5M H$_2$SO$_4$	0.17	(2.14)
$Rh^{III}(bpy)_3^{3+}$	-0.80	H$_2$O (pH 4.9)	0.02 (310nm)	(2.15)

2.2.1 Photolysis of simple ions in acid media

    This earliest as well as the simplest of all the methods involves photolysis of transition metal ions such as Cu^+, Fe^{2+}, Eu^{2+} and others in their CTTS (Charge Transfer To Solvent) bands in acidic media. Systems of this type for photoproduction of H_2 reported till date are collected in Table II. General features of this type of systems can be summarised as follows:

(i)   These are examples of reaction (2) of the C2 cycles;

(ii) CTTS absorption bands which are being irradiated are located in the UV region for majority of the TM ions. Quantum yields for

H_2 observed, however, are fairly high (0.1 - 0.8). Visible light photolysis of the CT bands in colored metal complexes of these ions, in few cases reported (e.g. $Cu(phen)^+$) however failed to produce H_2.

(iii) Mechanistically H_2 production in most cases is believed to go through metal hydride intermediates derived by reduction of the photo-electron initially produced

$$M^{n+} \xrightarrow{h\nu} M^{(n+1)+} + e^- \tag{17}$$

$$e^- + H^+ \longrightarrow H \tag{18}$$

$$M^{(n+1)+} + H \longrightarrow MH^{(n+1)+} \tag{19}$$

though only in few cases direct transient spectral evidence for the formation of MH• have been obtained, e.g. Cu photolysis. Photo-electron formation, however, has been shown in many cases.

(iv) Some of the metal ions that have been photolysed, e.g. Cr^{2+}, V^{2+}, Eu^{2+} are thermodynamically capable of reducing H_2O in the dark in dilute acids. H_2 production photochemically using UV light in these systems, thus, is not energy storing. It has been known for quite some time that TM ions such as Eu^{2+}, Cr^{2+} and V^{2+} get oxidised in H_2O and vigorously evolve H_2 upon introduction of redox catalysts such as a platinised Pt wire

$$2\, Eu^{2+} + 2\, H^+ \xrightarrow[\text{dark}]{Pt} 2\, Eu^{3+} + H_2 \tag{20}$$

(v) Except for the case of Ce^{3+} and Fe^{2+}, all these systems are sacrificial (non cyclic) in that, under optimal conditions, they liberate stoichiometric yields of H_2 and stop. Even in the case of Fe^{2+} and Ce^{3+}, regeneration of the starting materials require photo-lysis in the UV region. Also instability and complexation/polymeri-sation problems require utilization of very acidic conditions.

(vi) Rh (III) bipyridyls photolysis system is unique in that, in the presence of suitable electron-donors such as triethanolamine, the Rh (III) complex acts as a photocatalyst involved in a two-electron (Rh (III)$\leftrightarrow$Rh (I)) cycle. Also the H_2 yields are maximal in the neutral pH region (pH 5.8).

2.2.2 Photolysis of metal hydrides

H_2 production through C3 or C4 cycles of Balzani provide an efficient method. Metal hydrides from which photoproduction of H_2 has been observed are collected in Table III. As can be noted in the third column, the metal hydride photolysis require UV photons in majority of cases. The only case where visible light has proved to be effective is that of $[Rh^I(bridge)_4H]^{3+}$ of Gray et. al., but, as with metal ions photolysis, the medium photolysis is again concentrated acid. The systems are non-cyclic, with H_2 yields stoichiometric to that of the complex being photolysed. In majority of the metal hydrides that have been studied, their preparation processes suggest that the metal complexes serve more as a reversible H_2-carrier than those which take part in a true redox cycle involving H^+ reduction.

Table III: H_2 Evolution via Metal Hydride Photolysis

Metal Hydride	Medium	$\cdot\Gamma_{H_2}(\lambda\ nm)$	Reference
Co: $Co(phen)(PR_3)H_2^+$			(2.16)
Ir: $H\ Ir(PF_3)_4$	C_6H_6		(2.17)
$IrCl(PPH_3)_3H_2$	C_6H_6		(2.18)
$Ir(CO)(Cl)(H_2)(PPh_3)_3$	C_6H_6		(2.18)
Ru: $Ru(CO)(H_2)(PPh_3)_3$	C_6H_6		(2.19)
Mo: $Mo(\eta^5C_5H_5)H_2$	C_6H_6	0.1 (366nm)	(2.20,2.21)
$Mo_2\ Cl_8^{2-}$	3M HCl	0.14 (254nm)	(2.13)
$(Mo_2Cl_8H)^{3-}$			
$[Mo_2(SO_4)_4]^{4-}$	10^{-3}-5M H_2SO_4	0.18-0.19 (254nm)(515nm)	(2.13,2.14)
$Mo_2(aq)^{4+}$	1M CF_3SO_3H	0.035 (254nm)	(2.13)
W : $W(\eta^5C_5H_5)H_2$			(2.21)
Rh: $[Rh_2(bridge)_4H]^{3+}$	12M HCl	0.004 (546nm)	(2.22)

2.2.3 $\underline{H_2 \text{ production via dye-sensitized redox reactions}}$

These are the first examples of "in vitro" systems which produce large amounts of H_2 by H^+ reduction in water using visible light. The essential ingredients of systems of this type are a sensitizer (S), an electron relay (A) and a redox catalyst. Excitation of the sensitizer induces as electron-transfer:

$$S + A \underset{}{\overset{h\nu}{\rightleftharpoons}} \quad S^+ + A^- \tag{21}$$

which is followed by the catalytic step

$$A^- + H_2O \xrightarrow{\text{cat.}} \quad A + OH^- + \tfrac{1}{2} H_2 \uparrow \tag{22}$$

leading to H_2 generation. The back conversion of S^+ into S may be achieved by sacrificing a donor D added to the solution through irreversible oxidation

$$D + S^+ \longrightarrow D^+ + S \tag{23}$$

Alternatively, as mentioned in section (i), one can obtain these products via a reductive cycle:

$$S + D \underset{}{\overset{h\nu}{\rightleftharpoons}} \quad S^- + D^+ \tag{24}$$

$$S^- + A \longrightarrow S + A^- \tag{25}$$

$$A^- + H_2O \xrightarrow{\text{cat.}} A + OH^- + \tfrac{1}{2} H_2 \tag{26}$$

Examples of electron donors, sensitizers, acceptor relays and redox catalysts which have been found to be quite efficient are listed in Table IV. Before we discuss the detailed photochemistry of selected systems of this type, it is worth summarizing salient features of this type of photoredox reactions.

(i)   The threshold wavelengths for driving a reaction of this type depend on the nature of the sensitizer and this can extend into the red (600-700 nm) with porphyrins as photosensitizers.

(ii)   Isotopic studies have shown that the H_2 produced is not through a simple photosensitized dehydrogenation of the electron donors. Rather they are true water reduction processes.

(iii) Compared to the systems mentioned in the earlier part of this section, these reactions can be carried out in water and around neutral pH.

Table IV. Dye Sensitized Redox Reactions Leading to H_2 Evolution

Electron Donor (sacrificial)	Sensitizer	Acceptor	Redox Catalyst
EDTA	Bipy,phen.Complexes (2.23-2.32) of Ru,Cr,... $(Ru(bpy)_3^{2+})$	Viologens (MV^{2+})	Enzymes H_2-ase, N_2-ase
Cysteine, H_2S			
TEA & other amines	Acridine Dyes (2.32-2.37) Proflavine,Acridine Yellow	V^{3+}, Cr^{3+}, En^{3+}	Metal Oxides PtO_2, IrO_2, Al_2O_3
Ascorbate Eu^{2+}	Porphyrins (2.38-2.44) Zn, Ru, Mg (Chl)	$Rh(bpy)_3^{3+}$	Colloidal Pt, Au, Ag particles
	Phthalocyanines	$Co^{II}L_5$ Complexes	
	Flavines Deazaflavin		

(iv)   Mechanistically H_2 production can be visualised as in cycle Cl'. The general idea is to reduce an acceptor relay A whose redox potential is below that H_2/H^+ and generate hydrogen from A^- subsequently through the aid of redox catalysts. Early examples of redox catalysts are the enzymes hydrogenases and nitrogenases and recently these have been replaced by colloidal dispersions of Pt, Au and Ag. The redox catalysis as such will be reviewed later.

(v)   The turnover numbers for S, A and the catalyst systems composed of various components of Table IV have been impressively high. By the very design of the system, with sacrificial donors, the turnover with respect to the donor is $\leq$ 1.

(vi) The type of donors that have been successfully employed are not many. Positive results have been obtained only with the same 3 or 4 donors. With this restricted choice, it is rather difficult to generalise the conditions that these donors must satisfy. Obviously for cyclic water decomposition, the oxidised donors should be transformed back to D with concomitant O_2 production

$$2D^+ + H_2O \longrightarrow 2D + 2H^+ + \frac{1}{2}O_2 \qquad (27)$$

(vii) In three of the systems that have been examined in detail, the H_2 yields have been found to be very much pH dependent. While the low yields of H_2 at high pH with acceptors such as MV^{2+} can be attributed to thermodynamics, the drop in H_2 yields at low pH apparently is due to inefficiency of donors in reaction (23). The pH ranges where maximum yields of H_2 are observed roughly correlate with the pk_a of ground state protonation equilibria of these donors.

(viii) In systems where the sensitizer undergoes a reductive cycle (reactions (24)-(26), under suitable conditions the reduced sensitizer (S^-) itself can play the role of the acceptor relay. The basic requirements are $E_o(S/S^-) < E_o(H^+/H_2)$ and the presence of a very efficient redox catalyst to compete with reverse of reaction (24). In cases of sensitizers Proflavine, $Ru(bpy)_3^{2+}$, PhCOPh   with colloidal Pt-catalysts, H_2 production from reduced S has indeed been observed in two component systems.

2.2.4 Photochemistry of selected redox systems for H_2 evolution

H_2-evolving systems with $Ru(bpy)_3^{2+}$ or Proflavine as photosensitizers and Methylviologen (MV^{2+}) or $Ru(bpy)_3^{3+}$ acceptors have become very popular as prototypes for probing other suitable candidates for the D,S,A relays and it is useful to digress on the photochemistry of these systems. Both $Ru(bpy)_3^{2+}$ and Proflavine sensitized schemes have been the subject of detailed flash photolysis studies.

Let us consider the $Ru(bpy)_3^{2+}$ sensitized reduction of MV^{2+}. Excitation of $Ru(bpy)_3^{2+}$ with a 530 nm laser pulse leads to the formation of MLCT excited state with a lifetime of 620 nsec. in water. In the presence of electron acceptors like MV^{2+}, the excited state is oxidatively quenched yielding $Ru(bpy)_3^{3+}$ and MV^+.

$$Ru(bpy)_3^{2+} + MV^{2+} \xrightarrow{h\nu} Ru(bpy)_3^{3+} + MV^+ \qquad (28)$$

Kinetic analysis of this process can be performed by monitoring the characteristic absorption of MV^+ at 395, 600 nm, bleaching and subsequent recovery of $Ru(bpy)_3^{2+}$ ground state at 452 nm and the luminescence of $Ru(bpy)_3^{2+*}$ at 615 nm. Oscilloscope traces illustrating the time course of these events are presented in Figure 7. The signal at 605 nm grow concomitantly with the rapid decay of the luminescence at 615 nm, indicating that the quenching of the excited states leads to the formation of products as described by the equation (28). As the forward reaction is endoergic by 1.7eV with respect to the ground state $Ru(bpy)_3^{2+}$, the products back react spontaneously at a high rate (lowest trace of Figure 7). Quantitative studies at different MV^{2+} concentrations yield the rate constants for forward and reverse electron-transfers as $k_{28} = 2.4 \times 10^9 M^{-1} sec^{-1}$. The yield of the redox products in the quenching process is about 30%.

REDUCTION OF METHYLVIOLOGEN

BY EXCITED $Ru(bipy)_3^{2+}$

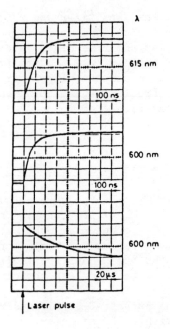

Fig. 7   Oscilloscope Traces Obtained from the 530 nm Laser
          Photolysis of Aqueous Solution of $Ru(bpy)_3^{2+}$ (4×10^{-5} m)
          and Methylviologen (10^{-2} M)

As indicated earlier, the back reaction can be prevented by adding a third component such as triethanolamine (TEOA) which is capable of reducing $Ru(bpy)_3^{3+}$ to the 2(+) state.

$$Ru(bpy)_3^{3+} + R_2\text{-}N\text{-}CH_2\text{-}CH_2\text{-}OH \longrightarrow Ru(bpy)_3^{2+} + R_2\text{-}N^+\text{-}CH_2\text{-}CH_2\text{-}OH \cdots$$

$$\text{(29)}$$

With increasing amounts of TEOA, reaction (29) removes $Ru(bpy)_3^{3+}$ in competition with reaction (28) leaving part of MV$^+$ formed as stable products.

INFLUENCE OF TEOA CONCENTRATION ON THE MV$^+$ DECAY KINETICS

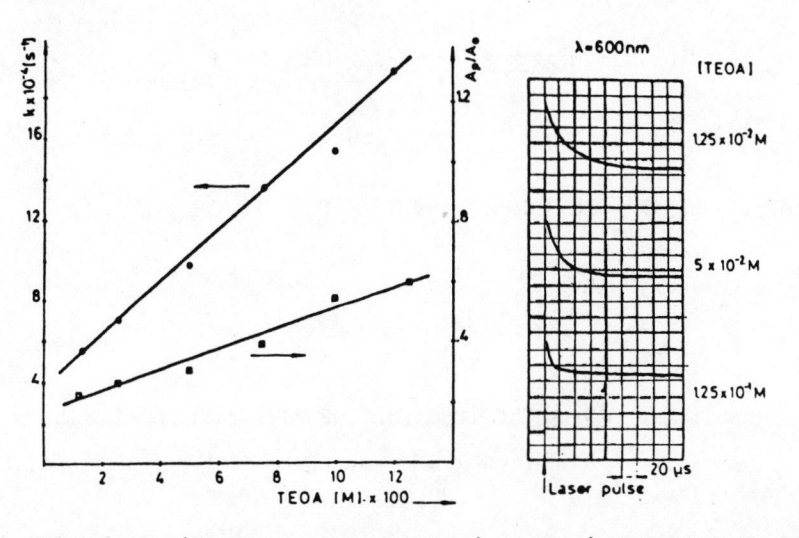

Fig. 8   Right: Oscilloscope traces showing the influence of added TEOA concentration on the decay kinetics of MV$^+$. radical monitored at 600 nm.

Left: Plot of ratio of transient absorbances obtained in the plateau region as well as rate of MV$^+$. decay versus TEOA concentration.

Figure 8 shows laser traces on the influence of TEOA concentration on MV$^+$ decay kinetics at 600 nm. Also included is a plot of the fraction of MV$^+$ that escapes recombination (reaction 28) as a function of the donor concentration. Steady state photolysis of TEOA/Ru(bpy)$_3^{2+}$/MV^{2+} show that the yields of MV$^+$ produced are crucially dependent on the donor concentration and also the pH of the solution. A rationale for the pH effect can be provided through examination of MV$^+$ absorption at 600 nm at various pH (Figure 9). In acid media (pH 5) the signal returns completely to the baseline indicating that all of the MV$^+$ formed in reaction (28) is reoxidised to MV^{2+}. A fractional decay is observed at neutral pH (pH 7) as was shown in Figure 9.

At alkaline pH (pH 9) the signal does not decay at all. On the contrary, the first and rapid rise of absorbance due to formation of MV^+ <u>via</u> reaction (28) is followed by a second and slower growth indicating additional MV^+ production. These observations can be explained in terms of an acid-base equilibrium of the TEOA cation and redox capabilities of the acid-base form of TEOA cation

$$R_2\text{--}N^+\text{--}CH_2\text{--}CH_2\text{--}OH \rightleftharpoons R_2\text{--}N\text{--}CH_2\text{--}CH\text{--}OH + H^+ \tag{30}$$

$$R_2\text{--}N\text{--}CH_2\text{--}CH\text{--}OH + MV^{2+} \longrightarrow R_2\text{--}N\text{--}CH_2\text{--}CHO + MV^+ + H^+ \tag{31}$$

$$R_2\text{--}N^+\text{--}CH_2\text{--}CH_2\text{--}OH + MV^+ \longrightarrow R_2\,N\,CH_2\text{--}CH_2\text{--}OH + MV^{2+} \tag{32}$$

EFFECT OF pH ON THE MV^+ KINETICS

01 M TEOA, 4×10^{-5} M $Ru(bipy)_3^{2+}$, 10^{-2} M MV^{++}

λ = 600 nm

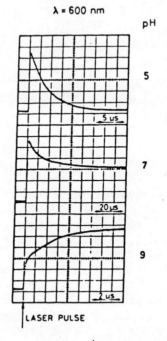

Fig. 9 Effect of pH on the MV^+ Radical Decay Kinetics.

At acid pH, in the absence of deprotonation of the cation, re-action (29) and (31) removes all MV^+. At neutral pH, reaction (29), (30) occur simultaneously and at higher pH$_+$(pH $>$ pKa) reactions (29), (30), (31) but not (32). Thus the net $[MV^+]$ yields per pulse in-creases from zero at pH $\sim$ 3 to almost twice the initial (redox yields of reaction (28) at higher pH values. One observes essentially simi-lar behaviour with EDTA as the electron donor.

The H_2 yields (derived from MV^+ via reaction 33) obtained in the presence of redox catalysts

$$2MV^+ + 2H^+ \xrightarrow{\text{cat.}} MV^{2+} + H_2 \tag{33}$$

such as PtO_2 or colloidal Pt show a similar pH dependence. The H_2 yields are maximal around pH 4-5 and fall off on either side of pH. The falloff in H_2 yields at pH $>$ 6.0 presumably arises from the thermodynamical inability of MV^+ (redox potential -0.44V pH independ-ant) to reduce water at these higher pH values. The reaction (33) is thermoneutral at pH $\sim$ 7.6.

Lehn et al. recently reported efficient H_2 production on visible light photolysis of TEOA/Ru(bpy)$_3^{2+}$/Rh(bpy)$_3^{3+}$ in aqueous solutions in the presence of K_4PtCl_6. The H_2 yields similarly show a pH depend-ence with maximal yields around pH 8.0 (pK$_a$ (TEOA) = 7.3). The mechanism involved is again oxidative quenching process, though there has been some controversy as to whether or not Rh(bpy)$_3^{3+}$ serves as an one-electron acceptor relay like MV^{2+} or as a two-electron relay undergoing Rh(III) $\leftrightarrow$ Rh(I) cycle. Most of the porphyrin-sensitized H_2 production reactions reported also involve oxidative cycles.

Under circumstances where the sensitizer is efficiently reduced by the donor in the excited state, (in comparison to oxidative quench-ing by the acceptor relay), then one obtains the same products via a reductive cycle. The proflavine-sensitized (also ZnTMPyP sensi-tized under high donor concentration) reduction of MV^{2+} with EDTA has been shown to involve a reductive cycle

$$^3P* + EDTA \longrightarrow P\overline{.} + EDTA^+ \tag{34}$$

$$P\overline{.} + H^+ \longrightarrow PH. \tag{35}$$

$$P\overline{.} + MV^{2+} \longrightarrow P + MV^+ \tag{36}$$

$$2PH. \longrightarrow P + PH_2 \tag{37}$$

2.3 Redox Systems for O_2-Evolution from Water

Production of a molecule of oxygen (O_2) from water via photoredox decomposition is, mechanistically, a more difficult process involving a concerted 4-electron transfer:

$$2H_2O \longrightarrow O_2 + 4H^+ + 4e- \tag{38}$$

and progress in this area consequently has been slow. Most of the known photoredox reactions involve one-electron transfer and any oxidation process other than a full 4-e⁻ transfer generates free radicals such as OH. or HO_2. Though there is no real need to avoid production of such free radicals with simple transition metal ions, with organic reactants the radical addition reactions invariably lead to total loss of the reactants.

To provide some insight into the mechanism of O_2 production from water, it is useful to examine chemical systems which evolve O_2 in dark chemical reactions. Since the early work of A.A. Noyes, several chemical oxidants have been postulated to oxidize water to molecular oxygen. These include Co^{III} (2.45-2.47), Mn^{IV} (2.48-2.54), Cu^{III} (2.45) Ni^{III}_{aq} (2.45) tellurates, (2.55-2.58) bipyridyl, phenanthroline complexes of the type $M(LL)_3^{3+}$ where M = Ru(III), Os(III) and Fe(III), and MV^{IV} porphyrins (2.59-2.64)

$$4D^+ + 2H_2O \longrightarrow 4D + 4H^+ + O_2 \tag{39}$$

Transition metal ions such as Ce^{4+} and Fe^{3+} in $HClO_4$ media oxidize water 2.67-2-69 to O_2 upon photolysis in the UV region. At least with the former, the reaction is distinctly photoassisted, though it does not occur in the dark in the absence of redox catalysts. The individual steps leading to the overall chemistry as shown by equation (39) has been the subject of numerous investigations. Studies with Fe^{III}, Co^{III} illustrate typical operative mechanisms: The first step invariably is one-electron oxidation leading to either free OH. radical or to an OH. adduct

$$M^{2+} + OH^- \longrightarrow M^{2+} + OH. \text{ or } (MOH)^{2+} \tag{40}$$

Following the (OH.) adduct formation, formation of a molecule of oxygen can occur through either one of the following mechanisms.

Mechanism I (through intermediatory formation of H_2O_2)

$$(MOH)^{2+} \longrightarrow M^{2+} + OH. \tag{41}$$

$$(MOH)^{2+} + OH \longrightarrow M^{2+} + H_2O_2 \tag{42}$$

or

$$OH \cdot + OH \cdot \longrightarrow H_2O_2 \tag{43}$$

along with

$$OH \cdot + H_2O_2 \longrightarrow HO_2 \cdot + H_2O \tag{44}$$

$$HO_2 + Fe^{3+} \longrightarrow Fe^{2+} + H^+ + O_2 \tag{45}$$

The production of H_2O_2 involves 2-e⁻ oxidation of OH^- radicals and subsequent metal-ions catalysed decomposition of H_2O_2 to O_2 involve 2 more electrons.

Mechanism II (through peroxo or 𝝁-oxo dimeric intermediates)

$$Co^{III} + OH^- \xrightarrow{H_2O} \left[(H_2O)_4 Co(OH)\right]_2^{4+} \quad (\text{𝝁-oxo dimer}) \tag{46}$$

$$\left[(H_2O)_4 - Co \overset{\overset{\displaystyle H}{\overset{\displaystyle |}{O}}}{\underset{\underset{\displaystyle H}{\underset{\displaystyle |}{O}}}{\diamond}} Co - (H_2O)_4 \right]^{4+}$$

This mechanism involves formation of binuclear complexes (𝝁-oxo complexes) which undergo intramolecular conversion to the peroxo dimer

$$\left[(H_2O)_4 Co \overset{\overset{\displaystyle H}{\overset{\displaystyle |}{O}}}{\underset{\underset{\displaystyle H}{\underset{\displaystyle |}{O}}}{\diamond}} Co(H_2O)_4 \right]^{4+} \quad --- \quad \left[(H_2O)_4 Co \overset{\overset{\displaystyle H}{\overset{\displaystyle |}{O}}}{\underset{\underset{\displaystyle H}{\underset{\displaystyle |}{O}}}{\diamond}} Co(H_2O)_4 \right]^{2+} \quad ..(47)$$

and then hydrolysis of the peroxo complex. There has been some discussion in the literature as to whether the bridging oxygen in the complexes of the type[2.68] L_5 Co-O-O-CoL$_5$ $^{5+}$ a superoxide ion, and that the oxidation of water to O_2 should involve peroxo dimers rather than 𝝁-oxo dimers. Based on his studies, with binuclear bipyridyl complexes of Mn(IV), Calvin[2.52] proposed 𝝁-oxo dimers as probable intermediates.

$$Mn^{IV} \overset{O}{\underset{O}{\diagup\diagdown}} Mn^{IV} \quad \rightleftarrows \quad Mn^{III} \overset{\cdot O}{\underset{O}{\diamond}} Mn^{III} \qquad Mn^{II} \overset{O}{\underset{O}{\Vert}} Mn^{II} \quad ..(48)$$

However, the similarity in bond strengths of the peroxide linkage to molecular O_2, the ease with which the known μ-peroxo Cobalt complexes liberate O_2 (in contrast to μ-oxo bipyridyl Mn dimers) on photolysis, kinetic barrierrs on μ-oxo to peroxo dimer conversions led Sawyer et al.[2.49-2.51] to suggest peroxo binuclear complexes as the most probable intermediates. More studies with model compounds are needed to elucidate this point. Various mechanisms proposed for water oxidations are variations of these two principal types.

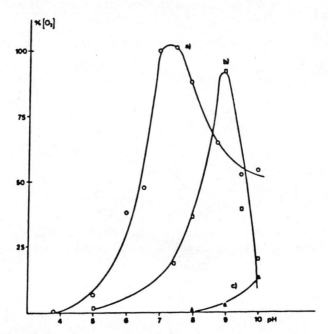

Fig. 10 Oxygen Yields Obtained in the Thermal Reduction   of
Fe(bpy)$_3^{3+}$ in Water as a Function of pH.   a) RuO_2 X H_2O
Powder (30mg/100ml), b) Colloidal RuO_2 (4 mg/100 ml),
c) Catalyst Free Solution.

Recent studies[2.58,2.69-2.70] have shown that the O_2 yields in reaction (39) are strongly pH dependent, both for the non-redox catalysed reductions as well as for the reaction redox-catalysed. Figure 10 presents[2.69] typical O_2-yields curves for the reduction of Fe(bpy)$_3^{3+}$ in H_2O in the presence of RuO_2 as redox catalysts.

with hydrated RuO_2 as the catalyst, curve (a), the reduction yields stoichiometric amounts of O_2 at $pH \sim 7.0$. With $Fe(bpy)_3^{3+}/$ $Fe(bpy)_3^{2+}$ ($E_o = 0.98V$) the energy losses connected with overvoltages for O_2 evolution in reaction (39) can be as low as 200 mV with RuO_2 catalysts.

    Figure 11 presents similar pH dependences of O_2 yields with colloidal Co^{2+} salts as redox catalysts, reported recently by Shafirovich et al.[2.72]   Presumably these redox catalysts assisted water oxidations to oxygen do not go through the same type of intermediates.

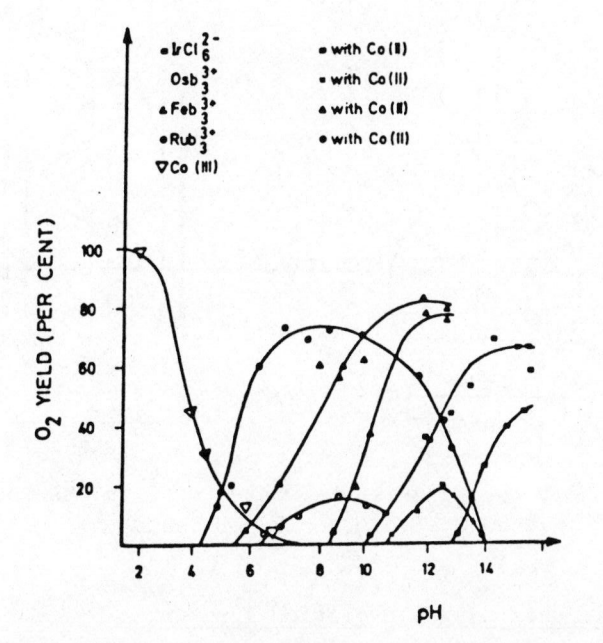

Fig. 11   Oxygen Yields as a Function of pH for Catalytic and Non-Catalytic Water Oxidation by One Electron Oxidants

    One of the major problems in studies of O_2 production from water has been analytical determination of the O_2 generated. Due to its very high solubility, up to saturation levels, O_2 tends to remain in solution and chromatographic determination of gas samples over the solution often prove inadequate. Valenty[2.65] has suggested use of a double-column for direct injection of the aqueous sample into the GLC. Detection of small quantities of O_2 in situ with the aid of a dark oxygen electrode (with the reaction mixture standing over the

electrode membrane) has been subject to controversy. In our studies,
we have adopted a flow technique where the O_2 generated (by chemical
or photochemical reaction) is flushed over and carried with a carrier
gas into a O_2-meter of End-o-Mess (<u>Figure 12</u>). Duncan et al.[2.66]
have described recently a method of quantitative trapping and detection with alkaline pyrogallol.

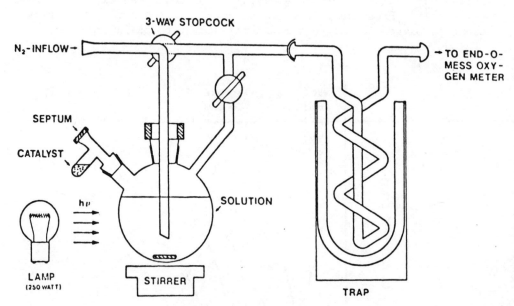

Fig. 12   Scheme of the Experimental Setup Used in the Thermal and
          Light-Induced O_2 Evolution Experiments

2.3.1   Photo-induced oxygen evolution from water

Table V lists various types of schemes that have been examined
to date, for light-induced oxygen evolution from water. Photolysis
of transition metal ions, as for the case for H_2-production, requires
UV-photons. Only for the Ce^{4+} system, the quantum yields are high
($\phi \sim 0.1$) but the reaction is one of photo-assisted and not energy
storing. Oxy anions such as VO_2^+, MnO_4^-, $S_2O_8^{--}$ are also known to liberate O_2 on photolysis but the sources of oxygen in these systems are
not clear. As mentioned earlier, while photolysis of the superoxo
dimeric complex of Co^{III} yield stoichiometric yields of O_2, photolysis of μ-oxo bridged binuclear Mn(IV) bipyridyls yield inefficiently
H_2O_2 at most. Jacobs, Schumacher and others[2.76-2.78] have reported
recently on the utility of Ag^+-ions exchanged Y-Zeolites for photodecomposition of water. Photolysis have been reported to yield O_2

with concomittant reduction of Ag^+ to the metal

$$2Ag^+ + 2Y^- + H_2O \xrightarrow{h\nu} 2Ag + 2YOH + \tfrac{1}{2}O_2 \tag{49}$$

Thermal treatment of the reduced zeolites at $873^{\circ}K$ leads to disappearance of hydroxyl groups and regeneration of Ag^+.

$$Ag + YH \xrightarrow{873^{\circ}K} Ag^+ + Y^- + \tfrac{1}{2}H_2 \tag{50}$$

Table V. Various Schemes for O_2 Evolution from Water Via Photolysis

a) metal ion photolysis[2.67-2.69,2.73] Ce^{4+}, Fe^{3+}, Mn^{4+},....

b) Oxy anions photolysis[2.74-2.75] VO_2^+, MnO_4^-

c) Photolysis on Zeolites[2.76-2.78]   Ag/Y- zeolite

d) binuclear μ-oxo, peroxo complexes[2.52-2.54,2.70]

e) Dye sensitized redox reactions with/without redox catalysts[2.71,2.72,2.79]

Sensitizer relay	Acceptor (Sacrificial)	Redox Catalyst
$Ru(bpy)_3^{2+}$	$Co^{III}(NH_3)_5X^{2+}$	RuO_2 (powder & colloidal)
	$Co^{III}(C_2O_4)_3^{3-}$	Ni
	Tl^{3+}, $S_2O_8^{--}$	
	Ag^+-crown	

Finally, analogous to the sacrificial systems for light-induced H_2 evolution, sacrificial systems have been developed for O_2 evolution. These systems consist of a photosensitizer ($Ru(bpy)_3^{2+}$), an irreversible electron acceptor and a redox catalyst and operates on the following scheme:

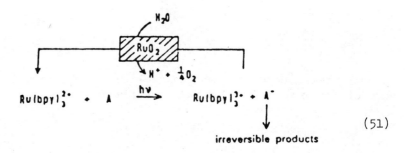

$$(51)$$

All the acceptors which are known to work in this scheme undergo irreversible side reactions following reduction e.q. disproportionation or decomposition. Thus, photolysis of $Ru(bpy)_3^{2+}$ with any of the acceptors listed in Table V in slightly acidic conditions leads to irreversible formation of $Ru(bpy)_3^{3+}$. In the presence of suitable redox catalysts, $Ru(bpy)_3^{3+}$ efficiently oxidizes H_2O to O_2, according to equation (52)

$$4\ Ru(bpy)_3^{3+} + 2H_2O \longrightarrow 4\ Ru(bpy)_3^{2+} + O_2 + 4H^+ \qquad (52)$$

Among the various acceptors that have been explored, $[Co(NH_3)_5 Cl]^{2+}$ complex has proved to be the most successful one in terms of O_2 yield. With $Co(NM_3)_5Br^{2+}$, $Co(C_2O_4)_3^{3-}$ etc., the O_2 yields are low, due to reduction of $Ru(bpy)_3^{3+}$ by photogenerated products Br^- and $C_2O_4^{--}$. Figure 13 presents pH dependence of O_2 yields obtained in the photolysis of $Co(NH_3)_5Cl^{2+}$ with $Ru(bpy)_3^{2+}$ and RuO_2. In Table VI we have listed typical O_2 yields obtained for other donors such as Fe^{3+}. Due to hydrolysis problems, studies with Tl^{3+} had to be carried out at pH 2.0, and O_2 yields are necessarily low under such acidic conditions. However, as will be shown later, the $Ru(bpy)_3^{3+}$ photogenerated under such conditions can be used to oxidize H_2O to O_2 in a photo-electrochemical cell with RuO_2-coated electrodes.

3.   STABILIZATION OF REDOX INTERMEDIATES THROUGH THE USE OF MULTI-
     PHASE SYSTEMS

     Consider the light-induced electron transfer reaction:

$$A + D \xrightarrow{\ h\nu\ } A^- + D^+ \qquad (53)$$

In this system light can be regarded as an electron pump which drives the reaction against a positive gradient of free energy change. The initial conversion efficiency, i.e. the ratio of free energy of

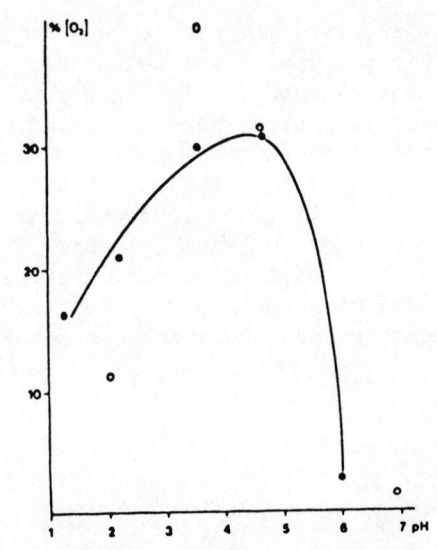

Fig. 13 Oxygen Yields Obtained in the Photolysis of Ru(bpy)$_3^{2+}$
(10^{-4}M) with (Co (NH$_3$)$_5$ Cl) Cl$_2$ (10^{-2}M) as a Function of
pH. O$_2$ Yields are for 3 hr Photolysis with a 150W Projector
Lamp.   0: RuO$_2$ powder (30mg/100ml)
            0: Colloidal RuO$_2$ (4mg/100ml)

Table VI: Light Induced Evolution of O$_2$ from Water with Ru(bpy)$_3^{2+}$
as Sensitizer and RuO$_2$ as Redox Catalyst

Electron Acceptor	Conc. (m/l.)	pH	RuO$_2$ form	Variant	t(h)	O$_2$ (ml)
TlCl$_3$	8×10^{-3}	1.6	Powder	A	0.5	0.3
			Colloid	A	0.33	0.4
[Co(NH$_3$)$_5$Cl]$^{2+}$	10^{-2}	4.8	Powder	A	1.5	1.2
			Colloid	B	3	1.2
[Co(NH$_3$)$_5$Br]$^{2+}$	10^{-2}	4.8	Powder	B	9	3.0
			Colloid	B	9	3.2
[Co(C$_2$O$_4$)$_3$]$^{3-}$	10^{-2}	4.8	Powder	A	1.5	0.3
			Colloid	A	1.5	0.4

Variant A: 50 ml of the solution with 20 mg of RuO$_2$ powder or 0.3 mg or RuO$_2$
colloid; Variant B: 150 ml of the solution with 50 mg of RuO$_2$ powder or 1.8
mg of RuO$_2$ colloid. pH 4.8 solutions are buffered with acetate buffer.

reaction over light energy input can be made impressively high ($>$ 80%) if suitable donor acceptor pairs are selected. However, a common feature of systems that are endoergic in the forward direction is their ability to back react at a diffusion-controlled rate in homogeneous solution. This leads to rapid thermal dissipation of chemical energy. One method whereby inhibition of the undesirable back transfer of electrons may be achieved is to employ heterogeneous (multiphase) systems such as micellar or membrane systems. Figure 14 illustrates structural features of various multiphase systems that can be employed: micelles, vesicles and microemulsions. Studies have also been carried out with other multiphase systems such as monolayers, multilayers and in polyelectrolytes.

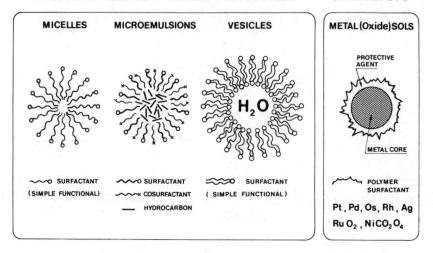

Fig. 14   Structural Features of Various Multiphase System and Redox Catalysts Used in Light-Induced Charge Separation Studies

3.1 Micelles

Micelles are surfactant assemblies that form spontaneously in aqueous solution[3.1-3.2] above a critical concentration (CMC). They are of approximately spherical structure (radius 15 to 30 A), the

polar head groups being exposed to the aqueous bulk phase while the
hydrocarbon tails protrude in the interior. A schematic illustration
of such a surfactant assembly is given in Figure 15. An important
feature of micellar solutions is their microheterogeneous character.
The apolar interior of the aggregates is distinguished from the aqueous phase through its hydrocarbon-like character allowing for the
solubilisation of hydrophobic species. Of crucial importance for
light energy conversion is the electrical double layer formed around
ionic micelles.

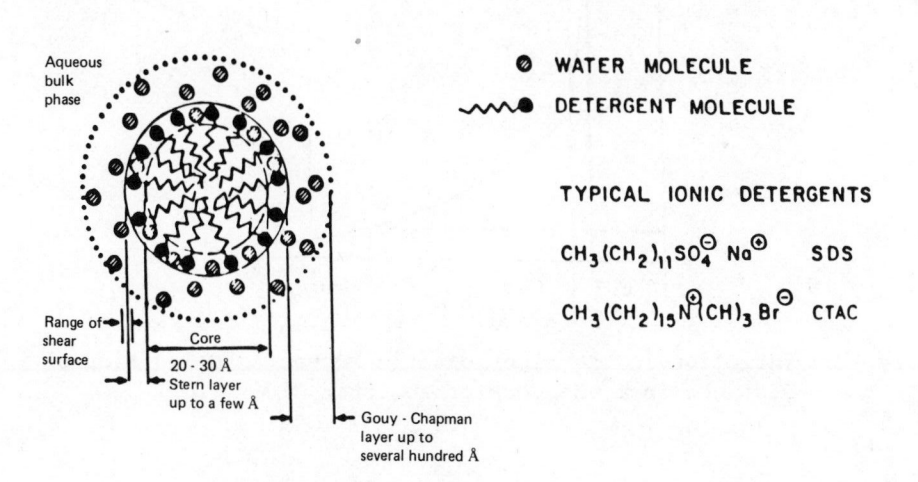

WATER MOLECULE

DETERGENT MOLECULE

TYPICAL IONIC DETERGENTS

$CH_3(CH_2)_{11}SO_4^{\ominus} \ Na^{\oplus}$ SDS

$CH_3(CH_2)_{15}N^{\oplus}(CH)_3 \ Br^{\ominus}$ CTAC

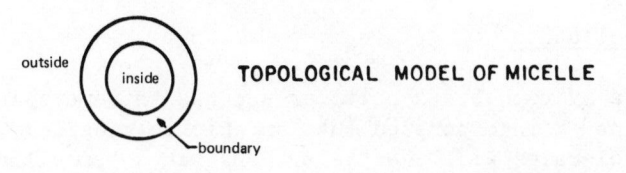

TOPOLOGICAL MODEL OF MICELLE

Fig. 15   Model of an Ionic Micelle.

This gives rise to surface potentials frequently exceeding values
of 150 mV,[3.3-3.4], Figure 16. This charged lipid-water interface provides a microscopic barrier for the prevention of thermal electron
back transfer. The role of the micellar double layer may thus be compared to that of the depletion or accumulation layer present at the
semiconductor/electrolyte interface. The following examples illustrate
this point:

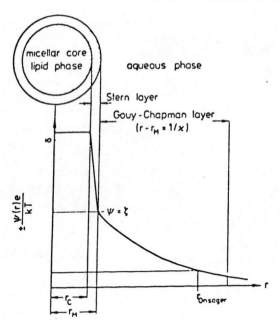

Fig. 16   Variation in the Electrostatic Forces as a Function of
           Distance from the Core of an Ionic Micelle.

3.1.1 Photoionisation

Consider a hydrophobic sensitizer such as tetramethylbenzidine[3.5]
or phenothiazine[3.6] incorporated into an anionic micelle. After excitation these molecules will eject electrons via a tunnelling mechanism[3.7-3.8] from the lipid into the aqueous phase where formation of
hydrated electrons (e^-_{aq}) occurs. The sensitizer cation will remain
within the micellar aggregate.

$$(S)_M \xrightarrow{\ h\nu\ } (S^+)_M + e^-_{aq} \qquad\qquad (54)$$

The negative surface potential of the anionic micelles prevents the
approach of hydrated electrons and sensitizer cations, e.g. holes,
Figure 17. The preferred pathway of reaction of e^-_{aq} is here hydrogen
formation $(2e^-_{aq} \longrightarrow H_2 + 2OH^-)$. Hence, with such a system,
the prevention of electron-hole recombination is achieved. While in
pure aqueous solution the lifetime of the sensitizer cation is only

several microseconds - due to the diffusion controlled back reaction
with e$^-_{aq}$ - the lifetime in anionic micellar systems can be as long
as days or weeks.

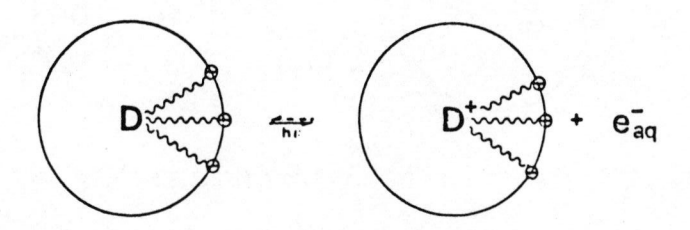

Fig. 17   Model of the Photoionisation of a Solute Solubilised in an
          Ionic Micelle.

3.1.2 Light induced electron transfer in the micelle

     Consider next a situation where two molecules i.e. the sensitizer
and the electron relay are incorporated in an anionic micelle. Light
quanta are used to excite one of the probes and promote electron
transfer from D to A, Figure 18. A radical ion pair is thereby produced within the aggregate. In an anionic micelle A$^-$ is clearly de-
stabilized with respect to the aqueous bulk solution. Therefore,once
it reaches the surface, it will be ejected into the water. Converse-
ly, the cation radical is electrostatically more stable in the micellar than in the aqueous phase and will thus remain associated with
the surfactant aggregate. Once A$^-$ and D$^+$ are separated, their dif-
fusional re-encounter will be obstructed by the ultrathin barrier of
the micellar double layer. The conjecture which evolves from these
considerations is that ionic micelles may be used as mediators to
achieve light-induced charge separation. The efficacy of such a process will crucially depend on the relative rates of A$^-$ ejection and
intramicellar back transfer of electrons from A$^-$ to D$^+$. The latter
reaction is thermodynamically favourable and can in principle occur
very rapidly. The rate of ejection of an anion from a negatively
chargedmicelle on the other hand is expected to depend critically on

the degree of hydrophobic interaction of A⁻ with the aggregated surfactant molecules.

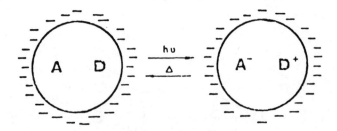

ELECTROSTATIC INTERACTION ENERGY

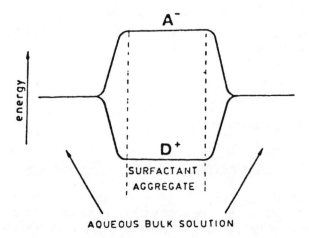

Fig. 18   Influence of Micellar Surface Charge on the Light-Induced
          Charge Separation.

          Using this principle, light induced charge separation has been
achieved for a number of donor/acceptor pairs. Noteworthy are the
examples where pyrene served as an electron acceptor and dimethyl
aniline as a donor[3.9-3.10] as well as the photoinduced reduction of
duroquinone by chlorophyll-a.[3.11] Experimental data obtained for
the latter system are presented in Figure 19. The upper two traces
represent absorption versus time curves from micellar solutions containing Chl-a alone. They reflect formation and decay of chl-a
triplet states. The time course of both the 685 and 465 nm absorption is drastically affected when DQ is co-solubilized with chl-a in
the micelles. In particular, a long-term bleaching of chl-a absorption is noted which can be attributed to chl-a⁺ cation radical

formation via:

$$^3 \text{chl-a}^* + DQ \longrightarrow \text{chl-a}^+ + DQ^- \tag{55}$$

Apparently, a large fraction of chl-a$^+$ escapes from geminite recombination with DQ$^-$ inside the aggregate. This must be due to the efficient ejection of DQ$^-$ from the micellar into the aqueous phase as was suggested above in our model considerations. Once in the aqueous phase DQ$^-$ cannot return into its native micelle since it is electrostatically rejected from the micellar surface. Hence, it will undergo disproportionation into durohydroquinone and DQ. The cation radical chl-a$^+$ is slowly reconverted into chlorophyll-a as indicated by the bleaching results.

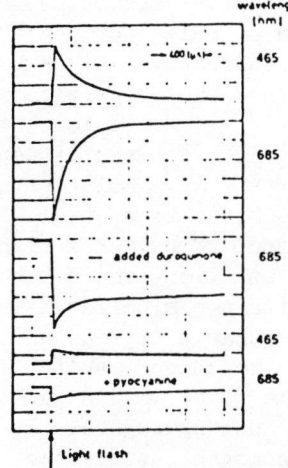

Fig. 19   Oscilloscope Traces Obtained in the 694 nm Laser Photolysis
          of Chlorophyll-a (3 X 10^{-5}M) in NALS (0.1M) Micelles. From
          the top: traces (1) and (2): decay of the chlorophyll-a
          triplets in the absence of additives; traces (3) and (4):
          chlorophyll cation radical decay in the presence of 3X10^{-3}M
          duroquinone; trace (5): chlorophyll cation decay in the pre-
          scence of 3X10^{-3}M DQ and 8X10^{-5}M Pycocyanine.

The effect of pycocyanine (PC$^+$) observed supports the suggested mechanism. PC$^+$ is present in the aqueous phase and is reduced by DQ$^-$. The neutral PC produced can enter the micelle and reduce in turn chl-a$^+$. Thus, the fraction of chl-a initially removed from the sphere of observation is replenished rapidly via this sequence of redox reactions. Hence, no long-term bleaching of chl-a is expected to occur as indeed is observed. Thus the lifetime of intermediates produced by the photoredox event could be prolonged by several orders of magnitude through employment of suitable micellar systems instead of the homogeneous solutions.

3.1.3 Solubilisation and spatial separation of reactants in micelles

The most common application of surfactants is as solubilising agents for a wide variety of solutes which are otherwise insoluble in water. In the redox pair D and A, if the hydrophobicity of either the donor or the acceptor is high, it provides an inherent selectivity in the solubilisation. With one of the reactants solubilised in an ionic micelle and the other randomly dispersed in the bulk water, the photoredox reaction is subject to strong electrostatic effects and in some cases also changes in the dimensionality of the reaction[3.12]. There has been several photochemical and pulse radiolytic studies based on this concept. Water in soluble photosensitizers such as metalloporphyrins[3.12] (including chlorophyll-a) have been solubilised in ionic and non-ionic micelles and photosensitized reduction of methylviologen relay investigated.[3.13-3.14] Variations of these include labelling of micelles with long-chain derivatives of sensitizers or acceptors such as $Ru(bpy)_3^{2+}$ and MV^{2+}.[3.15-3.19] Solutes involved in redox reactions in micellar media are subject to spatial separation and statistical effects of solubilisation. Consider a light driven redox reaction in a simple surfactant solution, where both donor and acceptor molecules are associated with the micelles. The situation is frequently encountered where the exchange of reactants between different micelles is much slower than the reaction itself. This implies that the reactive interactions take place within isolated groups of a few molecules contained in the host aggregate. It is evident that in such a case the rate laws of homogeneous systems are no longer applicable, since they are based on the concept of reactive coupling between the whole ensemble of dissolved species. In the specific case where a photo-excited acceptor (A*) abstracts an electron from a ground state donor (D):

$$A^* \;+\; D \;\xrightarrow{\;k_q\;}\; A^- \;+\; D^+ \qquad\qquad\qquad (56)$$

the following elementary processes should be taken into account:(cf. Figure 20)

a) Nonreactive deactivation of A* in micelles without D association

b) Reactive deactivation of A* in micelles with i D associations.

c) Intramicellar back reaction

$$A^* \cdot D \longrightarrow A^- \cdot D^*$$

a) NONREACTIVE DEACTIVATION OF A^* IN MICELLES WITHOUT D ASSOCIATION

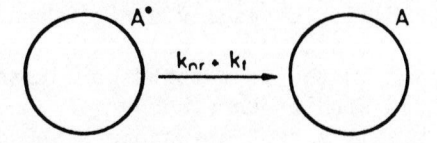

b) REACTIVE DEACTIVATION OF A^* IN MICELLES WITH i D ASSOCIATIONS

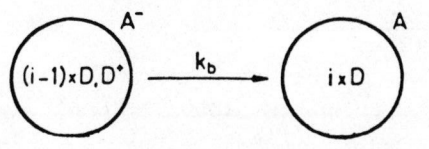

c) INTRAMICELLAR BACK REACTION

Fig. 20 Kinetic Features of Elementary Processes in the Intra-micellar Electrons Transfer Reaction

Recently, we have presented a kinetic model in which it was assumed that the donor distribution over the micelles follows Poisson's law, i.e.

$$\frac{M_i}{M} = \frac{m^i}{i!} e^{-m} \tag{57}$$

where M_i is the number of micelles having i donor associations, M is the total number of micelles and m is the average occupancy. Also the specific rate kq·(expressed in s^{-1}) was supposed to increase linearly with the number of D associations per micelle. From this treatment one obtains the following multiexponential time rate law for the decay of excited acceptor after excitation with a light pulse

$$[A^*](t) = [A^*](0) \exp\left\{ -(k_f + k_{nr})t + m\left[\exp(-k_q t)-1\right] \right\} \tag{58}$$

where $[A^*](0)$ stands for the excited acceptor concentration immediately after the laser pulse and $k_f + k_{nr}$ are the rate constants for radiative and nonradiative deactivation respectively competing with electron transfer quenching. Meanwhile, the validity of equation (58) has been demonstrated for various cases including the light induced reduction of $Ru(bipy)_3^{2+}$ by N-methylphenothiazine.[3,20]

From equation (58) it is readily derived that micelles enhance the rate of electron transfer reaction with respect to homogeneous solutions which is due to the fact that the reactants are brought in close proximity. Thus studies on a number of donor/acceptor pairs have shown that in micelles containing one A and one D molecule each, the average time required for the electron transfer is only 100-200 ns depending on micellar size. However, it is also noted from equation (58) that the yield of redox products may actually be smaller in micelles as compared to homogeneous solutions as the presence of both electron donor and acceptor is required in one micelle for electron transfer to occur. It is one of the advantages of functional surfactant systems that these statistical factors play no longer any role.

3.1.4 Functional micellar systems

Further progress in the development of molecular assemblies capable of converting light energy into chemical energy was made by designing and synthesizing surfactants with suitable functionality. These are distinguished from simple surfactants by the fact that the micelle itself participates in the redox events. One can visualise functional micellar assemblies constituted in various ways

a)   micelles with acceptor relays as counterions (transition metal ion micelles)

b)   micelles formed by crown ether surfactants

c)   micelles formed/labelled with long chain derivatives of sensitizers such as $Ru(bpy)_3^{2+}$ or acceptor relays of MV^{2+}

a)   <u>Redox reactions in transition metal ion micelles</u>:

The simplest type of functional surfactant is one where the counter ion is a participant in the redox reaction. For example, replacement of the Na^+ ions in sodiumlaurylsulfate (NaLS) by copper(II) yields assemblies in which Cu^{2+} ions constitute the counter ion atmosphere of the micelle, <u>Figure 21</u>. These may be photoreduced to the monovalent state by suitable donor molecules incorporated in the micellar interior. An illustrative example is that where D = N,N'-dimethyl 5,11-dihydroindolo 3,3-6 carbazole(DI). When dissolved in NaLS micelles, DI displays an intense fluorescence   and   the

fluorescence lifetime measured by laser techniques is 144 ns. Introduction of Cu^{2+} as counterion atmosphere induces a 300 fold decrease in the fluorescence yield and lifetime of DI. The detailed
laser analysis of this system showed that in $Cu(LS)_2$ micelles there
is an extremely rapid electron transfer from the excited singlet DI
to the Cu^{2+} ions. This process occurs in less than a nanosecond and
hence can compete efficiently with fluorescence and intersystem cross-
ing. This astonishing result must be attributed to a pronounced micellar enhancement of the rate of the transfer reaction. It is of
course a consequence of the fact that within such a functional surfactant unit regions with extremely high local concentrations of Cu^{2+}
prevail. (Theoretical estimates predict the counterion concentration
in the micellar Stern layer to be between 3 and 6 M).

PHOTOINDUCED ELECTRON TRANSFER IN FUNCTIONAL SURFACTANT SYSTEMS

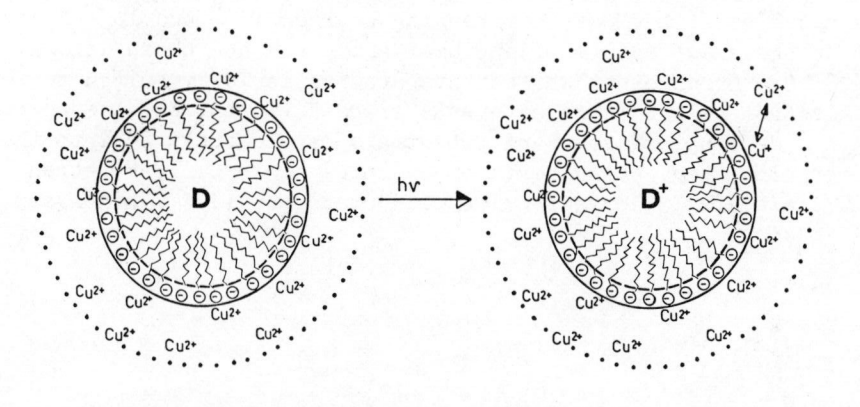

$$Cu^+ + Fe(CN)_6{}^{3-} \longrightarrow Cu^{2+} + Fe(CN)_6{}^{4-}$$

Fig. 21   Schematics of a Photoinduced Reduction of Cu^{2+} in Copper
          Lauryl Sulfate Micelles.

The significance of this functional organization becomes evident
also when the back reaction of Cu^+ and DI^+ is considered. Previous
studies have shown that the intramicellar electron transfer from Cu^+
to DI^+, though thermodynamically highly favourable, cannot compete
kinetically with the escape of the cuprous ion from its native

micelle into the aqueous phase. An efficient escape route if provided by the exchange with one of the Cu^{2+} ions present in high local
concentration in the Gouy-Chapman layer. By mere electrostatic arguments the latter ion is $exp(e\gamma/kT)$ times more likely to be absorbed
on the micellar surface than Cu^+. Once in the aqueous phase Cu^+ can
be used for a second redox process such as the reduction of $Fe(CN)_6^{3-}$.
The back reaction of the $Fe(CN)_6^{4-}$ with the oxidized donor, DI^+, is
prevented by the negatively charge micellar surface. Hence, such a
system is successful in storing light energy originally converted
into chemical energy during the photoredox process. Similar studies
include investigation[3.22] of the light-induced reduction of Eu^{3+}
ions by N-Methyl pheno thiazine and subsequent back transfer in
Eu(III) surfactant solutions.

b) <u>Micelles formed with crown ether surfactants</u>:

    Macrocyclic compounds with a paraffinic chain have first been
described by Cimquini et al.[3.23] In aqueous solution they form readily micelles with CMC values ranging between 10^{-5} and 10^{-3} M depending on the exact nature of the head group and the chain length.
[3.24-3.25] Crown ether surfactants are particularly suited to serve
as electron relays in redox events since different metal ions cover-
ing a wide range of standard potentials may be inserted into the
macrocyclic ring. Consider, for example, the crown 5-n-tetradecyl-1,
1,10-diaza-4,7,13,16-tetraoxi-cyclo-octadecane.(I) I binds strongly
Ag^+ ions[3.26]

the complexation constant being about 10^8. We shall now be concerned with the photoinduced of the Ag^+ derivative, i.e.

(II)

This surfactant possesses a CMC value of 3×10^{-4} M. The micellar aggregates formed are relatively large, the molecular weight being 6.3×10^6 as determined by quasielastic light scattering technique. A variety of sensitizers, when solubilized in aqueous micellar solutions of II were shown to be oxidatively bleached under simultaneous formation of Ag^0. Consider, for example, the cyanine dye:

(III)

Its fluorescence emission is totally quenched ($\emptyset_F < 10^{-6}$) in micellar solutions of II. Instead, under visible light illumination one observes a rapid bleaching of the cyanine and formation of a new stable absorption band with a maximum at λ = 415 nm. The spectrum can be attributed to silver atoms stabilized by the macrocyclic ligand. It appears that through insertion in the crown ether cavity Ag is prevented from aggregation and hence preserved in the monoatomic state. The speed of Ag^0 formation is so rapid that it cannot be resolved by ns-laser photolysis techniques. Obviously, this effect has to be attributed to the close proximity of photoactive donor and Ag^+ sites on the surface of the micelle. The photoinduced reduction of Ag^+ can be performed with a number of other donor chromophores such as porphyrins,

$Ru(bipy)_3^{2+}$ and pyrene. In all those cases no back reaction between oxidized donor and Ag^0 is observed. Apparently, through electrostatic interactions, the donor cation is ejected from the positively charge aggregates before back electron transfer can take place. Once in the aqueous phase, the reentry of D^+ in the micelle is prevented by the repulsive surface charge. We conclude that in the photoinduced reduction of Ag^+ to silver atoms the functional macrocyclic surfactant serves three important purposes: through high local concentration of the acceptor one accelerates drastically the forward rate - and hence the yield - of reaction; the Ag atom is stabilized by the macrocyclic ligand and finally, the back electron transfer prevented by the microscopic electrostatic barrier provided by the lipid/water interface.

c) Micelles with long chain derivatives of sensitizer or acceptor relays

One way of functionalising a redox system is to attach a long alkyl chain (surfactant) to the photosensitizer S or the acceptor relay A, and carry out the reaction either in the micelles formed by these or label the micelles formed by normal surfactants with these long chain derivatives. Following the report of photodecomposition of water on monolayer bound surfactant derivatives of $Ru(bpy)_3^{2+}$, there have been several studies[3.15,3.17,3.18,3.27-3.29] of redox reactions sensitized by surfactant derivatives of $Ru(bpy)_3^{2+}$ in micelles. The quenching of $Ru(C_{12}Bpy)(Bpy)_2^{2+}$ excited states reductively by amines such as dimethylaniline or oxidatively with viologens such as MV^{2+} have been shown to be strongly influenced by the presence of ionic and nonionic micelles. The charge separation is achieved by preferential adsorption/solubilisation of the oxidized Ru complex or the reduced viologen.

The fact that by suitable design of the system, micelles can be made to serve as a carrier for either the oxidized sensitized or the reduced relay has been exploited recently[3.30,3.31] to achieve significant charge separation with amphiphilic viologen derivatives. Consider, for example, the case where the derivative N-tetradecyl N-methylviologen is employed as an electron relay. In the oxidized state this compound exhibits a pronounced hydrophylic character while in the reduced state it has hydrophobic properties.

$$CH_3-N^+ \langle \rangle \langle \rangle N^+-(CH_2)_{13}-CH_3 \qquad (C_{14}MV^{2+}) \text{ hydrophylic}$$

$$\downarrow e^-$$

$$CH_3-N \langle \rangle \cdot \langle \rangle N^+-(CH_2)_{13}-CH_3 \qquad (C_{14}MV^+) \text{ hydrophobic}$$

In a solution containing $C_{14}MV^{2+}$ as an electron relay and the porphyrin ZnTMPyP^{4+}

$$R= -\langle \rangle \overset{+}{N}-CH_3 Cl^-$$

(or Ru(bpy)$_3^{2+}$) as a sensitizer light excitation of the latter will lead to formation of triplet states which subsequently transfer an electron to the viologen:

$$ZnTMPyP^{4+} + C_{14}MV^{2+} \underset{h\nu}{\overset{h\nu}{\rightleftharpoons}} C_{14}MV^+ + ZnTMPyP^{5+} \qquad (59)$$

The kinetics of this process was studied by laser photolysis technique. In <u>Figure 22</u> is shown the temporal behaviour of the absorbance of a solution containing 5×10^{-5}M ZnTMPyP^{4+} and 10^{-3}M C$_{14}$MV^{2+}. The concentration of C$_{14}$MV^{2+} is well below its CMC value, i.e. 7×10^{-3}M. The absorbance at 890 nm reflects the behaviour of the triplet state while that at 600 nm is characteristic for C$_{14}$ MV^{+} and ZnTMPyP^{5+}. The decay of the triplet state is accompanied by the formation of redox products (oscillogram b) which, however, undergo rapid back reaction. In marked contrast to this behaviour stand the results obtained in the presence of cationic micelles (cetyl trimethyl ammonium chloride). As is apparent from oscillogram c the forward electron transfer goes here to completion without subsequent intervention of the back reaction. The detailed kinetic analysis shows that CTAC micelles retard the reverse electron transfer by at least a factor of 1000.

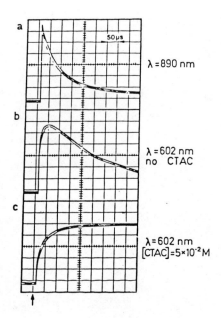

Fig. 22 Oscilloscope Traces Showing Light-Induced Reduction of
C$_{14}$MV^{2+} by the Triplet State of ZnTMPyP.
ZnTMPyP = 5×10^{-5}M; C$_{14}$MV^{2+} = 10^{-3}M.
a) triplet decay monitored at 890 nm, b) growth and decay of redox products at 602 nm in pure aqueous solution;
c) stabilisation of the redox products by the CTAC micelles.

This charge separation effect induced by the micelles is explained schematically in Figure 23. $C_{14}MV^{2+}$ due to its hydrophylic
character is mainly present in the aqueous phase and does not associate with the CTAC aggregates. The forward electron transfer will
therefore occur in water. In the reduced state the viologen relay
acquires hydrophobic properties. This leads to rapid solubilisation
in the CTAC assemblies. The oxidized porphyrin on the other hand is
prevented from approaching the micelles by the positive surface
charge. As the ζ-potential of CTAC is at least +100 mV[3.3] the probability of encounter with $ZnTMPyP^{5+}$ is smaller than $2x10^{-9}$. This
explains the effective micellar inhibition of the back reaction.

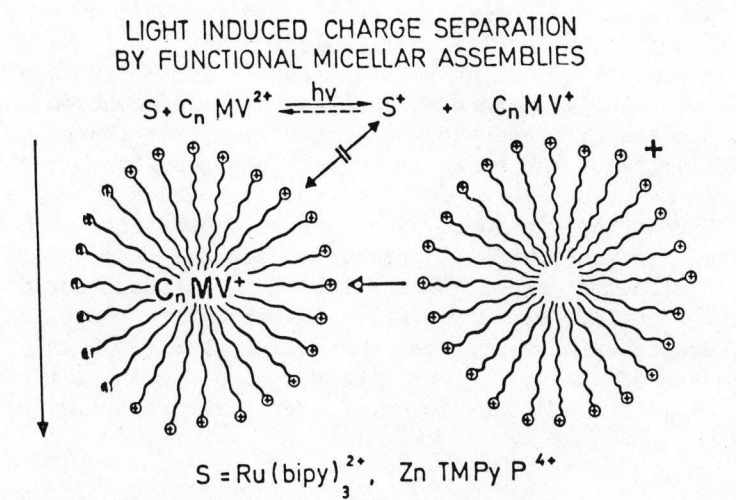

LIGHT INDUCED CHARGE SEPARATION
BY FUNCTIONAL MICELLAR ASSEMBLIES

$$S + C_n MV^{2+} \xrightarrow{\;h\nu\;} S^+ + C_n MV^+$$

$C_n MV^+$

$S = Ru(bipy)_3^{2+}, \; Zn \, TMPy \, P^{4+}$

Fig. 23 Schematics of Light-Induced Charge Separation by Functional
         Micellar Assemblies

3.2 Light-Induced Charge Separation in Vesicles

Earlier sections demonstrated how by suitable design of the system it is possible to achieve light-induced separation of charges in
energy storing redox reaction in aqueous micellar media. Similar studies have also been carried out in multiphase systems of unilamellar
vesicles and liposomes. These systems differ from simple micelles
in providing two aqueous phases (inner and outer) separated by a

bilayer formed by the lipid. In addition to vesicles and liposomes formed by biologically derived phospholipids, synthetic vesicles formed by surfactant derivatives such as dioctadecyl dimethyl ammonium chloride (DODAC) have been reported[3.3] recently. Vesicles, unlike micelles are static entities and can accomodate a substantial number of guest molecules per aggregate. As in micelles, these systems can organise donors and acceptors, lower ionisation potentials and more importantly through their interfaces or electrical double layer allow for some kinetic control of electron transfers.

The efficiency of light-induced charge separation and electron-transport through the bilayer lipid walls have been examined recently[3.34-3.39] for a variety of redox systems in lipid vesicles. Of particular interest has been the redox reactions sensitized by porphyrins such as chlorophyll-a and derivatives of $Ru(bpy)_3^{2+}$. Grätzel, Fendler and others[3.40-3.43] have investigated recently electron-transfer and charge-transfer phenomena in synthetic surfactant vesicles composed of DODAC and dihexadecyl phosphate (DHP). In one particular study, N-methyl phenothiazine (MPTH) and its long chain (C_{12}) analog were used as donors and a C_{18}-derivative of $Ru(bpy)_3^{2+}$ as the photoactivate acceptor. The models picture Ru-derivative to be anchored onto the surface of the vesicles and the MPTH molecules distributed in the hydrophobic bilayers of the DODAC vesicles. Laser photolysis studies identify three major pathways: for decay of photoredox products MPTH$^+$ and Ru$^+$. First there is rapid germinate recombination, secondly there is escape of some of MPTH$^+$ into inner aqueous water pools owing to spatial confinement and recombination occur at inner surface of the vesicles. Finally, a part of MPTH$^+$ escapes into the bulk aqueous phase where it survives for periods much longer (> 1 ms). By study of influence of added Cl$^-$ ions, the vesicles are shown to influence both the lifetime and the yields of the charge separated products.

3.3 Charge Separation Phenomena in other Multiphase Systems

In addition to micelles and vesicles, various other types of multiphase systems have been explored for achieving light-induced charge separations. These include microemulsions,[3.44-3.46] monolayer and multilayer assemblies[3.28,3.47-3.49], black lipid membranes[3.50] and polyelectrolytes[3.51-3.53]. Compared to micelles and vesicles, these systems are more complex in nature and obviously design and interpretation of experiments are more difficult. Preliminary results, however, appear very promising. However, for lack of space, we content ourselves citing the relevant studies and will not dwell in detail on these experiments.

4. REDOX CATALYSIS

4.1 Concept of Redox Catalysis

      In the previous section examples were presented illustrating
successful kinetic control of light-induced redox reactions by micellar and other multiphase systems. This achievement is, however, only
a first step on the way to conversion of solar light into chemical
energy. A crucial problem to be solved is the coupling of the photoredox events with catalytic steps leading to water decomposition. The
concept of redox catalysis was originally developed in 1938 by Wagner
and Traud.[4.1] To the solution are added finely dispersed catalytic
particles which serve as microelectrodes to afford water oxidation
or reduction selectively. Consider the hydrogen evolution step

$$2R^- \;+\; 2H_2O \;\xrightarrow{\text{cat.}}\; H_2 \uparrow + 2OH^- \;+\; 2R \tag{60}$$

Here the role of the particle is to couple the anodic oxidation of
the reduced relay with H_2-generation from water. The choice of the
catalytic material may be based on the same considerations which apply
for electrocatalytic reagents used on macroelectrodes: the exchange
current densities for the anodic and cathodic electron transfer steps
must be high. Colloidal platinum would then appear to be a suitable
candidate to mediate reaction (60). This fact was recognized already
at the end of the last century when numerous examples for the intervention of finely divided Pt in the process of water reduction by
agents such as Cr^{2+} and V^{2+} appeared in the german colloid literature.[4.2]

      Both for the oxygen-evolution and H_2-evolution reactions, there
are several advantages when one employs redox catalysts (Fig. 24).
They overcome the problems associated with free radicals generated in
discrete one-electron transfers. Even in biological systems, the
enzymes play essentially the role of mediators for multiselection
transfers. Thus redox catalysts can also be viewed as simple analogs
of enzymes. They provide selectivity and control over the gas production. Used in a flow-cell system, they enable gas production at a
given spot on a given time. The actual form of the redox catalyst
can be homogeneous (such as finely dispersed colloidal particles) or
heterogeneous (as pellets or beads or immobilised on a solid support
over which the solution flows). The colloidal particles have the
advantage of providing large surface area for catalysis but they suffer from recovery of the catalysts in recycling.

– <u>multi electron transfer catalysts</u>

$2A^- \cdot 2H_2O \longrightarrow 2A \cdot 2OH^- \cdot H_2\uparrow$ (2-e^-reduction)
$4D^+ \cdot 2H_2O \longrightarrow 4D \cdot 4H^+ \cdot O_2\uparrow$ (4-e^- oxidation)

– <u>selectivity/control</u> over gas production

(when and where)

– <u>homogeneous / heterogeneous</u>

– <u>efficiency</u> – <u>kinetically fast</u>

– <u>quantitative</u>

– <u>stability</u> over several cycles

Fig. 24   Principles of Redox Catalysis

The efficiency and overall performance of a given redox catalyst, as will be seen below, depend on several factors. Kinetically, re- action (60) on the catalyst surface should be very fast to compete with fast thermal back-electron-transfer reactions. Also the reac- tion should be quantitative as written. Furthermore, the catalyst should be stable and not susceptible to catalyst poisoning over ex- tensive cycles of the reactions such as (60).

4.2 <u>Redox Catalysis in the H_2-Evolution Reaction from Water</u>

Among the various acceptor relays that have been examined, methyl- viologen (MV^{2+}) has been found to be the most desirable one, due to its redox potential, high solubility in aqueous solutions and the ease of reduction. The coupling of chloroplasts with MV^{2+} relay and enzymes such as hydrogenase has been popular as a means of biophotoly- sis of water[4.3-4.4]

$$MV^{2+} + e^- \xrightarrow{\text{chloroplasts}}_{h\nu} MV^+ \qquad (61)$$

$$2MV^+ + 2H_2O \longrightarrow H_2 + 2OH^- + 2MV^{2+} \qquad (62)$$

The fact that this latter process can be catalyzed by Pt dispersions was discovered by Green and Stickland in 1934.[4.5] Earlier we re- viewed a number of photochemical systems that have been developed

in which MV^{2+} is reduced in a light driven electron transfer reaction
by a suitable sensitizer. The sensitizer cation undergoes a subsequent reaction with a third component which is irreversibly oxidized.
Such sacrificial systems served to optimize conditions for the light
induced hydrogen evolution. An illustrative example is the case
where $Ru(bipy)_3^{2+}$ serves as a sensitizer and EDTA as a sacrificial
electron donor.

Figure 25 illustrates the influence of the radius of the Pt
particle on the hydrogen evolution rates.[4.6-4.7] These particles
were stabilised by polyvinyl alcohol. Illuminations were carried
out with a XBO-450W Xe-lamp. One notices that a decrease of the
particle radius from 500 to 100 Å leads to a drastic augmentation
of the hydrogen evolution rate which is as high as 12 l/day/l solution for the smallest particle size. In fact, in the last case,
the bubbling of hydrogen gas occurring under illumination of the
solution is readily seen. In Figure 25 the common feature is the

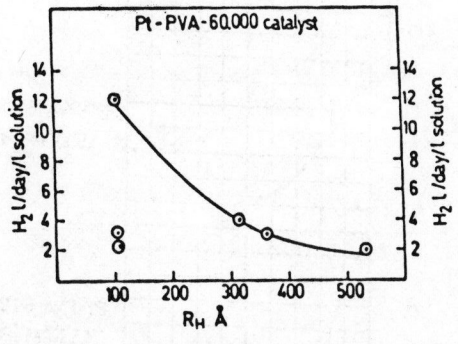

Fig. 25   Influence of the Size of the Colloidal Pt-PVA Particle on
the H_2 Evolution Rates in the System $Ru(bpy)_3^{3+}/MV^{2+}/EDTA$
at pH 4.7

Pt content of the solutions of different particle size, namely 3.5
mg Pt/25 cc. This figure also shows two extra points with 100 Å
particle size but with low Pt levels; 0.35 mg Pt/25 cc solution, the
upper one, and 0.25 mg Pt/25 cc solution, the lower one. Quantum
yield measurements indicate a stoichiometric relation between the
viologen reduced in the photoprocess and the amount of hydrogen produced. One might object to the high Pt levels (140 mg/ltr) required
to obtain these high yields. Such concentrations are intolerable
for practical systems. However, reduction in particle size to a
radius of only 15 A leads to a one hundred fold increase in the
activity of the catalyst. Hence, the same hydrogen output can now
be achieved with a Pt concentration of only 1.4 mg/ltr.

A further advantage of these finely divided platinum dispersions is that the solutions remain completely transparent even at high catalyst concentration. This allows to study directly the dynamics of the reaction of MV^+ with Pt particles by employing laser photolysis technique. The upper part of Figure 26 shows oscilloscope traces illustrating the temporal behaviour of the characteristic MV^+ absorbance at 602 nm in the absence and presence of catalyst. The upward deflection of the signal after the laser pulse is due to the formation of MV^+ via the photoredox process:

$$Ru(bipy)_3^{2+} + MV^{2+} \xrightarrow{h\nu} Ru(bipy)_3^{3+} + MV^+ \qquad (63)$$

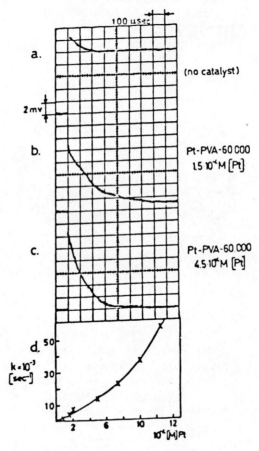

Fig. 26 Oscilloscope Traces Showing the Effect of the Pt–PVA Catalyst Concentration on the Behaviour of MV^+ Absorption at 600 nm (curve a,b,c). In (d), the rate constants for the MV^+ decay are plotted versus Pt concentration of the Pt–PVA catalyst used (pH 5).

In this case, no back reaction occurs since $Ru(bipy)_3^{3+}$ is reduced to the 2^+ state by EDTA. While in the absence of catalyst the MV^+ absorption is stable, addition of colloidal Pt induces a decay of the signal, the rate of which increases sharply with Pt concentration. From a fitting of the absorbance decay curves to an exponential time law, one obtains the rate constants which are plotted as a function of Pt concentration in the lower part of Figure 26. The ascent of the curve is steeper than linear, indicating that the reaction order is greater than one with respect to the Pt concentration. At a concentration of $10^{-4}M$ Pt in Pt-PVA-polymer, we observe a rate constant $k = 0.14 \times 10^3$ sec^{-1}. At the highest concentration of catalyst $(12.5 \times 10^{-4}M$ Pt content), the rate was 5.7×10^4 sec^{-1}. The lifetimes observed for MV^+ were shortened from about 7000 s to 15 sec, when the concentration of catalyst varied only by a factor of about 12. The high reaction rate of the platinum catalyst with the reduced viologen leading to H_2-formation depicted in Figure 26 are even exceeded by the rates obtained with 15 Å radius Pt particles. In this case the half lifetime of MV^{2+} in the presence of $10^{-4}M$ Pt (20 mg/ltr) is only ca. 20 μs. Taking into account that one particle contains 1200 Pt atoms, one derives a rate constant for reaction (7) of ca. $4 \times 10^{11}M^{-1}s^{-1}$ indicating that this process is essentially diffusion controlled. The very high catalytic activity of these Pt particles enables one to catalyse other H_2-producing reactions from reactive free radicals in competition to side reactions such as disproportionation. H_2 production via proflavine[4.8], Ph_2COH[4.9] and $Ru(bpy_3^+)$[4.10] illustrate this point.

These studies have since been extended[4.11] to preparation of colloidal Pt-particles by various other reduction schemes and with various protective agents. In recent work[4.11], 32 Å platinum particles have been produced in aqueous solutions by citrate reduction of hexachloro platinate. Carbowax 20M and styrene-maleic anhydride copolymer protected Pt-particles prepared this way were found to give outstanding stability and achieve high H_2 generation rates even at concentrations as low as 1.4 mg Pt/l. It should be pointed out that development of colloidal particles are not restricted to Pt alone. Recent studies of Henglein[4.12-4.13], Meisel[4.14-4.15] and others show that colloidal Ag and Au sols behave in a similar manner. In gamma-radiolysis studies in isopropanal media, these authors find, in addition to H_2 production, very efficient reduction of metal ions such as Tl^+, Cd^{2+}, Ag^+ catalysed in the presence of these colloidal sols.

These kinetic results are of great importance for the design of a cyclic water decomposition system. It appears that, by suitable choice of the catalyst, the conversion of A^- into A and simultaneous formation of hydrogen can be accomplished so rapidly that it can compete efficiently with the thermal back reaction. (In the photostationary state achieved under sunlight irradiation, the latter process, due to its bimolecular nature, should require at least several milliseconds).

4.3 Redox Catalysis in the O_2-Evolution Reaction from Water

If one envisages an economical process the formation of hydrogen from the reduced electron relay has to be coupled with water oxidation by the sensitizer cation. Earlier work in our laboratory has led to the discovery that certain noble metal oxides can serve as mediators in the oxygen generating reaction.[4.16] Amongst those RuO_2 has been most widely investigated.[4.17-4.18] Meanwhile a whole series of compounds such as Ce^{4+}, $Fe(bipy)_3^{3+}$ or $Ru(bipy)_3^{3+}$ have been identified which in the presence of colloidal or macrodisperse RuO_2 are capable to oxidize water with a stoichiometric yield. Other transition metal oxides such as cobalt or nickel oxide produce also oxygen in the presence of a suitable oxidizing agent but the yield is lower than stoichiometric.[4.19]

The mode of intervention of RuO_2 in the oxygen generation process may be best understood in terms of electrochemical concepts. The two coupled redox reactions occurring on the RuO_2 particle are depicted schematically in Figure 27. The particle serves as a microelectrode for which the theory of local elements may again be applied. The cathodic branch indicates the reduction of $Fe(bipy)_3^{3+}$ (or $Ru(bipy)_3^{3+}$) while the anodic branch corresponds to water oxidation. Two heterogeneous electron transfer reactions across the RuO_2/water interface are involved in the overall reaction. A decrease in the particle potential results in an increase of the rate of the cathodic electron transfer process, while that of the anodic reaction is retarded. Under steady state conditions, the RuO_2 will assume a potential E_p which is given by the intersection of the two current-voltage curves. At this potential a current i_R will flow which defines the overall reaction rate. It appears that in the case of oxidizing agents such as $Fe(bipy)_3^{3+}$, $Ru(bipy)_3^{3+}$, Ce^{4+}, or $ZnTMPyP^{5+}$ a driving force of ca. 150 mV for the water oxidation

$$4D^+ + 2H_2O \longrightarrow 4D + O_2 + 4H^+ \tag{64}$$

suffices to make the process occur rapidly and quantitatively.

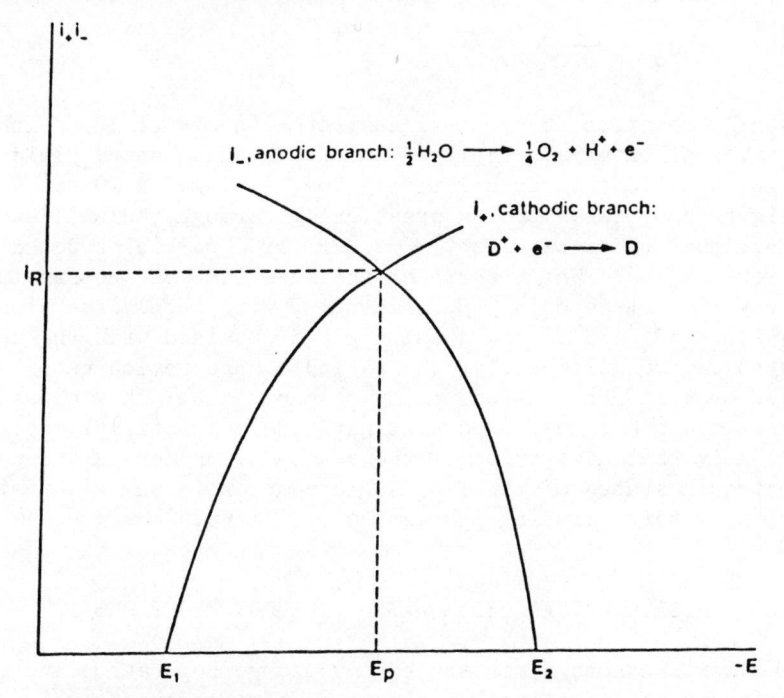

Fig. 27   Current-Voltage Diagram for Coupled Redox Processes Occuring on the RuO_2 Particle.
D^+ = Fe(bpy)$_3^{3+}$, Ru(bpy)$_3^{3+}$
E_p = potential of the particle; i_R = reaction current
       ($\rightarrow$ rate)
E_1 = E^0 + 0.059 log $[D^+]$ / $[D]$
E_2 = 1.23 - 0.059 log p_{O_2} $[H^+]$

4.4   Coupled Redox Catalysts for Water Decomposition

The case of the water oxidation by the Ru(bipy)$_3^{3+}$ complex is
particularly interesting as this species is formed in the redox reaction (51) which preceeds hydrogen generation from the reduced relay. One recognizes here the possibility to close the cycle of water
decomposition by visible light. In fact, irradiation of a solution
containing colloidal Pt (protected by a copolymer of maleic anhydride
and styrene) and macrodisperse RuO_2 as catalysts, apart from the
Ru(bipy)$_3^{2+}$ sensitizer and the MV^{2+} electron relay, produces the
two gases hydrogen and oxygen simultaneously.[4.20] One concludes
that in the combined system the catalysts preserve their specificity.
Also, the reaction of MV^+ with the Pt particles must be high enough
to be competitive with back electron transfer and the reduction of
oxygen

$$MV^+ + O_2 \longrightarrow O_2^- + MV^{2+} \tag{65}$$

The overall reaction scheme is illustrated in Figure 12. With the combination of catalysts employed initially the quantum yield of water splitting was found to be relatively low, i.e. 1.5×10^{-3}. However, this figure has been improved drastically through the utilisation of a cofunctional redox catalyst.[4,21] Here TiO_2 particles doped with RuO_2 were employed which serve at the same time as support for an extremely fine Pt deposit. Surprisingly, with such dispersions one achieves already 1/5 of the quantum yield obtained with the sacrificial systems. Significantly, the H_2 and O_2 production rates do not decrease over an irradiation period of many days. This shows that cross reactions are less important with the cofunctional catalyst. It is likely that adsorption of the reactants and/or participation of electronic states of the TiO_2 semiconductor in the redox events render the water splitting process so efficient.

5. PHOTOELECTROCHEMICAL CELLS BASED ON REDOX REACTIONS

Earlier sections demonstrated how charge separation and redox catalysis can be effectively employed in homogeneous solutions for the photodissociation of water, _via_ redox catalysis. Due to the explosive nature of the H_2-O_2 mixtures and also due to the prohibitive costs involved in the separation of the two gases, in systems for the complete photodissociation of water, it is desirable to design them in such a way that the two gaseous products are evolved in two different places with minimum mixing up. An attractive way of achieving this goal is through the employment of a photoelectrochemical cell (PEC). One possible design of a PEC is shown in Figure 28. In the cathode compartment, a light-driven redox process coupled to a Pt-catalyst leads to net reduction of water to H_2 with concomitant production of a strong oxidant. The redox equivalents of this strong oxidant are then carried through to the anode where with the assistance of an RuO_2-coated electrode, O_2 evolution is achieved. Recently there have been some interesting developments connected with the design of such PEC's.

Photoelectrochemical cells for light-induced H_2 evolution have been recently described by Kawai et al.,[5.1] and also by Durham et al., [5.2]. Kawai et al. describe operation of two types of PEC's.

Photocell I : TiO_2 / 0.2M NaOH // Phosphate buffer,MC^{2+}/metal

Photocell II : ZnTPP-Pt / 0.1M KCl,MV^{2+} // $Fe^{2+}(10^{-3}M)$ / Pt

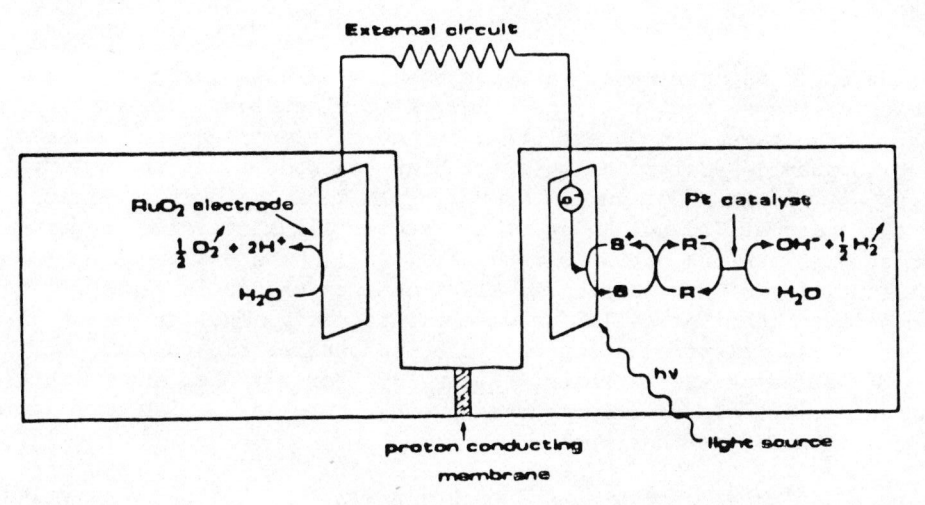

Fig. 28   Schematics of a Cell Device for the Cyclic Water Decomposi-
          tion by Visible Light.

Illumination of the TiO_2 in the photocell I with the band gap
light resulted in the production of O_2 at the TiO_2-surface and the
reduction of the MV^{2+} at the counter electrode. Addition of small
catalyst particles of $Pt-Al_2O_3$ or $Pd-Al_2O_3$ to the photocell resulted
in the reoxidation of MV^+ to give H_2.

Similarly, in the second photocell, by illumination of the ZnTPP-
coated Pt electrode immersed in the solution of MV^{2+}-KCl, net oxidation of Fe^{2+} to Fe^{3+} at the anode and MV^{2+} to MV^+ at the cathode was
achieved. In studies of Durham et al., the sacrificial system TEOA/
$Ru(bpy)_3^{2+}/MV^{2+}$ described earlier in section 2, has been adopted to
yield H_2 in the cathode compartment of a PEC:

$$Pt/Ru(bpy)_3^{2+}, MV^{2+}, TEOA, 0.1M \ NaCl, pH \ 9 \ // \ 1M \ HCl/Platinised \ Pt$$

Photoelectrochemical cells for light-induced oxygen evolution based
on irreversible oxidative quenching of the excited states of $Ru(bpy)_3^{2+}$
have also been described recently.[5.3-5.4] Meyer et al.,[5.3] photolysed $Ru(bpy)_3^{2+}$ in the presence of $Co(C_2O_4)_3^{3-}$ (reaction (64)) in the

cathode to generate $Ru(bpy)_3^{3+}$ irreversibly:

$$Ru(bpy)^{2+*} + Co(C_2O_4)_3^{3-} + 6H^+ \longrightarrow Ru(bpy)_3^{3+} + Co^{2+} + 3\ H_2C_2O_4$$

(64)

The $Ru(bpy)_3^{3+}$ so generated was then used to oxidise water to O_2 in the anode compartment with the aid of a Pt-electrode. Due to the high overvoltages for O_2 evolution on the Pt-electrode, the authors have to provide additional chemical bias (cathode solution: $Ru(bpy)_3^{2+}$ $Co(C_2O_4)_3^{3-}$ in 1M H_2SO_4 and 1M NaOH with 1M Na_2SO_4) to draw current of the order of 100 μA under steady state illumination, In our own studies carried out simultaneously, we[5.4] have examined the possible utility of RuO_2-coated Ti electrodes as anodes in similar photoelectrochemical cells for water photolysis. The low overvoltages on RuO_2 electrodes for O_2 evolution enables one to draw over 300 μA steady state currents. <u>Figure 29</u>, for example, shows the current-potential evolution curves during photolysis (with room light light) in a PEC of the type:

$$Pt/Ru(bpy)_3^{2+},\ S_2O_8^-,\ 0.1M\ Na_2SO_4, pH\ 4.7\ buffer//pH\ 4.7\ buffer, RuO_2-Ti$$

The performance characteristics of a few irreversible acceptor systems for water oxidation in such a PEC are shown in Table VII. Studies are currently underway to couple two such half-cells (one for for H_2 and one for O_2) in a photoelectrochemical cell to achieve cyclic photodissociation of water.

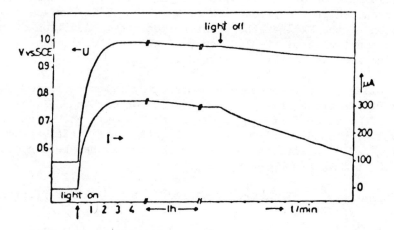

Fig. 29   Current and Potential Evolution During the Photolysis of $Ru(bpy)_3^{2+}$ / $S_2O_8^{--}$ System in a Photoelectrochemical Cell.

Table VII: Performance Characteristics of Various Redox Systems
             for the Oxidation of Water in the PEC.

Redox System	Illum. Source	Support Electrolyte*	Photo-Potentials Max, V	Photo-Currents Max, mA	Total Coulombs in First Hour
$Ru^{2+}/S_2O_8^{--}$	Room light	$1M\ Na_2SO_4$	0.980	0.325	1.10
	60 W	"	1.045	0.62	2.06
Ru^{2+}/Tl^{3+} ($1M\ H_2SO_4$)	250 W	$1M\ H_2SO_4$	1.03	0.32	
$Ru^{2+}/Co(NH_3)_5\ 2+$	60 W	$1M\ Na_2SO_4$	1.02	0.35	1.23

*Photolysis conditions used: Solutions: cathode: $\left[Ru(bpy)_3^{2+}\right] = 1.4 \times 10^{-4} M$;
$\left[Q\right] = 10^{-2}M$; support electrolytes as shown in the table.
Anode: 0.1M acetate buffer, pH 4.7. Illumination source was a regulated
(0-250W) t_ngten-halogen lamp used with a water filter and a 400 nm cutoff
glass filter. Potentials quoted are against Ag-Agcl(sat).

References

1.1)   J. Bolton, ed., Solar Power & Fuels, Academic Press, London,
        (1977).

1.2)   S. Claesson, ed., Photochemical Conversion and Storage of
        Solar Energy, Swedish National Energy Board Report, Stockholm,
        (1977).

1.3)   J. Bolton and D.O. Hall, Ann. Rev. Energy. 4:353 (1979).

1.4)   A. Harriman & J. Barber in "Photosynthesis in relation to
        model systems", J. Barber ed., Elsevier, Amsterdam (1979).

1.5)   N.J. Sutin, J. Photochem. 10:19 (1979)

1.6)   E. Schumacher, Chimia, 32:194 (1978)

1.7)   G. Porter & M.D. Archer, Interdisc. Sci. Rev., 1:119 (1976).

1.8)   G. Porter, Proc. Roy. Soc., (London) A 362:281 (1978).

1.9)   M. Calvin, Acc. Chem. Res. 11:369 (1978). Photochem. Photo-
        biol., 23:425 (1976).

1.10)   J. Bolton, Science, 202:705 (1978).

1.11)   V. Balzani, F. Boletta, M.T. Gandolfi and M. Maestri, Topics
        Current Chem., 75:1 (1978).

1.12)   M. Almgren, Photochem. Photobiol., 27:603 (1978)

1.13)   R.T. Ross, J. Chem. Phys., 45:1 (1966); ibid.,46:4590 (1967).

1.14)   R.T. Ross and M. Calvin, Biophys. J., 7:595 (1967).

1.15)   V. Balzani, L. Moggi, M.F. Manfrin, F. Boletta & M. Gleria,
        Science, 189:852 (1975).

1.16)   Light Induced Charge Separation in Biology and Chemistry,
        H. Gerischer and J.J. Katz, ed., Dahlem Conferences Life
        Sciences Research Report 12, Berlin (1979).

2.1)   D.D. Davis, K.L. Stevenson and C.R. Davis, J. Amer. Chem.
        Soc., 100:5344 (1978).

2.2)   G. Ferraudii, Inorg. Chem., 17:1370 (1978).

2.3)   K.L. Stevenson, D.M. Kaehr, D.D. Davis and C.R. Davis, Inorg.
        Chem., 19:781 (1980).

2.4)   P.R. Ryason, Solar Energy, 19:445 (1977).

2.5)   D.D. Davis, K.L. Stevenson and G.K. King, Inorg. Chem.,
        16:670 (1977).

2.6)   M. Brandys and G. Stein, J. Phys. Chem., 82:852 (1978).

2.7)   L.J. Heidt, M.G. Mullin, W.B. Martin and A.M.J. Beatty, J.
        Phys. Chem., 66:336 (1962).

2.8)   F.S. Dainton, E. Collinson and M.A. Malati, Trans. Farad. Soc.,
        55:2096 (1959).

2.9)   L.J. Heidt and A.F. McMillan, J. Amer. Chem. Soc., 76:2135
        (1954).

2.10)   B.V. Koryakin, T.S. Dzabiev and A.E. Shilov, Dokl. Akad. Nauk.
        SSSR, 229:614 (1976).

2.11)   D.R. Eaton and W. Stuart, J. Phys. Chem., 72:400 (1968).

2.12)   K.L. Stevenson and D.D. Davis, Inorg. Chem. Nucl. Chem. Lett
        12:905 (1976).

2.13)   W.C. Trogler, D.K. Erwin, G.L. Geoffroy and H.B. Gray, J.
        Amer. Chem. Soc:, 100:1160 (1978).

2.14)   D.K. Erwin, G.L. Geoffroy, H.B. Gray, G.S. Hammond, E.I.
        Solomon, W.C. Trogler and A.A. Zagars, J. Amer. Chem. Soc.,
        99:3620 (1977).

2.15)   K. Kalyanasundaram, Nouveau J. Chim., 3:511 (1979).

2.16)   A. Camus, C. Cocevar and G. Mestroni, J. Organometallic Chem.
        39:355 (1972).

2.17)   T. Knick, G. Sylvester, I.P. Kanan, Angew. Chem. Int'l Edn.,
        10:725 (1971).

2.18)   G.L. Geoffroy, G.S. Hammond and H.B. Gray, J. Amer. Chem.
        Soc., 97:3933 (1975).

2.19)   G.L. Geoffroy and M.G. Bradley, Inorg. Chem., 16:744 (1977).

2.20)   G.L. Geoffroy and M.G. Bradley, J. Organometallic Chem.,
        134:C27 (1977).

2.21) G.L. Geoffroy and M.G. Bradley, Inorg. Chem. 17:2410 (1978).

2.22) K.R. Mann, N.S. Lewis, V.M. Miskowski, D.K. Erwin, G.S. Hammond and H.B. Gray, J. Amer. Chem. Soc., 99:5525 (1977).

2.23) J.M. Lehn and J.P. Sauvage, Nouv. J. Chim., 1:449 (1977).

2.24) K. Kalyanasundaram, J. Kiwi and M. Grätzel, Helv. Chim. Acta, 61:2720 (1978).

2.25) A. Moradpur, E. Amouyal, P. Keller and H. Kagan, Nouv. J. Chim., 2:547 (1978).

2.26) K. Takuma, M. Kajiwara and T. Matsuo, Chem. Lett. 1199 (1977).

2.27) I. Okura, S. Nakamura, N.K. Thuan, K. Nakamura, J. Mol. Cat. 6:281 (1979).

2.28) G.M. Brown, S.F. Chan, C. Creutz, H.A. Schwarz and N. Sutin, J. Amer. Chem. Soc., 101:7638 (1979).

2.29) K. Kirsch, J.M. Lehn, J.P. Sauvage, Helv. Chim. Acta, 62:1345 (1979).

2.30) M. Gohn and N. Getoff, Z. Naturforschung, 34A:1135 (1979).

2.31) J. Kiwi and M. Grätzel, J. Amer. Chem. Soc., 101:7214 (1979).

2.32) K. Kalyanasundaram, J.C.S. Chem. Commun., 628 (1978).

2.33) B.V. Koryakin, T.S. Dzhabiev and A.E. Shilov, Dokl. Akad. Nauk, 233:620 (1977).

2.34) A.I. Krasna, Photochem. Photobiol., 29:267 (1979).

2.35) A.I. Krasna, Photochem. Photobiol., 31:75 (1980).

2.36) M.W.W. Adams, K.K. Rao and D.O. Hall, Photobiochem. Photobiophys., 1:33 (1979).

2.37) S. Markiewicz, M.S. Chan, R.H. Sparks, C.A. Evans and J.R. Bolton, Abstract International Conference on Photochemical Conversion and Storage of Solar Energy (1976).

2.38) A.A. Krasnovskii, V.V. Nikandro, G.P. Brin, I.N. Gogotov and V.P. Oshchepkov, Dokl. Akad. Nauk SSSR, 225:711 (1975).

2.39) K. Kalyanasundaram and G. Porter, Proc. Roy. Soc., A364:29 (1978).

2.40) T. Kawai, K. Tanimura and T. Sakata, Chem. Letts., 137 (1979).

2.41) I. Okura and N. Kim-Thuan, J. Mol. Cat., 6:227 (1979).

2.42) I. Okura and N. Kim-Thuan, J. Chem. Res.(S), 344 (1979).

2.43) I. Okura and Kim-Thuan, JCS Chem. Commun., 84 (1980).

2.44) M.P. Pileni, A.M. Braun, M. Grätzel, Photochem. Photobiol., 31:423 (1980).

2.45) M. Anbar and I. Pecht, J. Amer. Chem. Soc., 89, 2553 (1967).

2.46) M. Anbar and I. Pecht, Transac. Farad. Soc., 64:744 (1968).

2.47) V. Ya. Shafirovich and V.V. Strelets, Nouv. J. Chim., 2:199 (1978).

2.48) T.S. Dzhabiev, V. Ya. Shafirovich and A.E. Shilov, Reaction Kinetics Cat. Letts., 4-11 (1976).

2.49) M.E. Bodini and D.T. Sawyer, J. Amer. Chem. Soc., 98:8366 (1976).

2.50) K.D. Magers, C.G. Smith and D.T. Sawyer, Inorg. Chem., 17:575 (1978).

2.51) M.R. Morrison and D.T. Sawyer, Inorg. Chem., 17:333 (1978).

2.52) S.R. Cooper and M. Calvin, J. Amer. Chem. Soc., 99:6623 (1977).

2.53) M.R. Morrison and D.T. Sawyer, J. Amer. Chem. Soc., 99:257 (1977).

2.54) Y. Otsuji, K. Sawada, I. Morishita, Y. Taniguchi and K. Mizuno, Chem. Letts., 983 (1977).

2.55) G. Nord and O. Wernberg, J.C.S. Dalton, 866 (1972).

2.56) G. Nord and O. Wernberg, J.C.S. Dalton, 845 (1975).

2.57) V. Ya. Shafirovich, A.P. Moravskii, T.S. Dzhabiev and A.E. Shilov, Kinet. Kat., 18:509 (1977).

2.58) C. Creutz and N. Sutin, Proc. Nat'l Acad. Sci., 72:2858 (1975).

2.59) P. Loach and M. Calvin, Biochem., 2:361 (1963).

2.60) P. Loach and M. Calvin, Biochem. Biophys. Acta, 79:369 (1964).

2.61) I. Tabushi and S. Kujo, Tetrahedron Letts., 1577 (1974).

2.62) I. Tabushi and S. Kojo, ibid., 305 (1975).

2.63) G. Porter, Proc. Roy. Soc., A362:281 (1978).

2.64) A. Harriman and G. Porter, J.C.S. Faraday II, 75:1543 (1979).

2.65) S.J. Valenty, Anal. Chem., 50:669 (1978).

2.66) I.A. Duncan, A. Harriman and G. Porter, Anal. Chem., 51:2206 (1979).

2.67) L.J. Heidt and M.E. Smith, J. Amer. Chem. Soc., 70:2476 (1978).

2.68) M.G. Evans and N. Uri, Nature, 164:602 (1950).

2.69) G.V. Buxton, S.P. Wilford and R.J. Williams, J. Chem. Soc., 4957 (1962).

2.70) J.E. Barnes, J. Barrett, R.W. Brett and J. Brown, J. Inorg. Nucl. Chem., 30:2207 (1968).

2.71) K. Kalyanasundaram, O. Micic, E. Pramauro, and M. Grätzel, Helv. Chim. Acta, 62, 2432 (1979).

2.72) V. Ya. Shafirovich, N.K. Khannanov and V.V. Strelets, Nouv. J. Chim., 4:81 (1980).

2.73) T.S. Dzhabiev and Y.A. Shafirovich, Reaction Kinet. Cat. Letts., 4:419 (1976).

2.74) B.G. Jeliakova, S. Nakamma and H. Fukutoni, Bull. Chem. Soc. (Japan), 48:347 (1975).

2.75) V. Klanning and M.C.R. Symons, J. Chem. Soc., 3269 (1959).

2.76) P.A. Jacobs, J.B. Vytterhoeven and H.K. Beyer, J.C.S. Chem. Comm., 128 (1977).

2.77) S. Leutwyler and E. Schumacher, Chimia, 31:475 (1977).

2.78) S.M. Kuzimicki and E.M. Eyring, J. Amer. Chem. Soc., 100:6790 (1978).

2.79) J.M. Lehn, J.P. Sauvage and R. Ziessel, Nouv. J. Chim., 3:423 (1979).

3.1)   C. Tanford, "The Hydrophobic Effect; Formation of Micelles and Biological Membranes", Wiley-Interscience, New York, N.Y. (1973).

3.2)   J.H. Fendler and E.J. Fendler, "Catalysis in Micellar and Macromolecular Systems", Academic Press, New York (1975).

3.3)   M.S. Fernandez and P. Fromherz, J. Phys. Chem., 81:1755
        (1977).

3.4)   S. McLaughlin, in "Current Topics in Membranes and Trans-
        port", Vol. 9, p. 71 ff, Academic Press, New York (1977).

3.5)   S.A. Alkaitis and M. Grätzel, J. Amer. Chem. Soc., 98:3549
        (1976).

3.6)   S.A. Alkaitis, G. Beck and M. Grätzel, J. Amer. Chem. Soc.,
        97:5723 (1975).

3.7)   S.A. Alkaitis, M. Grätzel and A. Henglein, Ber. Bunsenges.
        Phys. Chem., 79:541 (1975).

3.8)   M. Grätzel, in "Micellization and Microemulsion", K.L. Mittal,
        ed., Plenum Press, Vol. 2, p. 831 (1977).

3.9)   B. Razem, M. Wong and J.K. Thomas, J. Amer. Chem. Soc.,
        100: 1679 (1978).

3.10)   Y. Waka, K. Hamamoto and N. Mataga, Chem. Phys. Lett.,53:242
        (1978).

3.11)   C. Wolff and M. Grätzel, Chem. Phys. Lett. 52:542 (1977).

3.12)   For a general review of micellar effects on photochemical
        reactions see, K. Kalyanasundaram, Chem. Soc. Rev. 7:435
        (1978); M. Grätzel, A. Braun and N.J. Turro, Angew. Chem.
        (1980) in press.

3.13)   A.A. Krasnovskii and A.N. Luganskaya, Dokl. Akad. Nauk SSSR,
        223:229 (1975).

3.14)   K. Kalyanasundaram and G. Porter, Proc. Roy. Soc. (London)
        A364:29 (1978) and references cited therein.

3.15)   K. Kalyanasundaram, J.C.S. Chem. Commun., 628 (1978).

3.16)   M.-P. Pileni, A.M. Braun and M. Grätzel, Photochem. Photo-
        biol., 31:423 (1980).

3.17)   R.H. Schmehl and D.G. Whitten, J. Amer. Chem. Soc., 102
        (1980).

3.18)   Y. Tsutsui, K. Takuma, T. Nishijima and T. Matsuo, Chem.
        Lett., 617 (1979).

3.19)   P.J. DeLaive, J.T. Lee, H. Abrina, H.W. Sprintschnik, T.J.
        Meyer and D.G. Whitten, Adv. Chem. Sci., 168:28 (1978).

3.20)   M. Maestri, P.P. Infelta and M. Grätzel, J. Chem. Phys.,
        69:1522 (1978).

3.21)   Y. Moroi, A.M. Braun and M. Grätzel, J. Amer. Chem. Soc.,
        101:567 (1979).

3.22)   Y. Moroi, P.P. Infelta and M. Grätzel, J. Amer. Chem. Soc.,
        101:573 (1979).

3.23)   M. Cinquini, F. Montanari, P. Tundo, J.C.S. Chem. Commun.
        393 (1975); J. LeMoigne, P. Gramain and J. Simon, J. Coll.
        Int. Sci., 60:564 (1977).

3.24)   Y. Moroi, E. Pramauro, M. Grätzel, E. Pelizzetti and P.
        Tundo, J. Coll. Int. Sci., 64:341 (1979).

3.25)   R. Humphry-Baker, M. Grätzel, P. Tundo and E. Pelizzetti,
        Angew. Chem. Int'l. Ed., 18:630 (1979).

3.26)   K. Monserrat, M. Grätzel and P. Tundo, J. Amer. Chem. Soc.,
        102 (1980) in press.

3.27)   P.J. DeLaive, D.G. Whitten and C. Giannotti, Adv. Chem. Sci.,
        173:236 (1979).

3.28)   D.G. Whitten, P.J. DeLaive, T.K. Foreman, J.-A. Mercer-Smith,
        R.H. Schmehl and C. Giannotti, in"Solar Energy Conversion
        and Storage",R.R. Hantala ed., Humana Press, (1979).

3.29)   T. Matsuo, K. Takuma, Y. Tsutsui and T. Nishijima, J. Coord.
        Chem., (1980) in press.

3.30)   P.-A. Brugger and M. Grätzel, J. Amer. Chem. Soc., 102:2461
        (1980).

3.31)   P.-A. Brugger, P.P. Infelta and M. Grätzel, J. Amer. Chem.
        Soc., 102 (1980) in press.

3.32)   T. Kunitake, J. Macromol. Sci. (Chem), A13:587 (1979) and
        references cited therein.

3.33)   J.H. Fendler, Acc. Chem. Res. (1980).

3.34)   M. Mengel, Biochem. Biophys. Acta, 430:459 (1976).

3.35)   W. Oettmeier, J.R. Norris and J.J. Katz, Z. Naturforschung,
        31C:163 (1976).

3.36)   S.S. Anderson, I.G. Lyle, R. Paterson, Nature, 259:147 (1976).

3.37)   Y. Toyoshima, M. Morino, H. Motoki and M. Sukigura, Nature,
        265:188 (1977).

3.38)   Y. Sudo and F. Toda, Chem. Lett., 1011 (1978).

3.39)   W.F. Ford, J.W. Otvos and M. Calvin, Proc. Nat'l Acad. Sci.,
        76:3590 (1979); Nature, 274:507 (1978).

3.40)   P.P. Infelta, M. Grätzel and J.H. Fendler, J. Amer. Chem.
        Soc., 102:1479 (1980).

3.41)   M.-P. Pileni, A.M. Braun and M. Grätzel, J. Phys. Chem.
        (1980) in press.

3.42)   M.-P. Pileni, Chem. Phys. Lett. (1980)

3.43)   J.R. Escabi-Perez, A. Romero, S. Lukac and J.H. Fendler,
        J. Amer. Chem. Soc., 101:2231 (1979).

3.44)   J.Kiwi and M. Grätzel, J. Amer. Chem. Soc., 100: 6314 (1978).

3.45)   J. Kiwi and M. Grätzel, J. Phys. Chem., 84 (1980) in press.

3.46)   C.E. Jones, C.A. Jones and R.A. Mackay, J. Phys. Chem.,
        83:805 (1979).

3.47)   H. Kuhn, Pure & Appl. Chem., 51:341 (1979) and references
        cited therein.

3.48)   A.F. Jansen and J.R. Bolton, J. Amer. Chem. Soc., 101:6342
        (1979) and references cited therein.

3.49)   F.H. Quina and D.G. Whitten, J. Amer. Chem. Soc., 99:877
        (1977).

3.50)   H. Ti Tien, Bilayer Lipid Membranes, Marcel Dekker, New York,
        (1974).

3.51)   D. Meisel and M.S. Matheson, J. Amer. Chem. Soc., 99:6577
        (1977).

3.52)   C.D. Jonah, M.S. Matheson and D. Meisel, J. Phys. Chem.,
        83:257 (1979).

3.53)   D. Meisel, J. Rabani, D. Meyerstein and M. Matheson,
        J. Phys. Chem., 82:985 (1978).

4.1)   C. Wagner and W. Traud, Z. Electrochem., 44:397 (1938).

4.2)   See for example, G. Bredig and R.Müller von Berneck, Ber.
        31:258 (1899).

4.3)   D.O. Hall, Fuel, 57:322 (1978).

4.4)   K.K. Rao and D.O. Hall, in "Photosynthesis in Relation to
        Model Systems, ed. J. Barber, Elsevier, Amsterdam (1979).

4.5)   D.E. Green and L.H. Stickland, Biochem. J., 28:898 (1934).

4.6)   J. Kiwi and M. Grätzel, Nature, 5733 (1979).

4.7)   J. Kiwi and M. Grätzel, J. Amer. Chem. Soc., 101:7214 (1979).

4.8)   K. Kalyanasundaram and M. Grätzel, JCS Chem. Commun. 1137
        (1979); K. Kalyanasundaram and D. Dung, J. Phys. Chem.,
        (1980) in press.

4.9)   C.K. Grätzel and M. Grätzel, J. Amer. Chem. Soc., 101:7741
        (1979)

4.10)   P.J. DeLaive, B.P. Sullivan, T.J. Meyer and D.G. Whitten,
        J. Amer. Chem. Soc., 101:4001 (1979).

4.11)   P.A. Brugger, P. Cuendet and M. Grätzel, J. Amer. Chem. Soc.,
        (1980) in press.

4.12)   A. Henglein, J. Phys. Chem., 83:2209 (1979); ibid. 83:2858
        (1979).

4.13)   A. Henglein, Ber. Bunsenges Phys. Chem., 84:253 (1980).

4.14)   D. Meisel, J. Amer. Chem. Soc., 101:6133 (1979).

4.15)   K. Kopple, D. Meyerstein and D. Meisel, J. Phys. Chem.,
        81:870 (1980).

4.16)   J. Kiwi and M. Grätzel, Angew. Chem. Int'l Ed., 17:860 (1978);
        ibid. 18:624 (1979); Chimia, 33:289 (1979).

4.17)   K. Kalyanasundaram, O. Micic, E. Promauro and M. Grätzel,
        Helv. Chim. Acta, 62:2432 (1979).

4.18)   J.M. Lehn, J.P. Sauvage and R. Ziessel, Nouveau J. Chim.,
        3:423 (1979).

4.19)   V. Ya. Shafirovich, M.K. Khannov and V.V. Strelets, Nouveau
        J. Chim., 4:81 (1980).

4.20)   K. Kalyanasundaram and M. Grätzel, Angew. Chem., Int. Ed.,
        18:701 (1979).

4.21)   J. Kiwi, E. Bogarello, E. Pelizzetti, M. Visca and M.Grätzel,
        Angew. Chem., Int. Ed. Engl., 19:0000 (1980).

5.1)   T. Kawai, K. Tanimura and T. Sakata, Chem. Letts., 137 (1979).

5.2)   B. Durham, W.J. Dressick and T.J. Meyer, J.C.S. Chem. Commun.
        381 (1979).

5.3)   D. Paul Riellemma, W.J. Dressick and T.J. Meyer, ibid. 247
        (1980).

5.4)   M. Neumann-Spallart, K. Kalyanasundaram, C. Grätzel and M.
        Grätzel, Helv. Chim. Acta., (1980) in press.

AUTHOR INDEX

Abrina, H., 382
Adams, M.W.W., 362
Albery, W.J., 313, 315, 316, 317, 320, 322, 324, 325, 334, 336, 341, 343, 347
Alkaisi, M.M., 112
Alkaitis, S.A, 378
Allen, C.W., 127
Allen, F.G., 72
Allison, J., 112
Almgrem, M., 350-351
Amamiya, T., 307
Amoyal, E., 362
Anbar, M., 368
Anderson, S.S., 392
Anderson, W.A., 68, 89, 98-99, 109, 112
Anderson, W.W., 126
Andrews, J.M., 73
Ang, P.G.P., 301
Antayposand, G.A., 184
Antypas, G.A., 172
Archer, M.D., 213, 313, 315, 317, 320, 322, 324, 341, 350
Archer, R.J., 72, 73
Arienzo, M., 160, 168, 192
Armantrout, G., 126
Arndt, W., 118-121, 152
Arnott, R., 125
Attalla, M.M., 72, 73
Augustine, F., 131, 156
Austin, R.G., 297

Backerra, S.C.M., 47, 90
Balzani, V., 350-352
Banerjee, A., 123

Barber, J., 350
Barber, S.J., 112
Bard, A.J., 245, 261, 263, 270, 278, 280, 283, 289
Barnes, J.E., 370, 373
Barnett, A.M., 118-119, 152
Baron, B., 156
Barrett, J., 370, 373
Bartlett, P.N., 324, 343, 347
Bawdet, Y., 90, 100-101
Beall, J., 158, 193-194
Beatty, A.M.J., 358
Beaulieu, R., 125, 158, 172, 186, 192-194
Beck, C., 378
Berg, R.S., 123
Berger, A., 129
Bernthsen, A., 328
Berry, W.B., 127
Beyer, H.K., 372-373
Bilger, G., 118, 119, 121, 152
Bloss, W.H., 118, 119, 121, 135, 139, 140, 146, 152
Bocarsly, A.B., 283, 289, 297
Boddy, P.J., 230
Bodini, M.E., 368, 370
Boer, K.W., 37
Boettcher, W., 340
Bogarello, E., 400
Bogus, K., 156
Boletta, F., 350
Bolton, J., 349-351, 362, 392
Bolton, J.M. (editor), 213
Bolts, J., 297
Bookbinder, D.C., 283, 289
Boudreaux, D.S., 289, 291
Bougnot, J., 117, 123, 135, 152

411

SUBJECT INDEX